AF389483

Advances in the Field of Electrical Machines and Drives

Advances in the Field of Electrical Machines and Drives

Editor

Athanasios Karlis

MDPI • Basel • Beijing • Wuhan • Barcelona • Belgrade • Manchester • Tokyo • Cluj • Tianjin

Editor
Athanasios Karlis
Department of Electrical and
Computer Engineering,
School of Engineering,
Democritus University of
Thrace Xanthi Greece,
Komotini, Greece

Editorial Office
MDPI
St. Alban-Anlage 66
4052 Basel, Switzerland

This is a reprint of articles from the Special Issue published online in the open access journal *Energies* (ISSN 1996-1073) (available at: https://www.mdpi.com/journal/energies/special_issues/ Advances_the_Field_Electrical_Machines_Drives).

For citation purposes, cite each article independently as indicated on the article page online and as indicated below:

LastName, A.A.; LastName, B.B.; LastName, C.C. Article Title. *Journal Name* **Year**, *Volume Number*, Page Range.

ISBN 978-3-0365-4285-0 (Hbk)
ISBN 978-3-0365-4286-7 (PDF)

Cover image courtesy of Athanasios Karlis.

Contents

About the Editor

Athanasios Karlis

Athanasios D. Karlis (born in 1967), received his Dipl.-Eng. and PhD degrees from the Electrical & Computer Engineering Department of the Aristotle University of Thessaloniki, Greece, in 1991 and 1996, respectively. In 2000, he was made a Lecturer in the Department of Electrical & Computer Engineering at Democritus University of Thrace, Greece. In 2017, he was made an Associate Professor, and since 2019, he has been Director of the Electrical Machines Laboratory. His research interests are in diagnostics, the control of electrical machines and drives, renewable energies, and power production from small hydro-wind energy conversion and photovoltaics power systems. His research activity includes more than 60 papers in scientific journals, book chapters and international conference proceedings. As a visiting researcher, he traveled to the Helsinki University of Technology; the Centre for Rapid Transit Systems in the Department of Electrical & Computer Engineering of the Virginia Polytechnic Institute and State University (Virginia Tech); and the Friedrich-Alexander-Universität Erlangen-Nürnberg, Germany. He speaks Greek (native), English and German. He has supervised more than 60 final-year and 10 master's theses. He is currently supervising five PhD and two master's theses.

He has been a member of IEEE since 2000, and a senior member since 2012. He is a member of the Editorial Board of two international scientific journals. He was Chair of the Local Organizing Committee of the International Conference on Electrical Machines (ICEM 2018), which was held in Alexandroupoli, Greece, 2018. He served for 2 years as a member of the board of directors of the Hellenic Institute of Electric Vehicles (HEL.I.E.V.). In 2015, he received the IEEE Outstanding Branch Chapter Advisor Award for IEEE Region 8, and in 2016, the Outstanding Student Branch Chapter Advisor Award for IEEE Industry Applications Society.

Preface to "Advances in the Field of Electrical Machines and Drives"

Electrical machines and drives dominate our everyday lives. This is due to their numerous applications in industry, power production, home appliances, and transportation systems such as electric and hybrid electric vehicles, ships, and aircrafts. Their development follows rapid advances in science, engineering, and technology. Researchers around the world are extensively investigating electrical machines and drives because of their reliability, efficiency, performance, and fault-tolerant structure. In particular, there is a focus on the importance of utilizing these new trends in technology for energy saving and reducing greenhouse gas emissions. This Special Issue will provide the platform for researchers to present their recent work on advances in the field of electrical machines and drives, including special machines and their applications; new materials, including the insulation of electrical machines; new trends in diagnostics and condition monitoring; power electronics, control schemes, and algorithms for electrical drives; new topologies; and innovative applications.

Athanasios Karlis
Editor

Article

Current and Stray Flux Combined Analysis for the Automatic Detection of Rotor Faults in Soft-Started Induction Motors

Angela Navarro-Navarro [1], Israel Zamudio-Ramirez [1,2], Vicente Biot-Monterde [1], Roque A. Osornio-Rios [2] and Jose A. Antonino-Daviu [1,*]

[1] Instituto Tecnológico de la Energía, Universitat Politècnica de València (UPV), Camino de Vera s/n, 46022 Valencia, Spain; annana3@etsii.upv.es (A.N.-N.); iszara@doctor.upv.es (I.Z.-R.); vibiomon@die.upv.es (V.B.-M.)
[2] HSPdigital CA-Mecatronica Engineering Faculty, Autonomous University of Queretaro, San Juan del Rio 76806, Mexico; raosornio@hspdigital.org
* Correspondence: joanda@die.upv.es

Abstract: Induction motors (IMs) have been extensively used for driving a wide variety of processes in several industries. Their excellent performance, capabilities and robustness explain their extensive use in several industrial applications. However, despite their robustness, IMs are susceptible to failure, with broken rotor bars (BRB) being one of the potential faults. These types of faults usually occur due to the high current amplitude flowing in the bars during the starting transient. Currently, soft-starters have been used in order to reduce the negative effects and stresses developed during the starting. However, the addition of these devices makes the fault diagnosis a complex and sometimes erratic task, since the typical fault-related patterns evolutions are usually irregular, depending on particular aspects that may change according to the technology implemented by the soft-starter. This paper proposes a novel methodology for the automatic detection of BRB in IMs under the influence of soft-starters. The proposal relies on the combined analysis of current and stray flux signals by means of suitable indicators proposed here, and their fusion through a linear discriminant analysis (LDA). Finally, the LDA output is used to train a feed-forward neural network (FFNN) to automatically detect the severity of the failure, namely: a healthy motor, one broken rotor bar, and two broken rotor bars. The proposal is validated under a testbench consisting of a kinematic chain driven by a 1.1 kW IM and using four different models of soft-starters. The obtained results demonstrate the capabilities of the proposal, obtaining a correct classification rate (94.4% for the worst case).

Keywords: current signals; stray flux signals; LDA; automatic fault diagnosis; induction motor; broken rotor bars; soft-starters

---

1. Introduction

Squirrel cage induction motors (SQIM) are indispensable elements widely used as powertrain drives in an extensive variety of industrial applications, constituting approximately 89% of the power demanded in industrial plants [1]. Their high efficiency, low cost, easy maintenance, and robustness have allowed the proliferation of these types of machines as the main drives for many mechanisms and processes, namely, large capacity exhaust fans, driving lathe machines, crushers, oil extracting mills, blowers, pumps, compressors. When the application requires continuous starts and stops of the driving motor, break/damage may arise due to the thermo-mechanical stresses developed in the rotor bars, especially during the start-up transient of the machine [2]. Thus, squirrel cage induction motors are vulnerable to frequent starts/stops and/or excessive load torque variations, and eventually a failure can occur in the rotor bar or end ring of the rotor cage [2]. Although a motor with damage in the rotor cage can continue operating, their performance and lifetime are degraded, which eventually may lead to the shutdown of the involved processes, causing huge time and economical losses if pertinent maintenance actions are not taken. In this

Citation: Navarro-Navarro, A.; Zamudio-Ramirez, I.; Biot-Monterde, V.; Osornio-Rios, R.A.; Antonino-Daviu, J.A. Current and Stray Flux Combined Analysis for the Automatic Detection of Rotor Faults in Soft-Started Induction Motors. *Energies* **2022**, *15*, 2511. https://doi.org/10.3390/en15072511

Academic Editor: Mario Marchesoni

Received: 24 February 2022
Accepted: 28 March 2022
Published: 29 March 2022

Publisher's Note: MDPI stays neutral with regard to jurisdictional claims in published maps and institutional affiliations.

context, in order to prevent/reduce the negative effects raised by continuous starts/stops and transient states, different starting systems are used in the industry, such as the star-delta starting, starting via auto-transformer or starting via soft-starters [3]. These methods have allowed a reduction in the high-amplitude currents developed under the start-up transient by controlling some parameters, such as the voltage profile, current profile or torque profile. In this context, the inclusion of soft-starters has enabled control of the start-up current profile by means of power electronic circuits containing thyristors installed in the different power source phases of the motor supply line. Subsequently, by changing the thyristors conduction time, a variation on the RMS value can be achieved, which in turn modifies the starting current profile. Thus, many devices have the possibility to limit the current during the startup transient, but most of them only allow the setting of the voltage and the time. However, although these devices reduce the starting current, this does not prevent motor faults from occurring. In fact, the use of soft-starters amplifies certain harmonics and introduces other frequency components, which could certainly make the motor diagnostic more difficult, even leading to erratic final diagnosis [4,5]. In this regard, although systems and mechanisms have been developed to avoid/reduce rotor-related failures, they can still occur. Hence, several monitoring systems and fault diagnostic techniques have been developed in order to prevent the costs/side effects associated with unexpected failures. Thus, the aim is to detect the failure in its initial stages, before catastrophic or irreversible damage can occur. Most of these techniques and proposed methodologies have used relevant information extracted from different physical quantities, which can be measured by means of primary sensors. In this regard, some common physical magnitudes that have been reported in the available technical literature are, among others: vibration signals [5], partial discharges [6], current signals [7], and stray fluxes [8]. Each technique has provided satisfactory results for the diagnosis of certain types of faults. For example, some previous works have proposed different methodologies relying on a single magnitude for the automatic classification of broken rotor bars for motors started by direct online (DOL) methods. In [9], the authors analyzed the stray flux signals with a feedforward neural network (FFNN) for the automatic classification, reporting a 97% effectiveness. By its part, Rivera et al. [10] studied specific signatures and patterns associated with a fault condition from current signals by means of a time-frequency (t-f) map, achieving a 97.5% overall effectiveness. However, as pointed out in some papers [11], the analysis of a single magnitude may be suitable to detect certain faults, but not all, and even, when the technique has exhibited satisfactory results in certain cases, there are specific situations in which the appropriate technique can provide false indications. In this regard, Zamudio et al. [12] proposed the fusion of stray flux and current signals by means of a feedforward neural network (FFNN), achieving a 95% overall effectiveness. Other works have proposed the use of convolutional neural networks (CNN) for the same purposes, reporting an accuracy rate higher than 97% by analyzing the current demanded by the motor as a principal magnitude, however, the method is limited to motors started by direct online (DOL) methods [13,14]. Moreover, although soft starters are widely used in industry, the number of works in the literature about the automatic diagnosis of electric motor failures started with these types of devices is limited. In this regard, Pasqualotto et al. [15] proposed an automatic classification technique for diagnosing broken rotor bars by analyzing stray flux signals, and using a convolutional neural network (CNN) as a main classifier. They achieved 94.4% accuracy. However, the use of this kind of technique demands high computational resources, and an elevated number of samples in order to train the method, which, in most of the cases, is very difficult to achieve under practical terms. Hence, some authors have proposed electric machine models by collecting data under healthy and faulty conditions [16], by means of the hybrid finite element method (FEM)—an analytical model for reducing the simulation time.

Considering the above-mentioned statements, recent works have shown that the analysis of two different quantities at the same time can provide a more complete analysis and can avoid false indications [12]. The proposed systems combine stray-flux and current signal analysis. Thus, the analysis of the current at the start-up could avoid false indicators [17] and, in the case of the stray flux analysis, it has shown good results in motors started by soft-starters [18]. Unfortunately, to the authors' best knowledge, there is no methodology capable of automatically diagnosing broken rotor bar failures in induction motors started by means of soft-starters, a challenging task since the frequency components and control technologies implemented by soft-starter manufacturers modifies the start-up transient profile, leading to unpredictable fault pattern evolutions.

The main contribution of this paper is a novel methodology for the automatic detection and severity quantification of rotor faults in soft-started induction motors by means of current-stray flux signal fusion. The proposed methodology relies on a couple of indicators that are introduced here, and are used to merge highly relevant information from current and stray flux signals, which are captured during the motor start-up transient. These indicators are based on the time-frequency maps obtained by applying the short-time Fourier transform to the captured signals. Additionally, a linear discriminant analysis (LDA) is performed in order to combine all the information, and finally a feed-forward neural network (FFNN) is trained to implement an automatic fault diagnosis and fault severity classification. The stray flux signals are captured by means of a handmade coil-based sensor, which can be installed on the frame of the machine. The effectiveness of the proposed method is verified under an experimental testbench with a 1.1 kW induction motor and using four different industrial soft-starters.

2. Materials and Methods

2.1. Current Monitoring for Broken Rotor Bars

The diagnostic of failures in electric motors based on the analysis of the current signals is a technique that has been extensively used, and has been greatly accepted at an industrial level, mainly due to its capability to perform a remote diagnosis, since the current signals of the motor can be accessed remotely (e.g., from the control center), and the large number of faults that can be diagnosed with this technique [7]. In this regard, the conventional motor current signature analysis (MCSA) is currently one of the most used online approaches in industry for detecting faults in electric motors. The technique is based on the fact that under normal conditions, the current in the rotor bars induces a clockwise field rotating at $s \cdot f$. When a rotor fault exists, an additional reverse rotating field is produced in accordance to Fortescue's Theorem, giving rise to a frequency component at $-s \cdot f$. This subsequently generates the amplification of the well-known lower sideband harmonic (f_{LSH}) appearing in the line current, which is given by Equation (1).

$$f_{LSH} = f \cdot (1 - 2 \cdot s) \tag{1}$$

Furthermore, as reported in some investigations [19], it is possible to associate with this anticlockwise field a negative-sequence current system which causes a pulsating torque and a speed oscillation which, in turn, generates an additional frequency component of the air-gap flux density component (f_{USH}) given by Equation (2) [20].

$$f_{USH} = f \cdot (1 + 2 \cdot s) \tag{2}$$

Hence, the MCSA mainly relies in the magnitude evaluation of these specific frequency components, which are amplified when the motor operates under a fault condition.

2.2. Stray Flux Signals Analysis

Magnetic flux analysis for the condition monitoring of electric machines has been found to be an excellent alternative to conventional techniques that have been widely used in the industry, such as MCSA. This fact can be mainly attributed to the various advantages that the analysis of magnetic flux signals provides over other approaches, among others: it has proven to be efficient and reliable in cases where conventional methods produce false indications (i.e., rotor axial air ducts, rotor magnetic anisotropy, low frequency load oscillations, etc.) [2,21], it is a non-invasive technique [19], very low cost sensors are required (e.g., handmade coil-based sensors) [22], flexibility and simplicity of installation of the available sensors [9]. In this regard, several previous pieces of research have been devoted to investigating and analyzing magnetic stray flux signals for the fault detection and diagnosis of electric motors under two main approaches: air-gap flux analysis, and stray flux analysis [23,24]. Since the stray flux-based methods are non-invasive (i.e., the sensors may be installed outside the frame of the machine), online methods without motor disassembly can be adopted, allowing for the development of online test methods.

In order to analyze the magnetic stray flux signals in an induction motor, two main components may be distinguished: axial stray flux, and radial stray flux [25]. These signals are known to be modified when the electric motor is working under a fault condition, hence introducing the amplification of some frequency components according to the fault [26]. Besides, as shown in some papers, as in [27], axial and radial stray flux can be captured by using a proper sensor installed in the vicinity of the motor frame. In order to illustrate this, Figures 1a,b show the presumed circulation of the radial and axial magnetic field lines, respectively. Additionally, depending on the location of the coil-sensor, different stray flux components are acquired, as shown in Figure 1. Thus, axial stray flux is measured at position A, radial stray flux at position C, and, at position B, both axial and radial stray flux are measured.

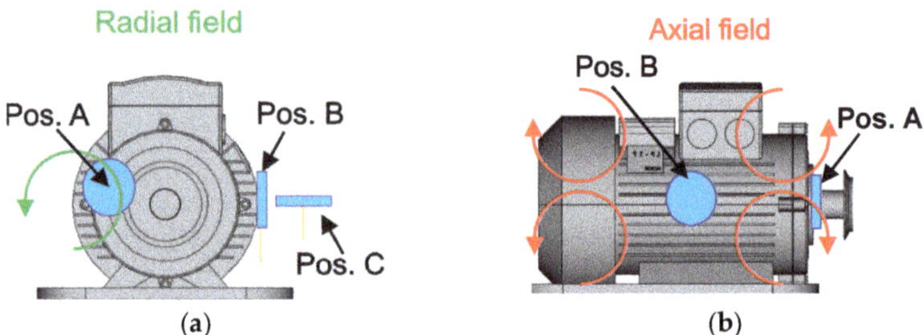

Figure 1. Stray flux components (**a**) Radial. (**b**) Axial.

2.3. Theoretical Fault-Related Frequency Evolution during Start-Up

Several research papers have proven that some faults modify the magnetic flux in the vicinity of the motor frame (magnetic stray flux), which may yield the amplification of some fault-related harmonics at specific frequencies [28]. Hence, as it has been pointed out in some works, for the case when the motor operates under rotor bar damages, the following harmonics may be observed to be amplified in the Fourier spectrum of the stray flux and current signals:

- Frequency components of axial nature (f_{axial}) [19]. These frequency components may be observed at one and three times the slip frequency, given by Equation (3) and Equation (4), respectively, and can be observed in the axial stray flux signals.

$$f_{axial} = s \cdot f \tag{3}$$

$$f_{axial} = 3 \cdot s \cdot f \tag{4}$$

where s = slip and f = power supply frequency.
- Sideband harmonics (f_{SH}), mainly observed in the radial stray flux signals [26,29]. These frequency components can be estimated by Equation (5).

$$f_{SH} = f \cdot (1 \pm 2 \cdot s) \tag{5}$$

In this regard, by analyzing Equations (1)–(4), it can be discerned that the harmonics related to rotor bar faults depend on the motor slip. Hence, it is expected to detect their evolution during the start-up transient, (when the motor speeds up) as the slip changes from a maximum value of 1 (i.e., at motor standstill), and approaches to 0 (when the motor reaches steady state). According to Equations (3)–(5), the theoretical evolution generated by the amplification of the fault-related frequency components produces very well-known patterns, which can be observed by means of a time-frequency map. Figure 2 shows the expected evolution of the different fault-related frequency components given by (3)–(5), according to the stray flux component (i.e., radial stray flux, axial stray flux) for a direct online (DOL) start.

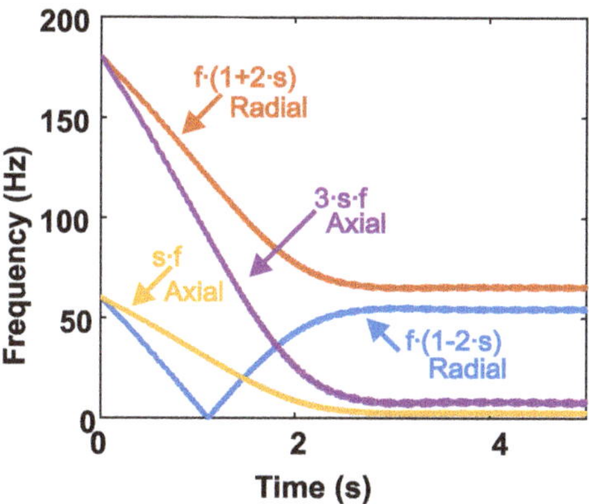

Figure 2. Evolutions of fault components related with broken rotor bars in the magnetic stray flux and current signals during the startup transient.

2.4. Linear Discriminant Analysis

The linear discriminant analysis (LDA) is a technique that has been widely used as a dimensionality reduction pre-processing stage for classification purposes. The main idea is to collect a new set of features, in which the maximization of the data separability is achieved for a certain number of considered classes. Hence, projecting a dataset onto a lower d-dimensional space with appropriate class-separability avoids overfitting, and helps reduce the computational burden [30]. The mathematical procedure to perform LDA can be found in [31].

2.5. Artificial Neural Network

Artificial Neural Networks (ANNs) are models employed to solve classification and pattern recognition problems [32]. Over the great diversity of ANN architectures that can be found in the literature, the feed-forward neural network (FFNN) structure is an excellent alternative for its application in automated final diagnosis schemes, since this type of ANN requires basic operations with a very low computational burden, has a simple and practical design, and generalizes well over the data on which it is trained. Hence, it can be easily incorporated into automated final diagnosis schemes that demand its implementation in programmable logic devices.

The most general structure of an FFNN is composed of different interconnected layers: one input layer (having n input neurons, I_{inputi}), one or more hidden layers, and one output layer (composed of k neurons, O_k), as shown in Figure 3a. The interconnection of the neurons in the different layers is performed by a series of weights (w_i), and biases (b_i) according to Equation (6). The mathematical model of each neuron (shown in the schematic of Figure 3b is given by Equation (6):

$$y = g\left(\sum_{i=1}^{n} w_i x_i + b_i\right) \tag{6}$$

where y is the output of the neuron, w_i the synaptic weights, x_i the inputs of the neuron, b_i the bias, and $g(\cdot)$ the activation function.

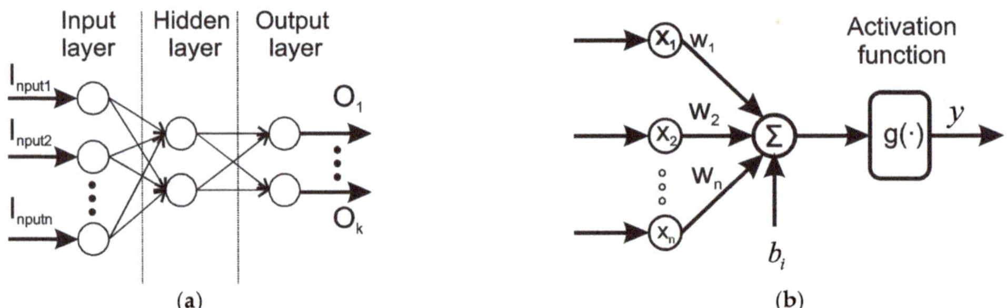

Figure 3. FFNN architecture: (**a**) general structure; (**b**) structure of a neuron.

2.6. Short-Time Fourier Transform

The short-time Fourier transform (STFT) is a well-known time-frequency (t-f) decomposition tool that allows the transformation of a time-domain signal into its time-frequency domain. A simple method that allows the STFT of a discrete signal to be obtained, is to dissect it into sliding windows that can overlap with each other, and then obtain the frequency content of each window, applying the fast Fourier transform (FFT). After computing the FFT of each window, a t-f map is obtained, containing the frequency content of the signal on different time intervals.

Mathematically, the STFT of a discrete-time signal, X_{STFT}, of length N can be computed by Equation (7):

$$X_{STFT}[k, l] = \sum_{n=0}^{N-1} x[n] \cdot w[k \cdot L - n] e^{-j(2\pi \cdot l \cdot \frac{n}{N})} \tag{7}$$

where $x[n]$ is the discrete-time signal, n is the time domain index, $l = 0, \ldots, N-1$, $k = 0, \ldots, [(N/L) - 1]$, $w[\cdot]$ is the applied windowing function and L determines the time separation among adjacent sections. Figure 4 shows the simplest way to obtain the time-frequency map of a signal by means of the STFT.

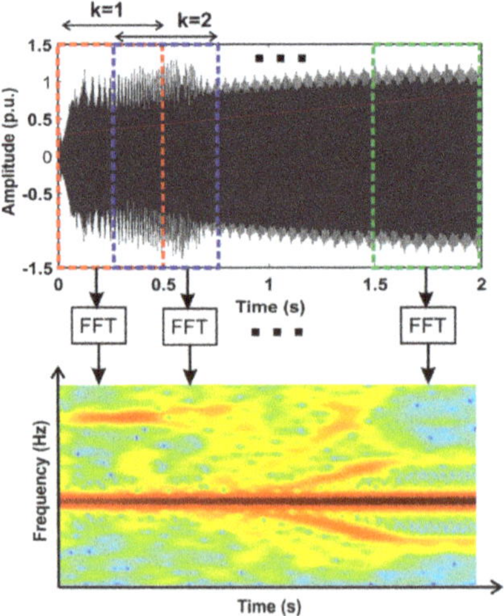

Figure 4. Short-time Fourier transform with overlap of a time domain signal.

3. Proposed Methodology

This section presents the proposed methodology for the automatic detection of broken rotor bars in soft-started induction motors under the start-up transient. When there is a failure in the motor, some harmonics are amplified in the stray flux and current spectrum. Hence, it is possible to observe their pattern evolution during the starting transient (since they are slip dependent) by means of a time-frequency map. These evolutions are identified by using a pair of indicators, which are the arithmetic mean and the maximum energy of specific t–f regions (depicted as shadowed areas in Figure 5) of the stray flux (RF_{ij}), and current signals (RC_{ij}). Such indicators can be computed by Equations (8)–(11), for the mean and maximum energy, respectively.

$$\text{mean}\left(RF_{ij}\right) = \frac{1}{N_{dp}}\left(\sum_{k=t_{initial}}^{t_{final}}\sum_{l=f_{initial}}^{f_{final}}\left(E_{k,l}\right)_{RF_{ij}}\right) \tag{8}$$

$$\text{mean}\left(RC_{ij}\right) = \frac{1}{N_{dp}}\left(\sum_{k=t_{initial}}^{t_{final}}\sum_{l=f_{initial}}^{f_{final}}\left(E_{k,l}\right)_{RC_{ij}}\right) \tag{9}$$

$$\text{max}\left(RF_{ij}\right) = \text{max}\left(\sum_{k=t_{initial}}^{t_{final}}\sum_{l=f_{initial}}^{f_{final}}\left(E_{k,l}\right)_{RF_{ij}}\right) \tag{10}$$

$$\text{max}\left(RC_{ij}\right) = \text{max}\left(\sum_{k=t_{initial}}^{t_{final}}\sum_{l=f_{initial}}^{f_{final}}\left(E_{k,l}\right)_{RC_{ij}}\right) \tag{11}$$

where $E_{k,l}$ is the normalized (over the fundamental frequency component) energy density at the (k, l) coordinate of the t–f map region under consideration (i.e., $RF_{i,j}$ or $RC_{i,j}$), $f_{initial}$ and f_{final} are, respectively, the initial and final frequency samples defining the considered t–f region, $t_{initial}$ and t_{final} are, respectively, the initial and final time samples defining the analyzed t–f region, and N_{dp} is equal to the total number of data points enclosed by the processed region.

Following the abovementioned definitions, the proposed methodology (depicted in Figure 5) is as follows:

- **Step 1.** Acquire current and magnetic flux signals (obtained in the vicinity of the motor frame, in axial + radial direction) simultaneously under the start-up transients by means of an oscilloscope, a magnetic flux sensor, and a current sensor. A coil-based sensor is the one used during the experimentation of this paper, which is described in detail in the next section.

- **Step 2.** Apply a time-frequency decomposition tool to obtain a time-frequency map of the captured stray flux and current signals; in this paper the STFT is computed by applying Equation (7). The STFT is selected since it can be easily implemented in programmable logic devices, and it allows a clear visualization of the fault components evolution.

- **Step 3.** Using Equations (8)–(11), compute the proposed indicators: $\text{mean}\left(RF_{ij}\right)$, $\text{mean}\left(RC_{ij}\right)$, $\text{max}\left(RF_{ij}\right)$, $\text{max}\left(RC_{ij}\right)$ in each region of interest, in order to characterize the fault-related pattern evolution for the stray flux and current signals. The regions of interest are obtained by dividing the t-f map into a grid of m rows by n columns. These regions of interest are in areas covered by the start-up transient. During the startup transient, the current reaches a high value in its amplitude, and it decreases until it arrives at steady state. Therefore, in using the envelope of the current signal, this part can be automatically isolated by setting a limit value (obtained through the last samples) equal to the maximum envelope amplitude at steady state. The intersection of this limit value with the time axis will be nearly at the end of the start-up transient.

- **Step 4.** Perform a feature reduction and fusion signals by applying an LDA, as described in Section 2.3. After that, a two-dimensional projection is obtained, in which the maximization of the data separability is achieved for the considered classes: healthy motor, one broken rotor bar, and two broken rotor bars. This projection allows the data clustering between the different fault severities to be observed, since the main projection axes are selected to be Feature 1 and Feature 2, respectively.
- **Step 5.** Perform an automatic classification of the motor condition status: healthy, one broken rotor bar, and two broken rotor bars, by means of the proposed indicators. For the purposes of this paper, an FFNN with hyperbolic tangent sigmoid and SoftMax activation functions in the hidden and output layers are used, respectively. The FFNN architecture is selected due to its simplicity, the low computation resources demanded by its calculation, and the ease of its implementation in hardware devices.

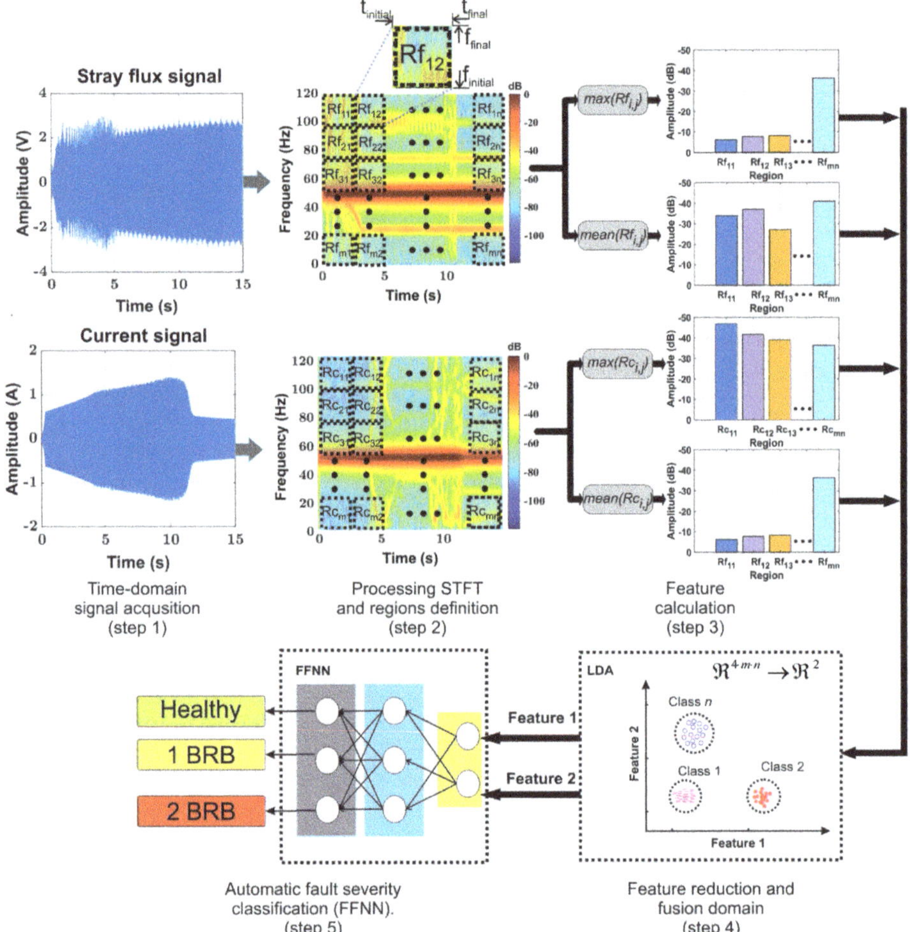

Figure 5. Proposed methodology flow-up.

4. Experimental Setup

Multiple experiments were carried out using a laboratory squirrel cage induction motor with 28 rotor bars, in order to validate the proposal. The main characteristics of the motor are listed in Table 1. The IM was driving a DC machine, which operated as a load. Hence, the load level was controlled by varying the excitation current of the DC machine. The experimentation testbench is shown in Figure 6. During the tests, two main physical magnitudes were captured: the magnetic stray flux, and the current demanded by the stator winding. To obtain the stray flux signals, a coil sensor is attached to the motor frame. This sensor was made in the laboratory, and it consists of an air coil with an internal diameter of 39 mm and an external diameter of 80 mm with 1000 turns (Figure 7 shows the dimensions and outline of the used sensor). Furthermore, in order to capture the current signals, a current clamp was installed in one of the stator power supply lines. Finally, an oscilloscope waveform recorder was used in order to acquire the current and stray flux signals for 30 s, and a sampling frequency of 5 kHz.

Table 1. Rated values and characteristics of the driving induction motor.

Power (kW)	1.1
Frequency (Hz)	50
Voltage (V)	400
Current (A)	2.4
Speed (rpm)	1440
Connection	Star
Number of Pole Pairs	2

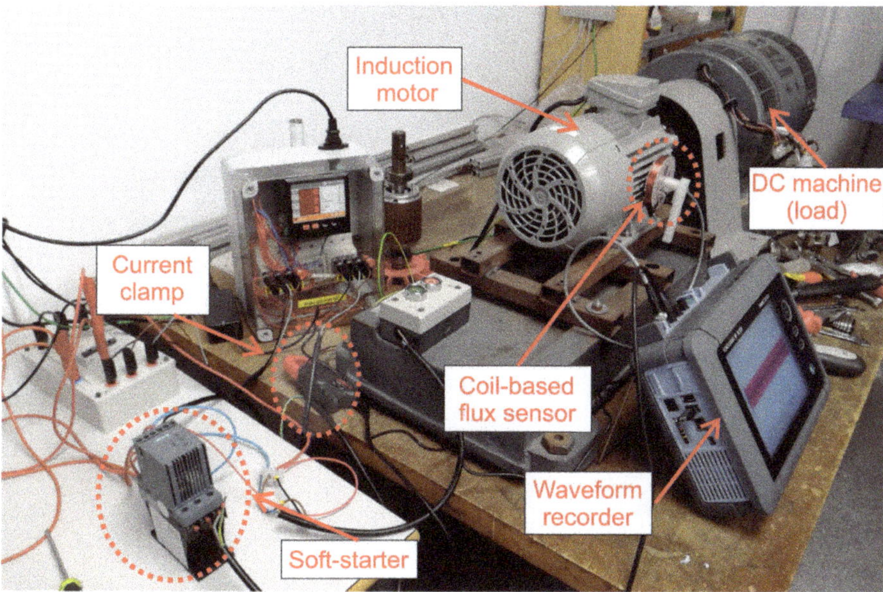

Figure 6. Experimental testbench.

Additionally, the machine was started by using four different soft-starters from different manufacturers (see Figure 8). The characteristics of the different soft-starters used are shown in Table 2. In this way, several tests were carried out using different levels of initial voltage/torque and different voltage ramp duration depending on the topology of the soft-starter, as described in Table 3. This allowed us to obtain several signals under different working conditions of the induction machine.

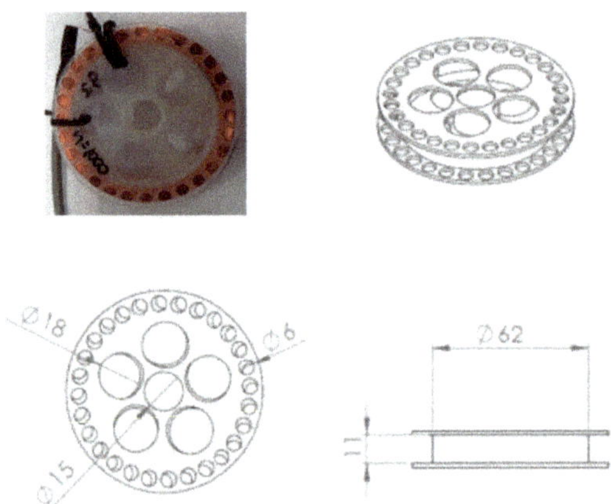

Figure 7. Shape and dimensions of the coil sensor used during the experiments.

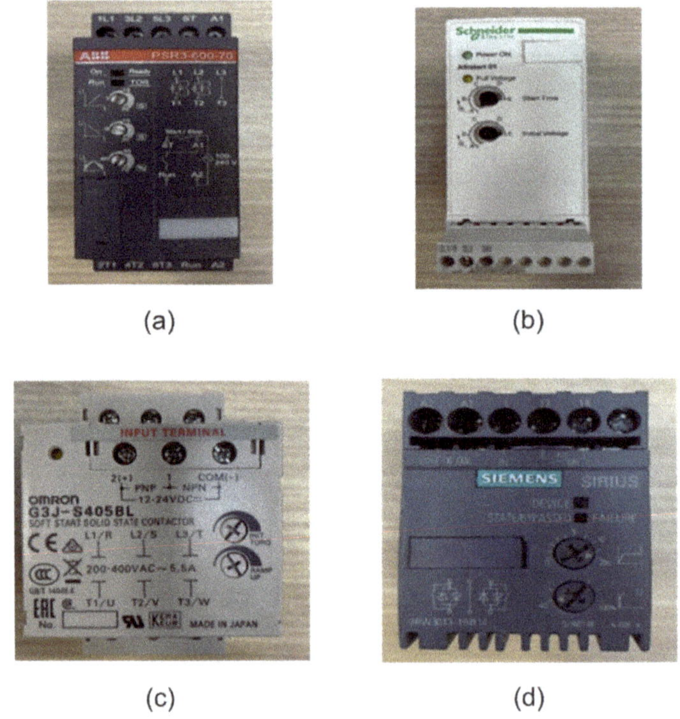

(a)

(b)

(c)

(d)

Figure 8. Tested soft-starters. (**a**) ABB PSR3-600-70. (**b**) Schneider. (**c**) Omron G3J-S405BL. (**d**) SIEMENS SIRIUS.

Table 2. Main characteristics of the industrial soft-starters used during experimentation.

Soft-Starter Model	Start-Up Transient Duration (s)	Initial Applied Voltage
Schneider	5	30%
Schneider	4	40, 50%
Schneider	3	55%
Schneider	2	67, 50%
Schneider	1	80%
ABB	20	40%
ABB	10	55%
ABB	1	70%
Omron	1	72%
Omron	12, 5	58%
Omron	25	44%
SIEMENS	0	100%
SIEMENS	5	70%
SIEMENS	10	50%
SIEMENS	20	40%

Table 3. Characteristics and rated values of the industrial soft-starters used during tests.

Manufacturer	Siemens	ABB	Omron	Schneider
Model type	3RW3013-1BB14	PSR3-600-70	G3J-S405BL	ATS01N109FT
Country of production	Germany	China	Japan	Germany
Number of controlled phases	2	2	3	1
Frequency (Hz)	50	50	50	50
Power (kW)	1.5	1.5	2.2	4
Voltage (V)	380–400	380–400	380–400	400
Maximum current (A)	3.6	3.9	5.5	9
Voltage ramp duration (s)	0–20	1–20	1–25	1–5

As a studied fault, broken rotor bars were induced in the induction motor by drilling a hole at the bar-end ring contact. Figure 9a shows the rotor in healthy conditions, before any hole was drilled. Besides, Figures 9b,c show the one and two induced broken rotor bars, respectively.

Figure 9. Rotor used during experimentation. (**a**) Healthy. (**b**) One broken bar. (**c**) Two broken bars.

5. Results and Discussion

This section shows and describes the results obtained during the experimentation by following the proposed methodology. Additionally, the fault-related harmonics pattern evolution during the start-up transient of the machine are evidenced and examined by means of time-frequency maps. Thereafter, the resulting decision regions modelled by the proposed FFNN architecture are presented. Finally, the effectiveness of the proposed methodology is presented.

Figures 10 and 11 show the results obtained when computing the STFT of the magnetic stray flux and current signals, respectively, for a healthy motor and a motor working under two broken rotor bars. These signals were captured during the experimentation, and they correspond to the Schneider starter when setting different starting times: 1 s, 2 s, 4 s. The STFT was obtained with a window size of 4096 samples, and a sliding size of 128 data points. When comparing the results shown there, it is relevant to note that due to the behavior established by each starter (defined by the technology used by the manufacturer), it is possible to appreciate a diversity of frequency components introduced by it, in such a way that different start-up times yield a different fault-related frequency evolution. Thus, for example, it is not possible to discern a clear difference (in terms of frequency content) between the healthy motor and the motor with two broken rotor bars in the stray flux signals when the starting time is set to 3 s, since similar t-f maps are obtained. However, if the start-up time is set to 4 s, it is possible to appreciate a very clear amplification of the rotor-related fault components (indicated with an arrow in Figure 10) in the t-f stray flux maps. In contrast, if this same analysis is carried out for current signals, it is more evident that there is a clear difference between a healthy motor and a motor with two broken rotor bars when the starting duration is 3 s. These results highlight the need and the relevance of merging the information provided by the magnetic stray flux and current signals simultaneously.

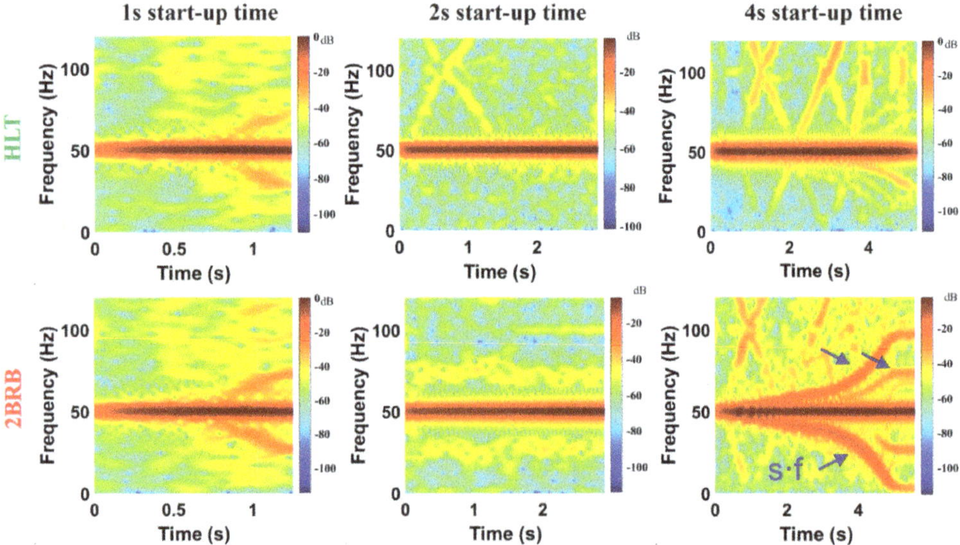

Figure 10. STFT t-f maps obtained by analyzing the stray flux signal of the IM for different start-up time transients when using Schneider soft-starter for a healthy motor, and two broken rotor bars.

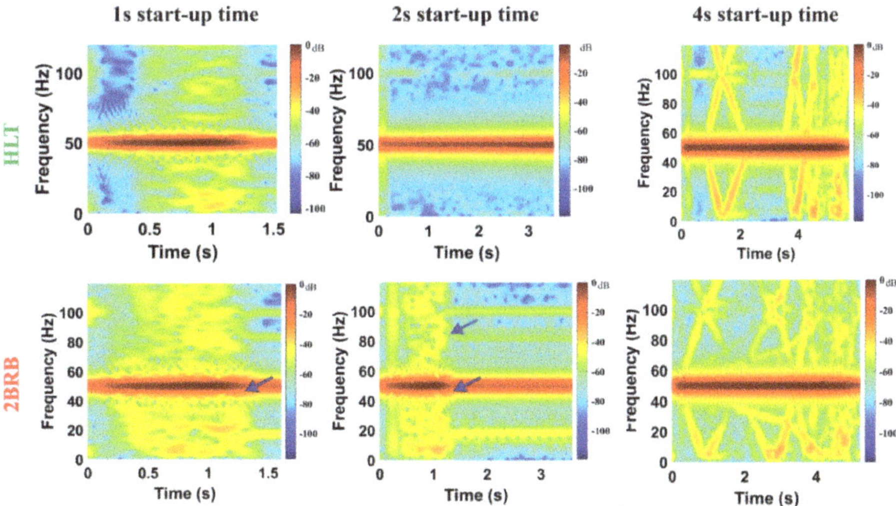

Figure 11. STFT t-f maps obtained by analyzing the current signal of the IM for different start-up time transients when using Schneider soft-starter for healthy motor, and one with two broken rotor bars.

On the other hand, the effectiveness, per evaluated class (HLT, 1BRB, 2BRB), is obtained through the calculation of the fault detection rate index (FDR). This index can be computed by dividing the number of correct classifications over the total number of samples in each class. In this regard, Figures 12–15 show the FDR obtained by evaluating the different soft-starters, and for the different machine healthiness states, i.e., a healthy motor (HLT), a motor with one broken rotor bar (1BRB), and a motor with two broken rotor bars. These results were obtained for an FFNN architecture with two neurons in the input layer, a hidden layer with two neurons, and three neurons in the output layer (corresponding to the three evaluated motor states). In addition, for the evaluation of the FFNN, 540 training samples were used for training, and another 180 different samples were used for validation, using each of the different models of soft-starters evaluated. As it can be observed in Figures 12–15, it is possible to discriminate and separate the different evaluated motor states by means of the proposed methodology, even in cases where variations in the fault-related pattern evolutions are introduced by alterations/perturbations according to the control components and mechanisms implemented by different soft-starters. Figure 12 shows that a high FDR is achieved, since only a few samples have been misclassified, reaching a 94.8% overall classification success. Similarly, Figure 13 confirms the excellent performance of the proposed methodology for the diagnosis and automatic classification of rotor faults when the ABB soft-starter is employed, since a total of 539 out of 540 samples have been classified correctly. In a similar fashion, Figures 14 and 15 show the classification results obtained when the proposed methodology is applied to the induction motor started by means of the Omron and Schneider soft-starter, respectively. In these Figures, it can be observed that the proposal is able to correctly classify among the different fault studied cases, since an overall FDR of 95% is achieved (only a few samples have been misclassified between healthy motor, and motor with one broken rotor bar). Additionally, the best results were obtained when evaluating the SIEMENS soft-starter, in which an FDR of 99.8% was achieved. On the other hand, the lowest performance was found when the motor was started using the ABB soft-starter, obtaining an FDR of 94.4%. This may be due to the diverse and erratic behavior of the slip changes experienced by the motor at start-up, introduced by the soft-starter, according to the manufacturer's technology.

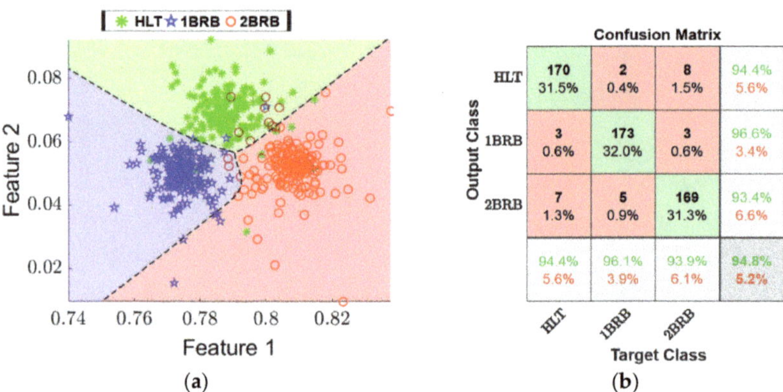

Figure 12. Classification results of the proposed methodology obtained for the Siemens soft-starter. (**a**) Resulting decision regions modelled by the proposed NN-based classifier over the 2-dimensional space; (**b**) Confusion matrix.

The performance for different t-f map grid sizes was evaluated using the overall efficiency reported by the FFNN, which is calculated dividing the samples that are correctly classified into the total number of samples. In this regard, Table 4 shows the results obtained for different time-frequency map sizes ranging from three rows by three columns, up to nine rows by nine columns. When comparing these results, it can be highlighted that the chosen grid size is highly relevant for the final diagnosis, and it has a significant impact on the efficiency achieved by the FFNN. Thus, the results show that the greater the size of the grid, the higher the efficiency achieved, however, a limit can be observed where the efficiency begins to decrease. This point is reached when the grid size is equal to nine rows by nine columns. Additionally, Table 4 shows that the best performance of the proposed methodology is achieved when the size of the t-f map grid is eight rows by eight columns.

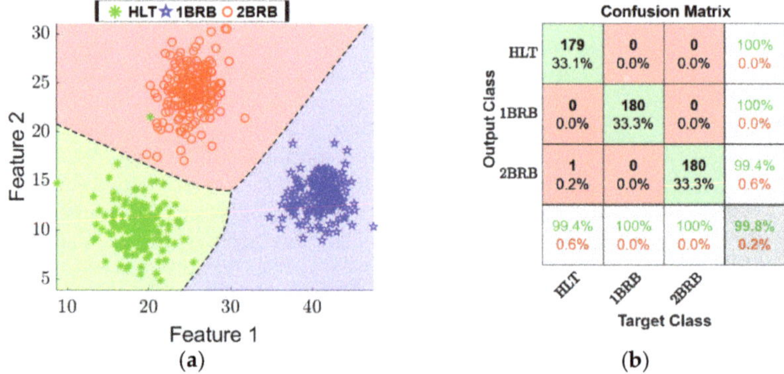

Figure 13. Classification results of the proposed methodology obtained for the ABB soft-starter. (**a**) Resulting decision regions modelled by the proposed NN-based classifier over the 2-dimensional space; (**b**) Confusion matrix.

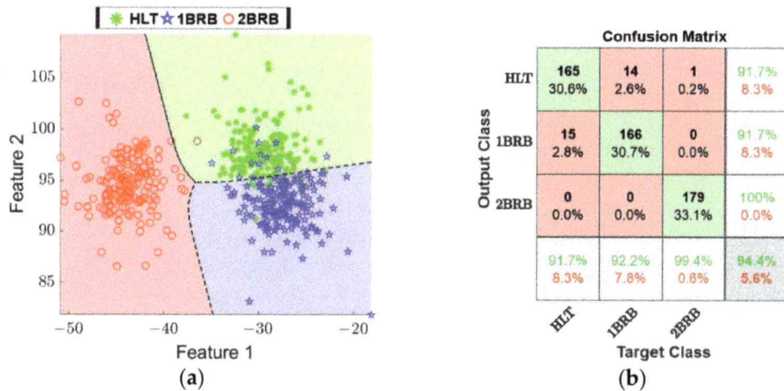

Figure 14. Classification results of the proposed methodology obtained for the Omron soft-starter. (**a**) Resulting decision regions modelled by the proposed NN-based classifier over the 2-dimensional space; (**b**) Confusion matrix.

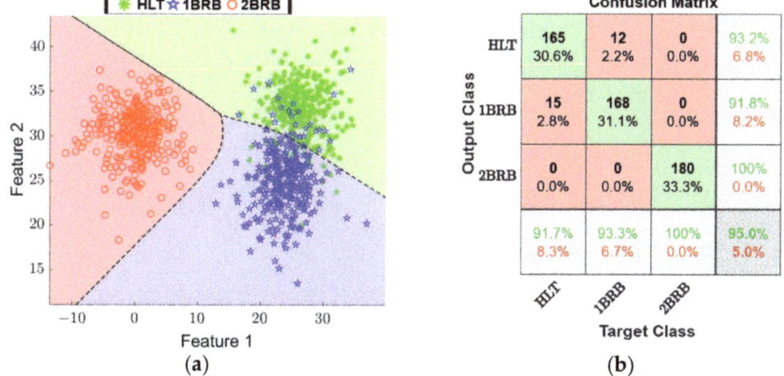

Figure 15. Classification results of the proposed methodology obtained for the Schneider soft-starter. (**a**) Resulting decision regions modelled by the proposed NN-based classifier over the 2-dimensional space; (**b**) Confusion matrix.

Table 4. Overall classification efficiency of the proposed method for different mesh sizes, and for the different soft-starters studied.

T-F Map Grid Size		Overall Classification Efficiency (%)			
Rows (m)	Columns (n)	ABB	Omron	Schneider	Siemens
3	3	83.6	72.8	74.6	79.2
4	4	96.3	74.1	84.8	79.4
5	5	96.4	83.6	86.9	87.3
6	6	97.5	83.6	90.4	93.8
7	7	99.1	85.2	91.1	94.1
8	8	99.8	94.4	95.0	94.8
9	9	98.1	91.4	92.0	94.8

Additionally, Table 5 summarizes the results obtained with the proposed methodology, and its comparison with the latest works reported in the literature for the automatic detection of broken bars in induction motors. The comparison includes the main techniques used in each proposed methodology, the technique applied to start the motor, the physical magnitude analyzed, and the accuracy rate achieved. As shown in Table 5, most of the works have been focused on the study of one unique magnitude to diagnose BRB faults, and many of them have reported the analysis under the transient regime of the current signals. For example, Martinez et al. [33] studied the current signals in order to provide an automated final diagnosis by means of an artificial neural network, reporting an overall effectiveness of 100%. However, they analysed only current signals, which may lead to a false diagnosis due to its intrinsic implications [21]. In contrast, Pasqualotto et al. [34] analysed stray flux signals, and proposed a methodology relying on a CNN in order to generate an automated final diagnosis. They achieved an accuracy rate of 66.7%, which in real terms is a very low efficiency that can potentially lead to a false indication. Furthermore, the use of a CNN as a main classifier may limit the viability of the method since a higher number of samples is required for training the CNN. In this regard, in order to avoid the high computational complexity for classification, in [35] the authors proposed a methodology based on the use of STFT with Gaussian and Kaiser windowing and an Otsu algorithm, achieving an accuracy rate of 100%. However, the proposed methodology is focused on DOL motor starters. In contrast, the proposed methodology is focused on the analysis of stray flux, and current signals, obtaining an overall effectiveness of 94.4%, being even higher in some of the soft starter models used in the experimentation as shown in Table 4.

Table 5. Comparison of different methodologies used in literature to detect broken bars in induction motors.

Reference	Methodology	Accuracy Rate	Start-Up Method	Signal Analyzed
Martinez et al. [33]	Homogeneity, kurtosis, ANN	100%	DOL	Current
Zamudio et al. [12]	STFT, FFNN	95%	DOL	Current and Stray Flux
Pasqualotto et al. [34]	CNN, STFT, data augmentation techniques	66.7%	DOL	Stray Flux
Pasqualotto et al. [15]	CNN, STFT, data augmentation techniques	94.4%	Soft-Starters	Stray Flux
Zamudio et al. [9]	STFT, FFNN	97%	DOL	Stray Flux
Lopez et al. [35]	Multi-STFT, Otsu Segmentation, Normal-distribution	100%	DOL	Current
Valtierra et al. [14]	STFT, CNN	100%	DOL	Current
Camarena et al. [36]	Wavelet Transform, Correlation Pearson	99%	DOL	Current
Ince et al. [13]	CNN, back-propagation (BP) algorithm	97.87%	DOL	Current
Rivera et al. [10]	Tooth-FFT, Pearson correlation	97.5%	DOL	Current
Proposed Approach	**STFT, FFNN, arithmetic mean and maximum value**	**94.4%**	**Soft-starter**	**Current and Stray Flux**

6. Conclusions

This paper has introduced a novel methodology for the automatic diagnosis of broken rotor bars in soft-started induction motors by means of the information fusion of current and stray flux signals. The proposed methodology relies on a pair of indicators proposed here. Such indicators are based on the arithmetic mean and maximum value, respectively, of specific regions from a time-frequency map; which are obtained by analyzing the current and stray flux signals captured during the start-up transient of the machine. These indicators are based on the fact that rotor faults yield the amplification of specific frequency components, which are found to be slip-dependent; hence, their evolution and amplification can be tracked by the proposed indicators. Additionally, as it can be observed in the results, it is very relevant to combine the information provided by the current and the stray flux signals, since the control mechanisms applied by each soft-starter manufacturer tend to modify the fault pattern evolutions, depending on the parameters used, according to the specific motor application and particular necessities of the final user. In this regard, the proposed methodology shows an excellent performance in the automatic classification among the studied faults, invariably to the soft-starter used, since an overall performance higher than 94.4% is achieved in any case. Finally, the obtained results show that, by means of the proposed methodology, it is possible to automatically discriminate among a healthy motor, a motor working under one broken rotor bar, and a motor working under two broken rotor bars. The proposal may find a great applicability under a vast array of applications demanding automated final diagnosis, especially those where the motor is constantly operated under starts/stops by means of a soft-starter.

Author Contributions: Conceptualization J.A.A.-D. and R.A.O.-R.; methodology, I.Z.-R., J.A.A.-D. and R.A.O.-R.; software, A.N.-N. and I.Z.-R.; validation, A.N.-N., I.Z.-R. and V.B.-M.; formal analysis, I.Z.-R. and A.N.-N.; investigation, A.N.-N., I.Z.-R. and V.B.-M.; resources, J.A.A.-D. and R.A.O.-R.; data curation, A.N.-N. and V.B.-M.; writing—original draft preparation, A.N.-N. and I.Z.-R.; writing—review and editing, I.Z.-R., J.A.A.-D. and R.A.O.-R.; visualization, I.Z.-R., J.A.A.-D. and R.A.O.-R.; supervision, J.A.A.-D. and R.A.O.-R.; project administration, J.A.A.-D. and R.A.O.-R.; funding acquisition, J.A.A.-D. and R.A.O.-R. All authors have read and agreed to the published version of the manuscript.

Funding: This work was supported by the Spanish 'Ministerio de Ciencia Innovación y Universidades' and FEDER program in the framework of the 'Proyectos de I+D de Generación de Conocimiento del Programa Estatal de Generación de Conocimiento y Fortalecimiento Científico y Tecnológico del Sistema de I+D+i, Subprograma Estatal de Generación de Conocimiento' (ref: PGC2018-095747-B-I00).

Acknowledgments: The authors would like to thank Consejo Nacional de Ciencia y Tecnología (CONACyT) under scholarship 652815.

Conflicts of Interest: The authors declare no conflict of interest.

References

1. Amezquita-Sanchez, J.P.; Valtierra-Rodriguez, M.; Perez-Ramirez, C.A.; Camarena-Martinez, D.; Garcia-Perez, A.; Romero-Troncoso, R.J. Fractal dimension and fuzzy logic systems for broken rotor bar detection in induction motors at start-up and steady-state regimes. *Meas. Sci. Technol.* **2017**, *28*, 75001. [CrossRef]
2. Park, Y.; Yang, C.; Kim, J.; Kim, H.; Lee, S.B.; Gyftakis, K.N.; Panagiotou, P.A.; Kia, S.H.; Capolino, G.-A. Stray flux monitoring for reliable detection of rotor faults under the influence of rotor axial air ducts. *IEEE Trans. Ind. Electron.* **2019**, *66*, 7561–7570. [CrossRef]
3. Larabee, B.; Pellegrino, B.; Flick, B. Induction motor starting methods and issues. In Proceedings of the Record of Conference Papers Industry Applications Society 52nd Annual Petroleum and Chemical Industry Conference, Denver, CO, USA, 12–14 September 2005; pp. 217–222. [CrossRef]
4. Corral-Hernandez, J.A.; Antonino-Daviu, J.; Pons-Llinares, J.; Climente-Alarcon, V.; Francés-Galiana, V. Transient-Based Rotor Cage Assessment in Induction Motors Operating with Soft Starters. *IEEE Trans. Ind. Appl.* **2015**, *51*, 3734–3742. [CrossRef]
5. Corral-Hernandez, J.A.; Antonino-Daviu, J. Startup-based rotor fault detection in soft-started induction motors for different soft-starter topologies. In Proceedings of the IECON 2016-42nd Annual Conference of the IEEE Industrial Electronics Society, Florence, Italy, 23–26 October 2016; pp. 6977–6982. [CrossRef]
6. Stone, Y.G.C.; Stranges, M.K.W.; Dunn, D.G. Common Questions on Partial Discharge Testing. *IEEE Ind. Appl. Mag.* **2016**, *22*, 14–19. [CrossRef]
7. Benbouzid, M.E.H. A review of induction motors signature analysis as a medium for faults detection. *IEEE Trans. Ind. Electron.* **2000**, *47*, 984–993. [CrossRef]
8. Capolino, A.; Romary, R.; Hénao, H.; Pusca, R. State of the art on stray flux analysis in faulted electrical machines. In Proceedings of the 2019 IEEE Workshop on Electrical Machines Design, Control and Diagnosis (WEMDCD), Athens, Greece, 22–23 April 2019; pp. 181–187. [CrossRef]
9. Zamudio-Ramírez, I.; Osornio-Ríos, R.A.; Antonino-Daviu, J.A.; Quijano-Lopez, A. Smart-sensor for the automatic detection of electromechanical faults in induction motors based on the transient stray flux analysis. *Sensors* **2020**, *20*, 1477. [CrossRef]
10. Rivera-Guillen, J.R.; de Santiago-Perez, J.J.; Amezquita-Sanchez, J.P.; Valtierra-Rodriguez, M.; Romero-Troncoso, R.J. Enhanced FFT-based method for incipient broken rotor bar detection in induction motors during the startup transient. *Measurement* **2018**, *124*, 277–285. [CrossRef]

11. Henao, H.; Capolino, G.-A.; Fernandez-Cabanas, M.; Filippetti, F.; Bruzzese, C.; Strangas, E.; Pusca, R.; Estima, J.; Riera-Guasp, M.; Hedayati-Kia, S. Trends in fault diagnosis for electrical machines: A review of diagnostic techniques. *IEEE Ind. Electron. Mag.* **2014**, *8*, 31–42. [CrossRef]

12. Zamudio-Ramirez, R.A.; Osornio-Rios, J.A.; Antonino-Daviu, J.A. Smart Sensor for Fault Detection in Induction Motors Based on the Combined Analysis of Stray-Flux and Current Signals: A Flexible, Robust Approach. *IEEE Ind. Appl. Mag.* **2022**, *28*, 56–66. [CrossRef]

13. Ince, T. Real-time broken rotor bar fault detection and classification by shallow 1D convolutional neural networks. *Electr. Eng.* **2019**, *101*, 599–608. [CrossRef]

14. Valtierra-Rodriguez, M.; Rivera-Guillen, J.R.; Basurto-Hurtado, J.A.; De-Santiago-Perez, J.J.; Granados-Lieberman, D.; Amezquita-Sanchez, J.P. Convolutional neural network and motor current signature analysis during the transient state for detection of broken rotor bars in induction motors. *Sensors* **2020**, *20*, 3721. [CrossRef] [PubMed]

15. Pasqualotto, D.; Navarro, A.N.; Zigliotto, M.; Antonino-Daviu, J.A.; Biot-Monterde, V. Fault Detection in Soft-started Induction Motors using Convolutional Neural Network Enhanced by Data Augmentation Techniques. In Proceedings of the IECON 2021—47th Annual Conference of the IEEE Industrial Electronics Society, Toronto, TO, Canada, 13–16 October 2021; Volume 2021. [CrossRef]

16. Asad, T.; Vaimann, A.; Belahcen, A.; Kallaste, A.; Rassõlkin, M.; Iqbal, M.N. The cluster computation-based hybrid fem–analytical model of induction motor for fault diagnostics. *Appl. Sci.* **2020**, *10*, 7572. [CrossRef]

17. Burnett, R.; Watson, J.F.; Elder, S. The detection and location of rotor faults within three phase induction motors. In Proceedings of the International Conference on Electrical Machines, Paris, France, 17–20 October 1994; pp. 288–293.

18. Biot-Monterde, V.; Navarro-Navarro, Á.; Antonino-Daviu, J.A.; Razik, H. Stray flux analysis for the detection and severity categorization of rotor failures in induction machines driven by soft-starters. *Energies* **2021**, *14*, 5757. [CrossRef]

19. Ceban, A.; Pusca, R.; Romary, R. Study of rotor faults in induction motors using external magnetic field analysis. *IEEE Trans. Ind. Electron.* **2012**, *59*, 2082–2093. [CrossRef]

20. Bellini, A.; Filippetti, F.; Franceschini, G.; Tassoni, C.; Kliman, G.B. Quantitative evaluation of induction motor broken bars by means of electrical signature analysis. *IEEE Trans. Ind. Appl.* **2001**, *37*, 1248–1255. [CrossRef]

21. Lee, S.B.; Shin, J.; Park, Y.; Kim, H.; Kim, J. Reliable Flux based Detection of Induction Motor Rotor Faults from the 5th Rotor Rotational Frequency Sideband. *IEEE Trans. Ind. Electron.* **2020**, *68*, 7874–7883. [CrossRef]

22. Jiang, C.; Li, S.; Habetler, T.G. A review of condition monitoring of induction motors based on stray flux. In Proceedings of the 2017 IEEE Energy Conversion Congress and Exposition (ECCE), Cincinnati, OH, USA, 1–5 October 2017; Volume 2017, pp. 5424–5430. [CrossRef]

23. Park, Y.; Lee, S.B.; Yun, J.; Sasic, M.; Stone, G.C. Air Gap Flux-Based Detection and Classification of Damper Bar and Field Winding Faults in Salient Pole Synchronous Motors. *IEEE Trans. Ind. Appl.* **2020**, *56*, 3506–3515. [CrossRef]

24. Henao, H.; Demian, C.; Capolino, G.A. A frequency-domain detection of stator winding faults in induction machines using an external flux sensor. *IEEE Trans. Ind. Appl.* **2003**, *39*, 1272–1279. [CrossRef]

25. Romary, R.; Roger, D.; Brudny, J.-F. Analytical computation of an AC machine external magnetic field. *EPJ Appl. Phys.* **2009**, *47*, 3. [CrossRef]

26. Bellini, A.; Concari, C.; Franceschini, G.; Tassoni, C.; Toscani, A. Vibrations, currents and stray flux signals to asses induction motors rotor conditions. In Proceedings of the IECON 2006-32nd Annual Conference on IEEE Industrial Electronics, Paris, France, 6–10 November 2006; Volume 2, pp. 4963–4968. [CrossRef]

27. Romary, R.; Pusca, R.; Lecointe, J.P.; Brudny, J.F. Electrical machines fault diagnosis by stray flux analysis. In Proceedings of the 2013 IEEE Workshop on Electrical Machines Design, Control and Diagnosis, WEMDCD 2013, Paris, France, 11–12 March 2013; pp. 247–256. [CrossRef]

28. Zamudio-Ramirez, I.; Osornio-Rios, R.A.A.; Antonino-Daviu, J.A.; Razik, H.; de Jesus Romero-Troncoso, R. Magnetic Flux Analysis for the Condition Monitoring of Electric Machines: A Review. *IEEE Trans. Ind. Inform.* **2021**, *3203*, 2895–2908. [CrossRef]

29. Goktas, T.; Arkan, M.; Mamis, M.S.; Akin, B. Separation of induction motor rotor faults and low frequency load oscillations through the radial leakage flux. In Proceedings of the 2017 IEEE Energy Conversion Congress and Exposition (ECCE), Cincinnati, OH, USA, 1–5 October 2017; Volume 2017, pp. 3165–3170. [CrossRef]

30. Song, X.; Liu, Z.; Yang, X.; Yang, J.; Qi, Y. Extended semi-supervised fuzzy learning method for nonlinear outliers via pattern discovery. *Appl. Soft Comput. J.* **2015**, *29*, 245–255. [CrossRef]

31. Jin, X.; Zhao, M.; Chow, T.W.S.; Pecht, M. Motor bearing fault diagnosis using trace ratio linear discriminant analysis. *IEEE Trans. Ind. Electron.* **2014**, *61*, 2441–2451. [CrossRef]

32. Camarena-Martinez, D.; Valtierra-Rodriguez, M.; Garcia-Perez, A.; Osornio-Rios, R.A.; Romero-Troncoso, R.D.J. Empirical mode decomposition and neural networks on FPGA for fault diagnosis in induction motors. *Sci. World J.* **2014**, *2014*, 908140. [CrossRef] [PubMed]

33. Martinez-Herrera, A.L.; Ferrucho-Alvarez, E.R.; Ledesma-Carrillo, L.M.; Mata-Chavez, R.I.; Lopez-Ramirez, M.; Cabal-Yepez, E. Multiple Fault Detection in Induction Motors through Homogeneity and Kurtosis Computation. *Energies* **2022**, *15*, 1541. [CrossRef]

34. Pasqualotto, D.; Navarro, A.N.; Zigliotto, M.; Antonino-Daviu, J.A. Automatic Detection of Rotor Faults in Induction Motors by Convolutional Neural Networks applied to Stray Flux Signals. In Proceedings of the 2021 22nd IEEE International Conference on Industrial Technology (ICIT), Valencia, Spain, 10–12 March 2021; Volume 2021, pp. 148–153. [CrossRef]
35. Lopez-Ramirez, M.; Ledesma-Carrillo, L.M.; Garcia-Guevara, F.M.; Munoz-Minjares, J.; Cabal-Yepez, E.; Villalobos-Pina, F.J. Automatic Early Broken-Rotor-Bar Detection and Classification Using Otsu Segmentation. *IEEE Access* **2020**, *8*, 112624–112632. [CrossRef]
36. Camarena-Martinez, D.; Perez-Ramirez, C.A.; Valtierra-Rodriguez, M.; Amezquita-Sanchez, J.P.; Romero-Troncoso, R.D.J. Synchrosqueezing transform-based methodology for broken rotor bars detection in induction motors. *Meas. J. Int. Meas. Confed.* **2016**, *90*, 519–525. [CrossRef]

Article

Design and Implementation of Input AC Filters and Predictive Control for Matrix-Converter Based PMSM Drive Systems

Tian-Hua Liu * and Jia-Han Li

Department of Electrical Engineering, National Taiwan University of Science and Technology, Taipei City 106335, Taiwan; m10807217@mail.ntust.edu.tw
* Correspondence: Liu@mail.ntust.edu.tw; Tel.: +886-2-2737-6678

Abstract: Matrix converters have many advantages, including high-efficiency, single-stage AC/AC energy conversion, bidirectional power flow, a near-unity input power factor, sinusoidal three-phase input currents, and sinusoidal three-phase output currents. However, matrix converters have 360 Hz voltage pulsations at the virtual DC-bus, which produce input harmonic currents and output harmonic currents, which cause unsatisfactory responses. To solve the problem of the input harmonic currents, a systematic design of an input three-phase current modulation method and an input three-phase AC filter that uses two different design methods are proposed. In addition, to improve dynamic responses, two predictive speed controllers are investigated and compared, and a predictive current controller is studied to reduce the output harmonic currents. A digital signal processor and an FPGA are used to execute the control algorithms. Several experimental results validate the theoretical analysis and show that the proposed methods effectively improve the power quality of the PMSM drive system and its input power-source quality.

Keywords: matrix-converter; input AC filter design; PMSM; predictive control

Citation: Liu, T.-H.; Li, J.-H. Design and Implementation of Input AC Filters and Predictive Control for Matrix-Converter Based PMSM Drive Systems. *Energies* **2022**, *15*, 748. https://doi.org/10.3390/en15030748

Academic Editor: Athanasios Karlis

Received: 24 December 2021
Accepted: 18 January 2022
Published: 20 January 2022

Publisher's Note: MDPI stays neutral with regard to jurisdictional claims in published maps and institutional affiliations.

1. Introduction

Matrix converters have simple and compact power circuits, which provide bidirectional power flow, sinusoidal input currents, sinusoidal output currents, a unity input power factor, and regeneration capabilities [1]. Recently, matrix converters have gained a lot of attention from researchers, and several have focused on improving the input currents for matrix converters. For example, Lei et al. proposed a damping control of matrix converter via modifying input reference currents by injecting damping signals. By using this method, the oscillations in input currents could be suppressed directly [2]. Sahoo et al. systematically designed an input filter for matrix converters by using an analytical estimation of root-mean-square current ripples. A step-by-step procedure was shown to determine the inductance parameter and capacitance parameter from the specifications of allowable total harmonic distortion in the input currents and voltages. In addition, a resistance parameter was determined to ensure that the filter had a minimum ohmic loss and a reasonable damping factor [3]. Orser et al. investigated using input filter capacitors as an energy storage device when the matrix converter was ridden through [4]. Dasgupta proposed a filter design for direct matrix converters, which focused on dynamic performance and reliable commutation [5]. Kume et al. studied an integrated filter, which reduced common-mode currents, and provided near sinusoidal output voltages. By using that integrated filter, the traditional R-L-C filter was eliminated [6]. Liu et al. investigated a modeling analysis and parameters design of an LC-filter. Those experimental results showed that the LC-filter-integrated quasi-Z-source network provided the necessary functions. As a result, the traditional input filter was eliminated [7].

In this paper, we propose two different approaches for designing the input R-L-C filter of a matrix converter. First, we use a step-by-step procedure to determine the inductance parameter, capacitance parameter, and resistance parameter. This proposed method has some advantages when it is compared to previous papers [3,4]. In the previously published paper [3], three equations with three coupling coefficients were used. As a result, a numeric solution obtained by using a computer simulation was required. To solve this problem, in this paper, we propose three equations in the first method. Each equation is related to only one or two parameters. As a result, the capacitance parameter, inductance parameter, and resistance can be directly solved by using the three equations and simple algebra. In addition, we use a transfer function to determine the required parameters of the R-L-C filter in the second method. After that, we compare the advantages and disadvantages of these two methods. Finally, experimental results demonstrate the effectiveness of the two different filter designs.

Besides dealing with input harmonic currents, the performance of the motor is important as well. Several researchers have concentrated on modulation methods and controller design of matrix converter-based permanent magnet synchronous motor (PMSM) drive systems. For example, Deng et al. proposed a direct torque control for matrix converter-based PMSM drive systems with minimized common-mode voltages [8]. Zhang et al. proposed a modified PI controller and a proportional resonant (PR) controller for matrix converters and then compared their steady-state tracking performance. However, the output of the matrix converter was connected to a three-phase resistor but not a three-phase AC motor [9]. Mubarok et al. implemented a matrix-converter-based IPMSM position control system, in which a model-free predictive current controller was used [10]. Siami et al. proposed a simplified finite control for matrix converter-based PMSM drive systems. By using that simplified method, the computation of the digital signal processor (DSP) was reduced [11]. Xia et al. investigated direct torque control of matrix converter-based PMSM drives by using duty cycles to reduce 30% of torque pulsations [12]. Khiem et al. proposed improving matrix converter-based PMSM drive systems by using an online detection and fault-tolerant switching strategy [13]. Friedli et al. compared the detailed performance of a three-phase AC-AC matrix converter-based PMM drive system and a DC-link voltage back-to-back converter-based PMSM drive system [14]. Furthermore, Di et al. investigated a novel predictive control method with an optimal switching sequence for a two-stage matrix converter [15]. Li et al. implemented a finite set model predictive control strategy for an indirect matrix converter [16]. Di et al. proposed a continuous control (predictive model control) strategy for an indirect matrix converter [17]. Dendouga designed second-order sliding-mode controllers for a direct matrix converter-based PMSM drive system [18]. Bu et al. designed output filters for a GaN-based matrix converter drive system [19]. Feng et al. investigated an improved model predictive control for matrix converters [20]. Orcioni et al. proposed a driving technique for a direct matrix converter based on a sigma-delta modulation technique [21]. Tuyen et al. implemented the space-vector modulation for an indirect matrix converter [22]. He et al. proposed a step-by-step design for a low-pass input filter for a single-stage converter [23].

However, these previous papers, which focused on matrix converter-based PMSM drive systems, only investigated one-step predictive control [10–13]. The main reason is that the DSP of a matrix converter-based drive system has to execute a four-step commutation, current-loop control, speed-loop control, and coordinate transformation. To fill this research gap, in our paper, an FPGA is used to execute the four-step commutation. In addition, a DSP is used to execute predictive current-control, one-step predictive speed-control, as well as two-step speed-control. Compared to traditional one-step predictive speed-control, the proposed two-step predictive speed-control provides more flexibility in determining the control input of the PMSM drive system and also has fewer steady-state errors than the one-step predictive control. To the authors' best knowledge, this comparison of the two methods to design the three-phase input filter of the matrix converter and the comparison of the two-step predictive speed-loop control and one-step predictive speed-loop control for a matrix converter PMSM drive system are original and have not been investigated by previously published papers [1–23]. These two methods and their comparison are the main contributions of this paper.

2. Indirect Matrix-Converter Control

Figure 1a shows the main circuit of the matrix converter. The indirect control of the matrix converter in this paper includes a three-phase input current control and a virtual inverter voltage control, which is shown in Figure 1b. The details are discussed as follows:

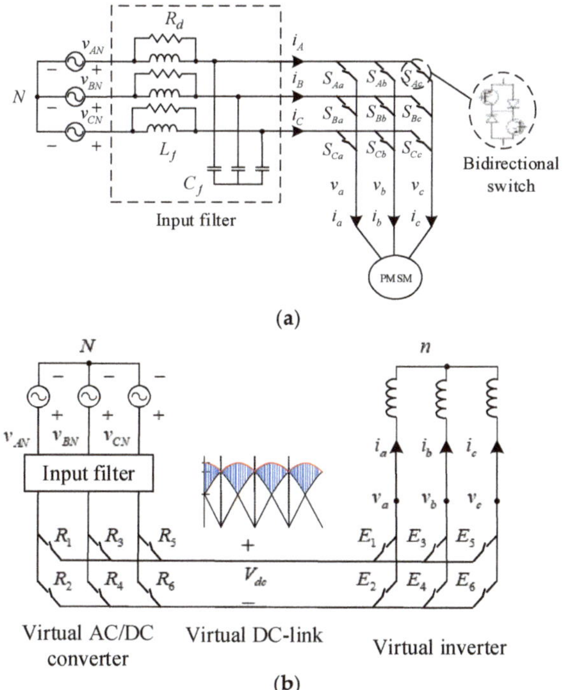

Figure 1. Matrix converter. (**a**) Main circuit, (**b**) equivalent circuit.

The relationship between input voltage and output voltage of the matrix converter in Figure 1a can be shown as the following:

$$\begin{bmatrix} v_a \\ v_b \\ v_c \end{bmatrix} = \begin{bmatrix} S_{Aa} & S_{Ab} & S_{Ac} \\ S_{Ba} & S_{Bb} & S_{Bc} \\ S_{Ca} & S_{Cb} & S_{Cc} \end{bmatrix} \begin{bmatrix} V_{AN} \\ V_{BN} \\ V_{CN} \end{bmatrix} \tag{1}$$

In addition, the switching states of the nine switches in Figure 1a and the equivalent switching states of the relative switches of the virtual AC/DC converter and inverter in Figure 1b are shown as follows:

$$\begin{bmatrix} S_{Aa} & S_{Ab} & S_{Ac} \\ S_{Ba} & S_{Bb} & S_{Bc} \\ S_{Ca} & S_{Cb} & S_{Cc} \end{bmatrix} = \begin{bmatrix} E_1 & E_2 \\ E_3 & E_4 \\ E_5 & E_6 \end{bmatrix} \begin{bmatrix} R_1 & R_3 & R_5 \\ R_2 & R_4 & R_6 \end{bmatrix} \tag{2}$$

2.1. Three-Phase Input Current Control

The desired three-phase input currents are shown as follows:

$$i_A = I_m \cos(\omega_g t) \tag{3}$$

$$i_B = I_m \cos\left(\omega_g t - \frac{2\pi}{3}\right) \tag{4}$$

and,

$$i_C = I_m \cos\left(\omega_g t + \frac{2\pi}{3}\right) \tag{5}$$

where i_A, i_B, and i_C are input three-phase currents from the input three-phase voltage source, I_m is the amplitude of the input three-phase currents, and ω_g is the angular frequency of the input three-phase voltages or currents. Figure 2 illustrates the space vector of the input current vector, which includes six sections based on the α-axis and β-axis coordinates. First, we assume the input current vector is I_{ref} and is between I_1 and I_6, which is shown in Figure 2. Then, the input current vector I_{ref} is expressed as follows:

$$\begin{aligned} I_{ref} &= \frac{t_\mu}{T_s} I_1 + \frac{t_v}{T_s} I_6 + \frac{t_0}{T_s} I_0 \\ &= d_\mu I_1 + d_v I_6 + d_0 I_0 \end{aligned} \tag{6}$$

and,

$$d_0 = \frac{t_0}{T_s} = 1 - (d_\mu + d_v) \tag{7}$$

where I_0 is the zero current vector; I_1 and I_6 are the active current vectors; t_μ, t_v, and t_0 are the time intervals of I_1, and I_0 individually; and d_u, d_v, and d_0 are the duty cycles of the current vectors I_1, I_6, and I_0 individually. In Figure 2, we can see that when the switches R_1 and R_6 from Figure 1 are turned on, the switching state of $I_1(A, C)$ is created. The other switching states can be expressed in the same way. Table 1 shows the relationship between the input current vectors and the switching states of the virtual AC/DC converter.

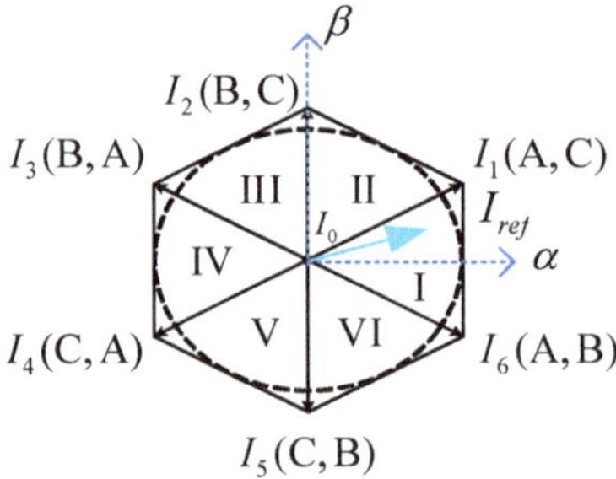

Figure 2. Input current vector.

Table 1. Relationship between input current vectors and switches.

Input Current Vectors	R_1	R_2	R_3	R_4	R_5	R_6
I_1	1	0	0	0	0	1
I_2	0	0	1	0	0	1
I_3	0	1	1	0	0	0
I_4	0	1	0	0	1	0
I_5	0	0	0	1	1	0
I_6	1	0	0	1	0	0
	1	1	0	0	0	0
I_0	0	0	1	1	0	0
	0	0	0	0	1	1

2.2. Output Voltage Control of Virtual Inverter

The desired three-phase output voltages are shown as follows:

$$v_a = V_{om}\cos(\omega_o t) \tag{8}$$

$$v_b = V_{om}\cos\left(\omega_o t - \frac{2\pi}{3}\right) \tag{9}$$

and

$$v_c = V_{om}\cos\left(\omega_o t + \frac{2\pi}{3}\right) \tag{10}$$

where v_a, v_b, and v_c are output three-phase voltages, and ω_o is the angular frequency of the output three-phase voltages of the virtual inverter. Figure 3 shows the eight space vectors of the output voltage vectors based on the α-axis and β-axis coordinates. In Figure 3, when the output voltage vector is v_{ref} and is located between V_1 and V_2, then the output voltage vector v_{ref} is shown as follows:

$$\begin{aligned} v_{ref} &= \tfrac{t_1}{T_s}V_1 + \tfrac{t_2}{T_s}V_2 + \tfrac{t_0}{T_s}V_0 \\ &= d_1 V_1 + d_2 V_2 + d_0 V_0 \end{aligned} \tag{11}$$

where V_0 is the zero voltage vector; V_1 and V_2 are the active voltage vectors; t_1, t_2, and t_0 are the time intervals of V_1, V_2, and V_0; and d_1, d_2, and d_0 are the duty cycles of the voltage vectors V_1, V_2, and V_0. The duty cycle of the zero vector is as follows:

$$d_0 = \frac{t_0}{T_s} = 1 - (d_1 + d_2) \tag{12}$$

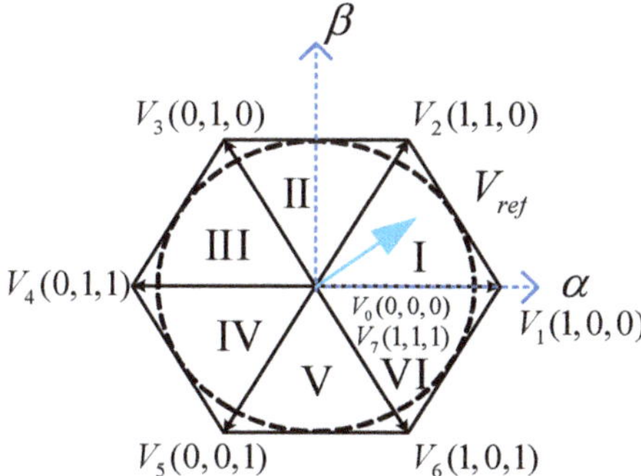

Figure 3. Output voltage vectors of virtual inverter.

The switching state of V_1 (1, 0, 0) means that when the switch of the virtual inverter E_1 turns on, E_2 turns off, and when E_4 turns on, E_3 turns off, and when E_6 turns on, E_5 turns off, all of which can be seen in Figure 1b. The other switching states can be expressed in the same way.

3. Input Filter Design

A systematic design procedure of the input filter for a matrix converter is described below.

3.1. Method 1

Several papers have investigated the optimal design method of input filters for AC/AC matrix converters [24]. In this paper, by using this systematic design procedure, the parameter λ_1 is used to determine the ratio of the input harmonic currents to the input fundamental currents. Then, the parameter λ_2 is used to determine the ratio of the input harmonic voltages to the input rms voltages. Finally, the λ_3 is used to determine the ratio of the power loss of the R-L-C filter to the rated output power of the matrix converter. The details are discussed as follows.

Step 1: Determine the input current harmonics

In the first step, the input harmonic currents are determined. Then, the input rms value of the fundamental currents of the matrix converter, which also includes i_A, i_B, and i_C, shown in Figure 1a, can be calculated as follows [3]:

$$I_{in1\_RMS} = \frac{3}{2} m_I \cdot m_V \cdot I_{o1\_RMS} \cdot \cos \phi_o \tag{13}$$

where I_{in1_RMS} is the input rms value of the fundamental currents of the matrix converter, m_I is the modulation index of the input current vectors of the matrix converter, m_V is the voltage modulation index of the virtual inverter, I_{o1_RMS} is the output rms value of the fundamental current of the matrix converter that includes i_a, i_b, and i_c shown in Figure 2a, and ϕ_o is the phase angle between the output voltages and output currents. In Equation (13), the output rms currents of the fundamental currents of the matrix converter, I_{o1_RMS}, can be expressed as follows:

$$I_{o1\_RMS} = \frac{P_o}{3V_{g\_RMS} \cos \phi_o} \tag{14}$$

where P_o is the rated output power of the matrix converter and V_{g_RMS} is the rms value of the fundamental voltages of the input voltage sources. After that, the value of parameter λ_1 can be determined by the designer and is shown as follows:

$$\lambda_1 = \frac{I_{gsw\_RMS}}{I_{in1\_RMS}} \tag{15}$$

where I_{gSW_RMS} is the rms value of the switching harmonic currents of the input currents, which includes i_A, i_B, and i_C, and I_{in1_RMS} is the input rms value of the fundamental currents of the input currents of the matrix converter. From Equation (15), we can obtain:

$$I_{gsw\_RMS} = \lambda_1 I_{in1\_RMS} \tag{16}$$

The ratio of the rms value of the switching harmonic currents of the input currents to the switching harmonic currents of the matrix converter, I_{insw_RMS}, which is related to I_{in1_RMS}, can be determined by the designer. Finally, the relationship between the input harmonic currents and output harmonic currents of the R-L-C filter can be shown as follows [6]:

$$\frac{I_{gSW\_RMS}}{I_{insw\_RMS}} = \frac{1}{\sqrt{1 + \dfrac{(1 - \omega_s^2 L_f C_f)^2 - 1}{(1 + \dfrac{\omega_s^2 L_f^2}{R_d^2})}}} \tag{17}$$

where ω_s is the switching frequency, and L_f, C_f, and R_d are the inductance, capacitance, and resistance of the R-L-C filter. In this paper, the parameters of L_f and R_d are directly obtained from Equations (17) and (21) by using an analytic method without a computer.

Step 2: Determine the input harmonic voltages

In the second step, the input harmonic voltages are determined. First, the ratio between the input harmonic voltages, V_{in_ripple}, and the input rms line voltages, $V_{in\_line\_rms}$, is described as follows:

$$\lambda_2 = \frac{V_{in\_ripple}}{V_{in\_line\_rms}} \tag{18}$$

where V_{in_ripple} is the input harmonic voltages and $V_{in\_line\_rms}$ is the input rms line voltages. After that, the capacitance of the input R-L-C filter, C_f, is shown as follows [7]:

$$C_f = \frac{I_o}{\pi \omega_s V_{ripple}} \sin \pi D \tag{19}$$

where D is the turned-on duty cycles of each switch in the matrix converter.

Step 3: Determine the ratio of the filter power loss to the rated output power

First, the λ_3 is determined by the designer to obtain the ratio of the filter power loss to the rated output power as follows:

$$\lambda_3 = \frac{P_{loss}}{P_o} \tag{20}$$

After doing some mathematical processes, one can obtain the following equation [6]:

$$\lambda_3 = \frac{I_{in1\_RMS}}{V_{g\_RMS}} \left(\frac{\omega_g^2 L_f^2 R_d}{\omega_g^2 L_f^2 + R_d} \right) \tag{21}$$

As a result, from Equations (16) and (21), one can obtain the unique solution of the L_f, which is the inductance of the R-L-C filter and the R_d, which is the resistance of the R-L-C filter. The single-phase equivalent R-L-C filter is shown in Figure 4, which includes the switching RMS voltages, the fundamental RMS voltages, the switching RMS currents, and the fundamental RMS currents.

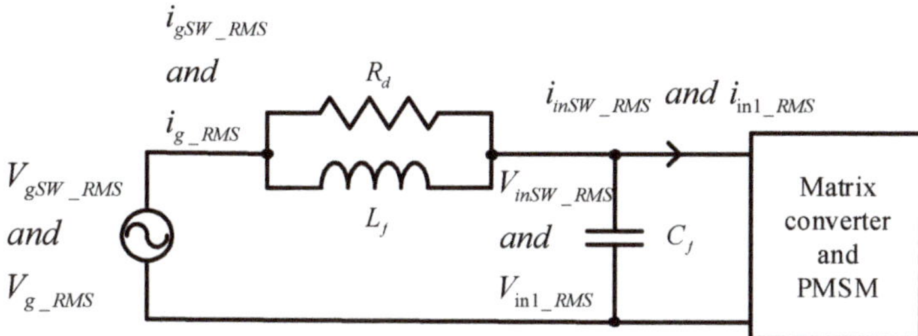

Figure 4. The single-phase equivalent circuit of the input filter for the matrix converter.

3.2. Method 2

The second method uses a frequency domain to design the R-L-C filter. The details are described below.

Generally speaking, the R-L-C filter resonant frequency is less than $1/3$ of the switching frequency of the matrix converter, and over 20 times greater than the fundamental frequency. The relationship is as follows:

$$20 f_g \leq f_{res} \leq \frac{1}{3} f_s \tag{22}$$

where f_g is the frequency of the input voltage source, f_{res} is the resonant frequency of the filter, and f_s is the switching frequency of the matrix converter. The capacitor C_f of the filter causes a phase shift between the input currents and input voltages. Therefore, the capacitor C_f has to be smaller than the allowed maximum capacitor that causes the allowed maximum phase angle θ_{max}. This relationship is expressed as follows [24]:

$$C_f < C_f{}^{max} \tag{23}$$

and,

$$C_f{}^{max} = \frac{I_{in1\_RMS}}{V_{g\_RMS} \omega_g} \tan \theta_{max} \tag{24}$$

where $C_f{}^{max}$ is the allowed maximum capacitor, and θ_{max} is the allowed maximum phase angle. The resonant frequency is then defined as:

$$f_{res} = \frac{1}{2\pi\sqrt{L_f C_f}} \tag{25}$$

From Equation (25), one can derive the following equation:

$$L_f = \frac{1}{C_f \cdot (2\pi f_{res})^2} \tag{26}$$

In this paper, the resistance is in parallel with the inductance. The transfer function of the second-order system is shown in Figure 5a and is as follows:

$$\frac{I_{g\_RMS}(s)}{I_{in1\_RMS}(s)} = \frac{s\frac{L_f}{R_d}+1}{s^2 L_f C_f + s\frac{L_f}{R_d}+1}$$
$$= \frac{s\frac{\omega_{res}}{Q_{res}}+\omega_{res}^2}{s^2+2\zeta\omega_{res}s+\omega_{res}^2} \tag{27}$$

In Equation (27), the related parameters are as follows:

$$\omega_{res} = \frac{1}{\sqrt{L_f C_f}} \tag{28}$$

$$Q_{res} = R_d\sqrt{\frac{C_f}{L_f}} \tag{29}$$

and,

$$\zeta = \frac{1}{2R_d}\sqrt{\frac{L_f}{C_f}} \tag{30}$$

where ω_{res} is the resonant frequency in rad/s, Q_{res} is the quality factor, and ζ is the damping ratio. Figure 5b illustrates the relationship of magnitude and frequency in a Bode diagram, and Figure 5c illustrates the relationship of the phase angle and frequency in a Bode diagram for a typical R-L-C filter.

Although the first method requires more complicated computation processes, it obtains the parameters of the R-L-C filter via the THD of the real currents and voltages. The second method quickly determines the parameters of the R-L-C filter; however, it is difficult to estimate the THD of the input current. The major reason is that the second method focuses on frequency-domain responses but not harmonic current or voltage constraints.

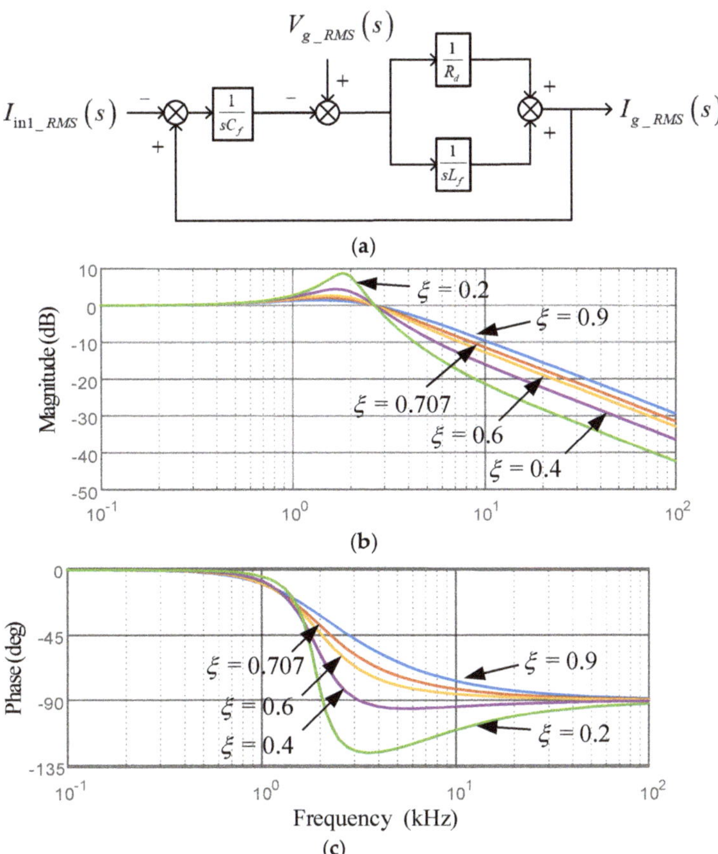

Figure 5. Three-phase R-L-C filter using method 2. (**a**) Equivalent diagram, (**b**) amplitude response, (**c**) phase response.

4. Predictive Controller Design

4.1. One-Step Predictive Speed Controller Design

The discrete form of the dynamic speed equation for a PMSM is shown as follows [25]:

$$\hat{\omega}_m(k+1) = e^{-\frac{B_m T_s}{J_m}} \omega_m(k) + \frac{1}{B_m}\left(1 - e^{-\frac{B_m T_s}{J_m}}\right) T_e(k) \tag{31}$$

where $\hat{\omega}_m(k+1)$ is the predictive speed at the $(k+1)$th sampling interval, T_s is the sampling interval of the speed-loop control, J_m is the inertia, B_m is the viscous coefficient, and T_e is the electromagnetic torque. The electromagnetic torque is calculated as follows:

$$T_e(k) = \frac{3}{2}\frac{P}{2}\lambda_m i_q(k) = K_T i_q(k) \tag{32}$$

where P is the pole number, λ_m is the flux linkage from the permanent magnet of each pole on the rotor, K_T is the torque constant, and $i_q(k)$ is the q-axis current. To simplify the dynamic speed equation of the PMSM, Equation (31) can be rewritten as follows:

$$\hat{\omega}_m(k+1) = a_m \omega_m(k) + b_m i_q(k) \tag{33}$$

The parameters a_m and b_m in Equation (33) are expressed as follows:

$$a_m = e^{-\frac{B_m T_s}{J_m}} \tag{34}$$

and,

$$b_m = \frac{K_T}{B_m}(1 - e^{-\frac{B_m T_s}{J_m}}) \tag{35}$$

From Equation (33), one can derive the following equation:

$$\omega_m(k) = a_m \omega_m(k-1) + b_m i_q(k-1) \tag{36}$$

Subtracting (36) from (33), one can obtain:

$$\Delta\omega_m(k+1) = a_m \Delta\omega_m(k) + b_m \Delta i_q(k) \tag{37}$$

and then the estimated speed of the $(k + 1)$th interval is:

$$\begin{aligned}\hat{\omega}_m(k+1) &= \omega_m(k) + \Delta\omega_m(k+1)\\ &= \omega_m(k) + a_m \Delta\omega_m(k) + b_m \Delta i_q(k)\end{aligned} \tag{38}$$

Then the performance index is defined as follows [25]:

$$J_s = (\omega_m^*(k+1) - \hat{\omega}_m(k+1))^2 + q\Delta i_q^2(k) \tag{39}$$

where $\widehat{\omega}_m(k+1)$ is the predictive speed at the $(k + 1)$th sampling interval and q is the weighting factor. By taking $\frac{\partial J_p(k)}{\partial i_q(k)} = 0$, one can obtain the following:

$$\Delta i_q^*(k) = \frac{b_s \omega_m^*(k+1) - a_s b_s \Delta\omega_m(k) - b_s \omega_m(k)}{b_s^2 + q} \tag{40}$$

Finally, the q-axis current command can be shown as follows:

$$i_q^*(k) = i_q(k-1) + \Delta i_q^*(k) \tag{41}$$

From Equations (40) and (41), one can obtain the block diagram of the one-step predictive speed control which is shown in Figure 6. In this paper, a more complicated two-step predictive speed control has been investigated and compared to the one-step predictive speed control, which is discussed in Equation (31) to (41). The particulars of the two-step predictive speed controller are as follows:

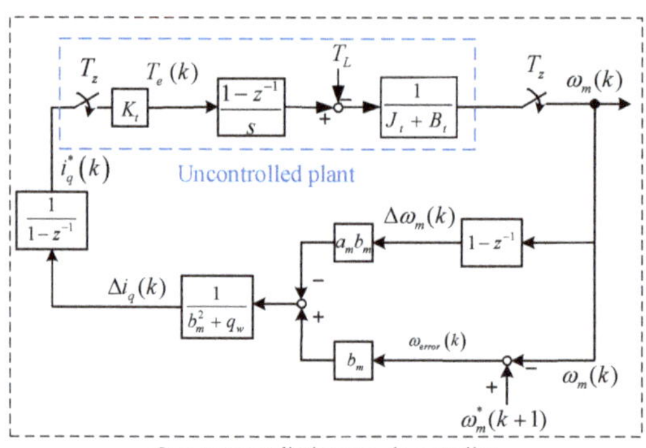

Figure 6. Block diagram of one-step predictive speed control.

4.2. Two-Step Predictive Speed Controller Design

For a deeper investigation, a two-step predictive speed controller is also investigated in this paper. When we discuss the two-step predictive speed controller, the predictive speed $\hat{\omega}_m(k+1)$ and the predictive speed $\hat{\omega}_m(k+2)$ are both considered.

First, from Equation (38), the predictive speed $\hat{\omega}_m(k+1)$ is developed. Then, one can develop the predictive speed $\hat{\omega}_m(k+2)$ as follows:

$$
\begin{aligned}
\Delta\omega_m(k+2) &= a_m\Delta\omega_m(k+1) + b_m\Delta i_q(k+1) \\
&= a_m[a_m\Delta\omega_m(k) + b_m\Delta i_q(k)] + b_m\Delta i_q(k+1) \\
&= a_m{}^2\Delta\omega_m(k) + a_m b_m\Delta i_q(k) + b_m\Delta i_q(k+1)
\end{aligned}
\tag{42}
$$

Then, the estimated $(k+2)$th speed is as follows:

$$
\begin{aligned}
\widehat{\omega}_m(k+2) &= \widehat{\omega}_m(k+1) + \Delta\omega_m(k+2) \\
&= \widehat{\omega}_m(k+1) + a_m{}^2\Delta\omega_m(k) + a_m b_m\Delta i_q(k) + b_m\Delta i_q(k+1) \\
&= \omega_m(k) + a_m\Delta\omega_m(k) + b_m\Delta i_q(k) + a_m{}^2\Delta\omega_m(k) + a_m b_m\Delta i_q(k) + b_m\Delta i_q(k+1)
\end{aligned}
\tag{43}
$$

Combing Equations (42) and (43), we can derive the following equation:

$$
\begin{aligned}
Y_{sm} &= \begin{bmatrix} \widehat{\omega}_m(k+1) \\ \widehat{\omega}_m(k+2) \end{bmatrix} \\
&= \begin{bmatrix} a_m & 1 \\ a_m + a_m{}^2 & 1 \end{bmatrix} \begin{bmatrix} \Delta\omega_m(k) \\ \omega_m(k) \end{bmatrix} + \begin{bmatrix} b_m & 0 \\ b_m(1+a_m) & b_m \end{bmatrix} \begin{bmatrix} \Delta i_q(k) \\ \Delta i_q(k+1) \end{bmatrix} \\
&= F_{sm}X_{sm} + \theta_{sm}\Delta U_{sm}
\end{aligned}
\tag{44}
$$

with,

$$
F_{sm} = \begin{bmatrix} a_m & 1 \\ a_m + a_m{}^2 & 1 \end{bmatrix}
\tag{45}
$$

and,

$$
\theta_{sm} = \begin{bmatrix} b_m & 0 \\ b_m(1+a_m) & b_m \end{bmatrix}
\tag{46}
$$

Next, the performance index is defined as follows:

$$
J_{sp} = [Y_{sm}{}^* - Y_{sm}]^T[Y_{sm}{}^* - Y_{sm}] + \Delta U_{sm}{}^T\eta\Delta U_{sm}
\tag{47}
$$

where $Y_{sm}{}^*$, η, and ΔU_{sm} are defined as follows:

$$
Y_{sm}{}^* = \begin{bmatrix} \omega_m{}^*(k+1) \\ \omega_m{}^*(k+2) \end{bmatrix}
\tag{48}
$$

In Equation (48), $Y_{sm}{}^*$ is the vector that includes the first-step speed command and the second-step speed command. The weighting matrix η is:

$$
\eta = \begin{bmatrix} \eta_w & 0 \\ 0 & \eta_w \end{bmatrix}
\tag{49}
$$

where η is the weighting matrix and η_w is the weighting factor for the first and second q-axis predictive difference-currents. The control input currents at sampling intervals k and $k+1$ are:

$$\Delta U_{sm} = \begin{bmatrix} \Delta i_q(k) \\ \Delta i_q(k+1) \end{bmatrix} \tag{50}$$

where ΔU_{sm} are the control input currents at sampling intervals k and $k+1$. After that, by taking the partial difference, one can obtain the following equation:

$$\frac{\partial J_{sp}}{\partial \Delta U_{sm}} = \left(-2\theta_{sm}{}^T(Y_m{}^* - F_{sm}X_{sm}(k)) + 2(\theta_{sm}{}^T\theta_{sm} + \eta)\Delta U_{sm}(k) \right) = 0 \tag{51}$$

From (51), the optimal predictive control input can be expressed as follows:

$$\Delta U_{sm} = \left(\theta_{sm}{}^T\theta_{sm} + \eta \right)^{-1} \left(\theta_{sm}{}^T Y_{sm}{}^*(k) - \theta_{sm}{}^T F_{sm}X_{sm}(k) \right) \tag{52}$$

The relative results are as follows:

$$\theta_{sm}{}^T\theta_{sm} + \eta = \begin{bmatrix} a_m^2 b_m^2 + 2a_m b_m^2 + 2b_m^2 + \eta_w & a_m b_m{}^2 + b_m{}^2 \\ a_m b_m{}^2 + b_m{}^2 & b_m{}^2 + \eta_w \end{bmatrix} \tag{53}$$

$$
\begin{aligned}
\theta_{sm}{}^T Y_m{}^* &= \begin{bmatrix} b_m & b_m(1+a_m) \\ 0 & b_m \end{bmatrix} \begin{bmatrix} \omega_m{}^*(k+1) \\ \omega_m{}^*(k+2) \end{bmatrix} \\
&= \begin{bmatrix} b_m\omega_m{}^*(k+1) + b_m(1+a_m)\omega_m{}^*(k+2) \\ b_m\omega_m{}^*(k+2) \end{bmatrix}
\end{aligned}
\tag{54}
$$

and,

$$
\begin{aligned}
\theta_{sm}{}^T F_{sm}X_{sm}(k) &= \begin{bmatrix} b_m & b_m(1+a_m) \\ 0 & b_m \end{bmatrix} \begin{bmatrix} a_m & 1 \\ a_m + a_m{}^2 & 1 \end{bmatrix} \begin{bmatrix} \Delta\omega_m(k) \\ \omega_m(k) \end{bmatrix} \\
&= \begin{bmatrix} [2(a_m b_m + a_m^2 b_m)\Delta\omega_m(k) + [b_m + b_m(1+a_m)]\omega_m(k) \\ b_m(a_m + a_m{}^2)\Delta\omega_m(k) + b_m\omega_m(k) \end{bmatrix}
\end{aligned}
\tag{55}
$$

Substituting (53), (54), and (55) into (52), one can derive the following equations:

$$
\begin{aligned}
\Delta U_{sm}(\mathrm{k}) = \begin{bmatrix} \Delta i_q(k) \\ \Delta i_q(k+1) \end{bmatrix} &= \frac{\begin{bmatrix} 2b_m{}^2 + \eta_w & -a_m b_m{}^2 - b_m^2 \\ -a_m b_m{}^2 - b_m^2 & a_m^2 b_m^2 + 2a_m^2 b_m^2 + 2b_s^2 + \eta_w \end{bmatrix}}{b_m{}^4 + \eta_w{}^2 + b_m^2\eta_w\left(a_m^2 + 2a_m + 3\right)} \\
\begin{bmatrix} b_m\omega_m{}^*(k+1) + b_m(1+a_m)\omega_m{}^*(k+2) - 2(a_m b_m + a_m^2 b_m)\Delta\omega_m(k) - [b_m + b_m(1+a_m)]\omega_m(k) \\ b_m\omega_m{}^*(k+2) - [b_m(a_m + a_m{}^2)]\Delta\omega_m(k) - b_m\omega_m(k) \end{bmatrix} \\
&= \frac{1}{Z}\begin{bmatrix} \Psi + \Theta \\ \Phi + N \end{bmatrix}
\end{aligned}
\tag{56}
$$

and,

$$Z = b_m{}^4 + \eta_w{}^2 + b_m^2\eta_w\left(a_m^2 + 2a_m + 3\right) \tag{57}$$

$$\Psi = \left(2b_m^2 + \eta_w\right)\left\{ b_m\omega_m{}^*(k+1) + b_m(1+a_m)\omega_m{}^*(k+2) \\ -2\left(b_m a_m + b_m a_m{}^2\right)\Delta\omega_m(k) - [b_m + b_m(1+a_m)]\omega_m(k)\right\} \tag{58}$$

$$\Theta = \left[-a_m b_m{}^2 - b_m^2\right]\left\{ b_m\omega_m{}^*(k+2) - b_m(a_m + a_m{}^2)\Delta\omega_m(k) - b_m\omega_m(k)\right\} \tag{59}$$

$$\Phi = \left[-a_m b_m{}^2 - b_m^2\right]\left\{ b_m\omega_m{}^*(k+1) + b_m(1+a_m)\omega_m{}^*(k+2) \\ -2\left(b_m a_m + b_m a_m{}^2\right)\Delta\omega_m(k) - [b_m + b_m(1+a_m)]\omega_m(k)\right\} \tag{60}$$

and,

$$N = \left[a_m{}^2 b_m{}^2 + 2b_m^2 + 2a_m b_m{}^2 + \eta_w\right]\left\{ b_m\omega_m{}^*(k+2) - b_m(a_m + a_m{}^2)\Delta\omega_m(k) - b_m\omega_m(k)\right\} \tag{61}$$

Next, the q-axis current command at the k-th sampling interval is as follows:

$$i_q^*(k) = i_q^*(k-1) + \Delta i_q(k) \tag{62}$$

By using the same method, the $(k+1)$th q-axis current command is shown in the following equation:

$$i_q^*(k+1) = i_q^*(k) + \Delta i_q(k+1) \tag{63}$$

The output q-axis current command sends out only one value for each sampling interval. As a result, we can combine Equations (62) and (63). Then, the final q-axis current using the two-step predictive speed controller is defined as follows:

$$i_q{}^*(k)_{-2step} = \rho i_q{}^*(k) + (1-\rho)i_q{}^*(k+1) \tag{64}$$

From Equation (56) to (64), we can obtain the block diagram of the two-step predictive speed control. Figure 7 shows the block diagram of two-step predictive speed control system.

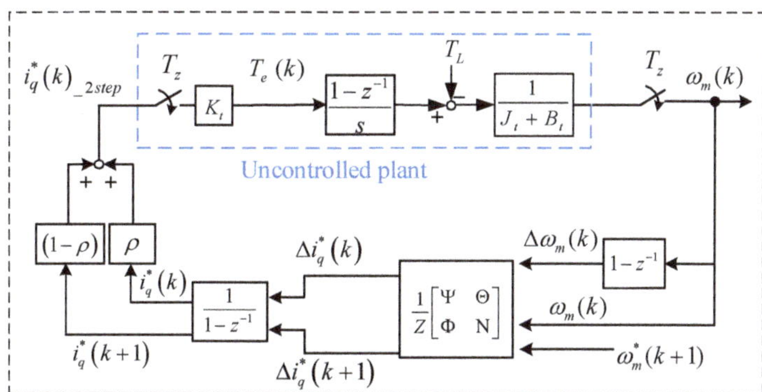

Figure 7. Block diagram of two-step predictive speed control.

5. Predictive Current Controller Design

The predictive current controller is developed by using a similar method as the predictive speed controller. First, the differential equations of the d-axis and q-axis currents are as follows:

$$\frac{d}{dt}i_d = \frac{1}{L_d}\left(v_d - r_s i_d + \omega_e L_q i_q\right) \tag{65}$$

and,

$$\frac{d}{dt}i_q = \frac{1}{L_q}\left(v_q - r_s i_q + \omega_e(L_d i_d + \lambda_m)\right) \tag{66}$$

By inserting these two zero-order hold devices into the d-q axis currents and then taking the z-transformation, we obtain the following equation:

$$\begin{bmatrix} i_d(k+1) \\ i_q(k+1) \end{bmatrix} = \begin{bmatrix} a_d & 0 \\ 0 & a_q \end{bmatrix}\begin{bmatrix} i_d(k) \\ i_q(k) \end{bmatrix} + \begin{bmatrix} b_d & 0 \\ 0 & b_q \end{bmatrix}\begin{bmatrix} v_d(k) + \omega_e(k)L_q i_q(k) \\ v_q(k) - \omega_e(k)(L_d i_d(k) + \lambda_m) \end{bmatrix} \tag{67}$$

The parameters a_d, a_q, b_d, and b_q are as follows:

$$a_d = e^{-\frac{r_s T_c}{L_d}} \tag{68}$$

$$b_d = \frac{1}{r_s}\left(1 - e^{-\frac{r_s}{L_d}T_c}\right) \tag{69}$$

$$a_q = e^{-\frac{r_s T_c}{L_q}} \tag{70}$$

and,

$$b_q = \frac{1}{r_s}\left(1 - e^{-\frac{r_s}{L_q}T_c}\right) \tag{71}$$

Next, we can define the control input $u_d(k)$ and $u_q(k)$ as the following two equations:

$$u_d(k) = v_d(k) + \omega_e(k)L_q i_q(k) \tag{72}$$

and,

$$u_q(k) = v_q(k) - \omega_e(k)(L_d i_d(k) + \lambda_m(k)) \tag{73}$$

Substituting Equations (72) and (73) into (67), we can obtain a new and simplified state-variable presentation equation as follows:

$$\begin{bmatrix} x_d(k+1) \\ x_q(k+1) \end{bmatrix} = \begin{bmatrix} a_d & 0 \\ 0 & a_q \end{bmatrix}\begin{bmatrix} x_d(k) \\ x_q(k) \end{bmatrix} + \begin{bmatrix} b_d & 0 \\ 0 & b_q \end{bmatrix}\begin{bmatrix} u_d(k) \\ u_q(k) \end{bmatrix} \tag{74}$$

where $x_d(k)$ is the new state variable of $i_d(k)$, and $x_q(k)$ is the new state variable of $i_q(k)$. Equation (74) can then be rewritten as the new state-variable vector presentation as the following equations:

$$X_{cm}(k+1) = A_{cm}X_{cm}(k) + B_{cm}U_{cm}(k) \tag{75}$$

and,

$$X_{cm}(k) = \begin{bmatrix} x_d(k) \\ x_q(k) \end{bmatrix} \tag{76}$$

$$A_{cm} = \begin{bmatrix} a_d & 0 \\ 0 & a_q \end{bmatrix} \tag{77}$$

$$B_{cm} = \begin{bmatrix} b_d & 0 \\ 0 & b_q \end{bmatrix} \tag{78}$$

and,

$$U_{cm}(k) = \begin{bmatrix} u_d(k) \\ u_q(k) \end{bmatrix} \tag{79}$$

The new output equation of the state-variable vector presentation is as follows:

$$Y_{cm} = \begin{bmatrix} y_d(k) \\ y_q(k) \end{bmatrix} = \begin{bmatrix} 1 & 0 \\ 0 & 1 \end{bmatrix}\begin{bmatrix} x_d(k) \\ x_q(k) \end{bmatrix} \tag{80}$$

Then, we can define the difference of the state variable as follows:

$$\begin{aligned} \Delta X_{cm}(k+1) &= X_{cm}(k+1) - X_{cm}(k) \\ &= A_{cm}\Delta X_{cm}(k) + B_{cm}\Delta U_{cm}(k) \end{aligned} \tag{81}$$

where $\Delta X_{cm}(k)$ is the difference of the state variables, and $\Delta U_{cm}(k)$ is the difference of the output variables. After that, we can define the augmented state variables as follows:

$$X_m(k+1) = \begin{bmatrix} \Delta X_{cm}(k+1) \\ Y_{cm}(k+1) \end{bmatrix} = A_m X_m(k) + B_m \Delta U_{cm}(k) \tag{82}$$

with:

$$A_m = \begin{bmatrix} A_{cm} & 0_{cm}^T \\ A_{cm} & 1 \end{bmatrix} \tag{83}$$

$$X_m(k) = \begin{bmatrix} \Delta X_{cm}(k) \\ Y_{cm}(k) \end{bmatrix} \tag{84}$$

and,

$$B_m = \begin{bmatrix} B_{cm} \\ B_{cm} \end{bmatrix} \tag{85}$$

Next, we can define the output $Y_m(k+1)$ of the augmented model as follows:

$$\begin{aligned} Y_m(k+1) &= \begin{bmatrix} 0_{cm}^T & 1 \end{bmatrix} X_m(k+1) \\ &= C_m X_m(k+1) \end{aligned} \tag{86}$$

The performance index of the current-loop controllers is defined as follows [26]:

$$J_c = (X_{cm}^*(k+1) - Y_m(k+1))^2 + q\Delta U_{cm}^2(k) \tag{87}$$

where q is the weighting factor. By taking the differential of the performance index to $\Delta U_{cm}(k)$, and then by assuming the result to be zero, one can derive the following optimal difference control input $\Delta U_{cm}^*(k)$ as follows:

$$\begin{aligned} \Delta U_{cm}^*(k) = & \left(B_{cm}^2 + q\right)^{-1} [B_{cm} X_{cm}^*(k+1) \\ & - A_{cm} B_{cm} \Delta X_{cm}(k) - B_{cm} Y_{cm}(k)] \end{aligned} \tag{88}$$

From Equation (88), the $\Delta v_d^*(k)$ and $\Delta v_q^*(k)$ can be expressed as the following equations:

$$\begin{aligned} \Delta v_d^*(k) = & \frac{b_d\left(i_d^*(k+1)-i_d(k)\right)-a_d b_d \Delta i_d(k)}{b_d^2+q} \\ & -\Delta \omega_e(k) L_q \Delta i_q(k) \end{aligned} \tag{89}$$

and,

$$\begin{aligned} \Delta v_q^*(k) = & \frac{b_q\left(i_q^*(k+1)-i_q(k)\right)-a_q b_q \Delta i_q(k)}{b_q^2+q} \\ & -\Delta \omega_e(k)\left(L_d \Delta i_d(k) + \lambda_m\right) \end{aligned} \tag{90}$$

Finally, from Equations (89) and (90), we can develop the block diagrams of the d-axis current controller and the q-axis current controller as in Figure 8a,b.

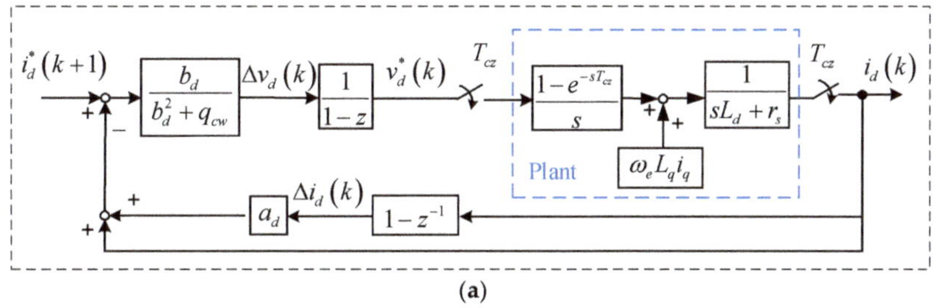

(a)

Figure 8. *Cont.*

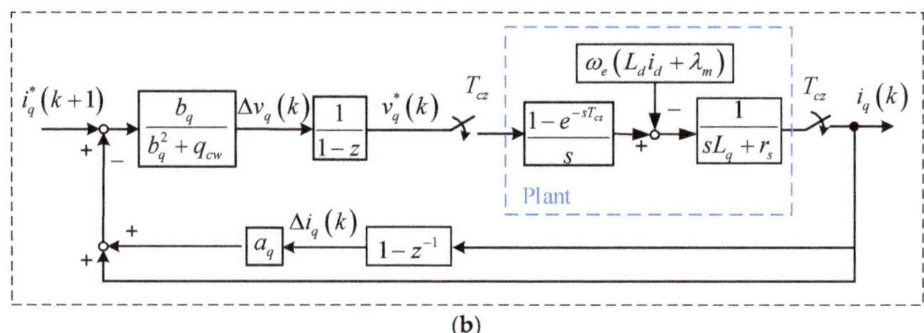

(b)

Figure 8. Block diagram of the predictive current control. (**a**) *d*-axis current control, (**b**) *q*-axis current control.

6. Implementation

A digital signal processor (DSP), type SH 7237 (manufactured by Renesas Electronics Corporation, Tokyo, Japan), and an FPGA, type 10M16SAU16917G (manufactured by Intel Corporation, Santa Clara, CA, USA), were used to execute the control algorithms. The sampling interval of the speed-loop was 1 ms, and the sampling interval of the current-loop was 100 μs. The switching frequency of the matrix converter was 10 kHz. The PMSM was an 8-pole motor, which had a 2000 r/min rated speed, 9.55 N·m of rated torque, 9 A of rated current, 0.58 Ω of stator resistance, 1.3 mH of *d*-axis inductance, 1.7 mH of *q*-axis inductance, 0.003 N·m·s/rad of inertia, and a 1.14 N·m/A torque constant. The three-phase input filter of method 1 had the following parameters: R_d = 15 Ω, L_f = 1.5 mH, and C_f = 6.8 μF. In addition, the three-phase input filter of method 2 had the following parameters: R_d = 22 Ω, L_f = 1.5 mH, and C_f = 4.7 μF. Figure 9a shows the software and hardware block diagrams of the control system. Figure 9b shows the circuits for a matrix converter, including a clamped circuit, some drivers, a matrix converter, an AC/DC power supply, and a control circuit.

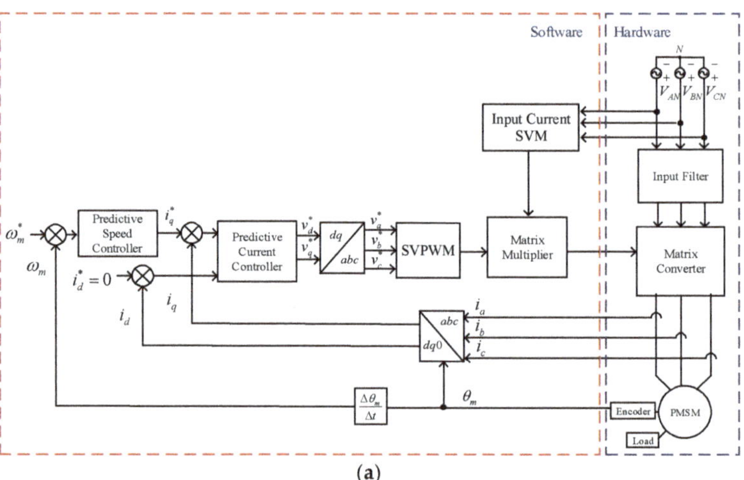

(a)

Figure 9. *Cont.*

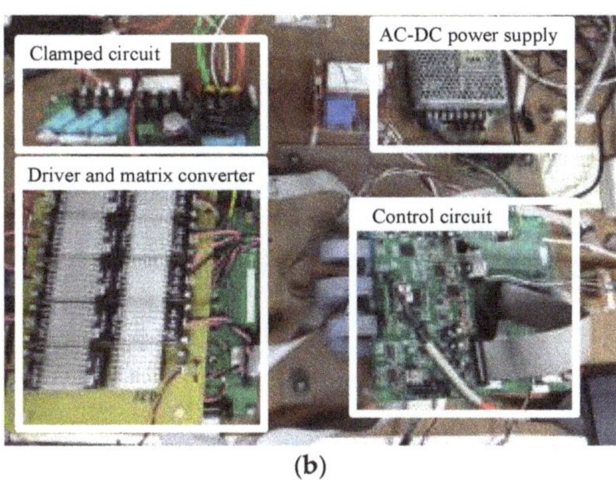

(b)

Figure 9. The implemented matrix-converter PMSM drive system. (**a**) Block diagram, (**b**) photo of matrix-converter.

7. Experimental Results

Several experimental results are shown here. Figure 10a demonstrates the measured input A-phase current of the matrix converter without using an input filter. The input A-phase current is close to a square-wave high-frequency PWM current, which has a 132.14% THD. Two simplified three-phase input AC filter design methods were proposed without using computer simulations. Figure 10c demonstrates the measured A-phase current using the proposed method 1 of the three-phase input AC filter, in which the parameters include $\lambda_1 = 0.11$, $\lambda_2 = 0.019$, and $\lambda_3 = 0.0006$. The measured A-phase current is nearly a sinusoidal waveform with a 9.55% THD. Figure 10e demonstrates the measured A-phase current using the proposed method 2 input three-phase AC filter, in which the parameters include $\omega_{res} = 12{,}570$ rad/s, $Q_{res} = 1.23$, and $\xi = 0.4$. The A-phase current using method 2 includes a 12.08% THD. As we can observe, the current without using an input filter has the highest THD. The major reason is that the high-frequency PWM current creates a lot of harmonic currents. The proposed method 1 of the three-phase input AC filter design provides lower THD than the proposed method 2. The major reason is that method 1 focuses on harmonic currents and voltages; method 2, however, focuses on frequency responses.

Figure 11a demonstrates the measured output currents of the a-phase matrix converter using a PI current controller, which produces an 11.91% THD. Figure 11c demonstrates the measured output currents of the a-phase matrix converter using a predictive current controller, which has a 9.25% THD, which is lower than the THD of the PI current controller.

Figure 12a illustrates the measured 300 r/min step-input speed responses by using a PI controller, a one-step predictive speed controller, and a two-step predictive speed controller. The PI controller is designed by pole assignment with two major poles $P_1 = -10.6 + j7.5$ and $P_2 = -10.6 - j7.5$. As we can observe, the PI controller provides the highest overshoot among the three different speed controllers. The one-step predictive speed controller, which chooses a weighting factor $q = 0.25$, has the quickest transient response when compared to the two-step predictive speed controller and the PI controller. The two-step predictive speed controller, which chooses a weighting factor $\eta = \begin{bmatrix} 0.25 & 0 \\ 0 & 0.25 \end{bmatrix}$ and $\rho = 0.5$, has the lowest overshoot but the slowest transient response when compared to the PI controller and the one-step predictive speed controller. Figure 12b illustrates the load disturbance responses at 300 r/min with a 2 N·m external load. The one-step predictive controller provides the smallest speed drop and the fastest recovery time relative to the PI controller and the two-step predictive controller. However, the two-step predictive controller provides fewer steady-state errors than the PI controller and the one-step predictive speed controller.

Figure 13a demonstrates the measured q-axis current response by using the PI controller, which has a higher overshoot and slower response than the one-step predictive current controller. Figure 13b demonstrates the measured q-axis current response by using the one-step predictive controller. As can be observed, the one-step predictive current controller performs better than the PI controller again, including faster transient responses and lower overshoot.

Table 2 shows the comparison of the input harmonic currents. Method 1 has fewer input harmonic currents than method 2. Table 3 shows the comparison of the speed responses. The one-step predictive speed controller provides quicker transient responses and quicker recovery time. However, the two-step predictive speed controller provides fewer speed drops and smaller overshoots than any other controller.

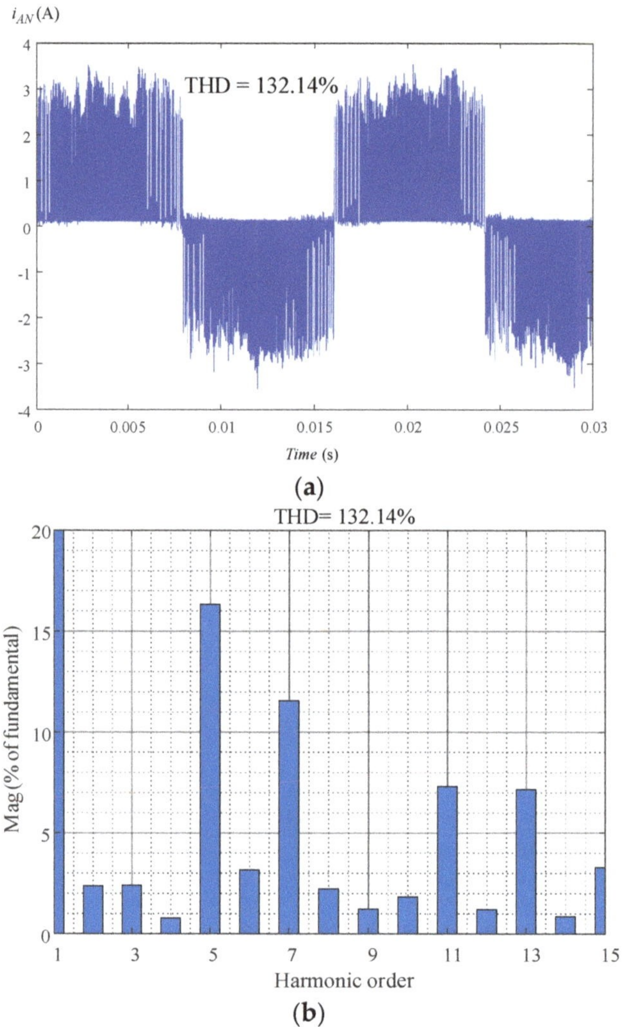

Figure 10. *Cont.*

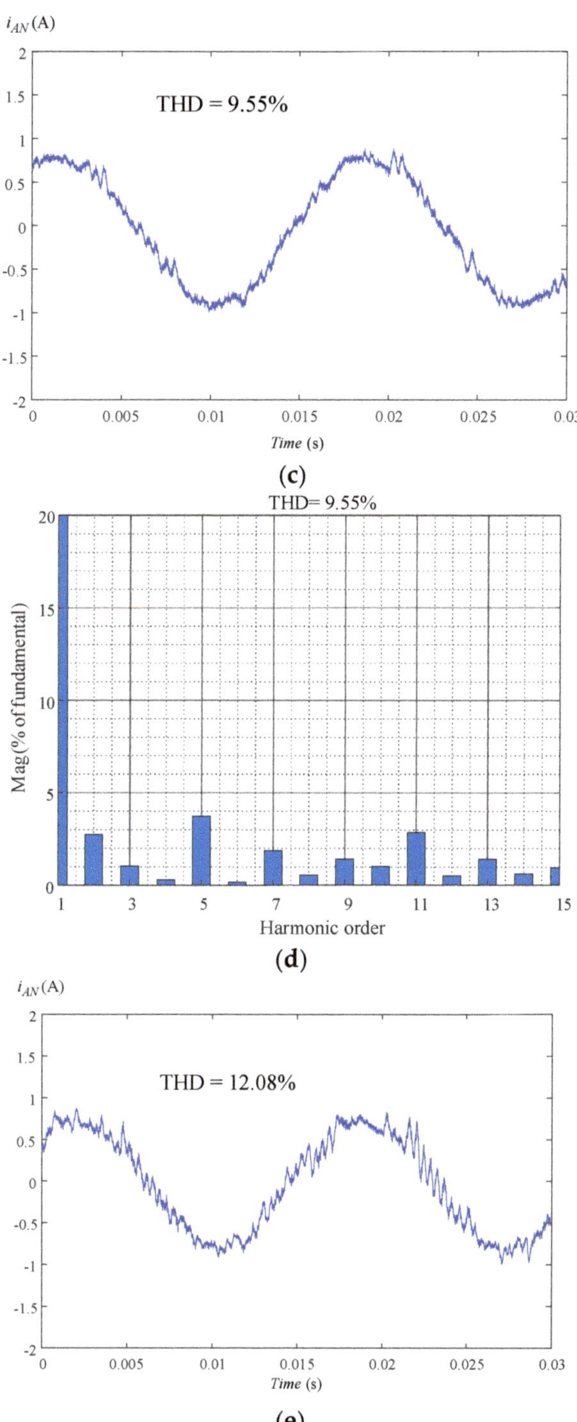

Figure 10. *Cont.*

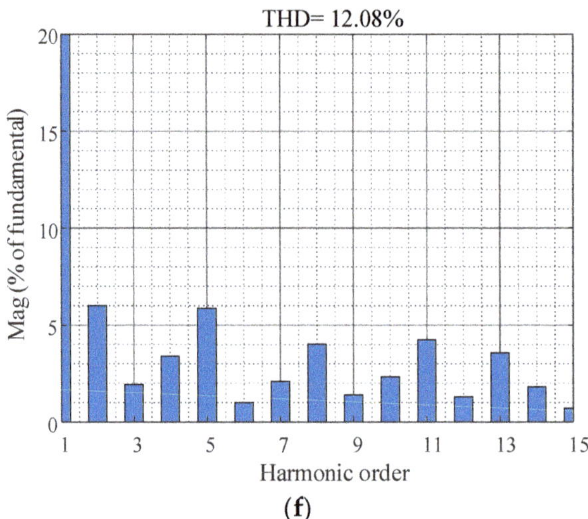

(**f**)

Figure 10. Measured input waveforms (**a**) i_{AN} without filter, (**b**) THD without filter, (**c**) i_{AN} with method 1. (**d**) THD method 1, (**e**) i_{AN} with method 2, (**f**) THD with method 2.

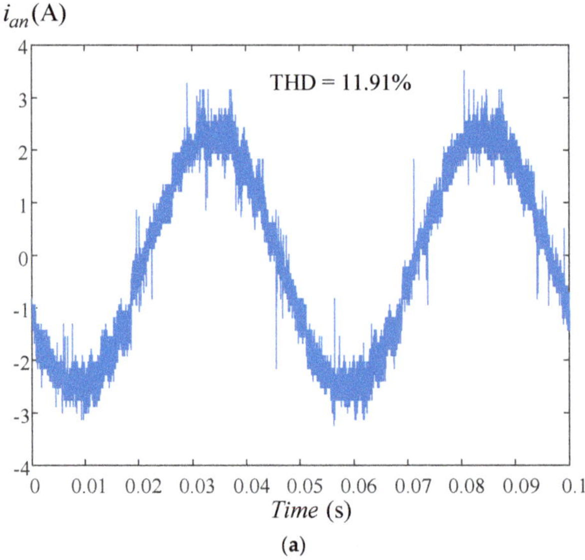

(**a**)

Figure 11. *Cont.*

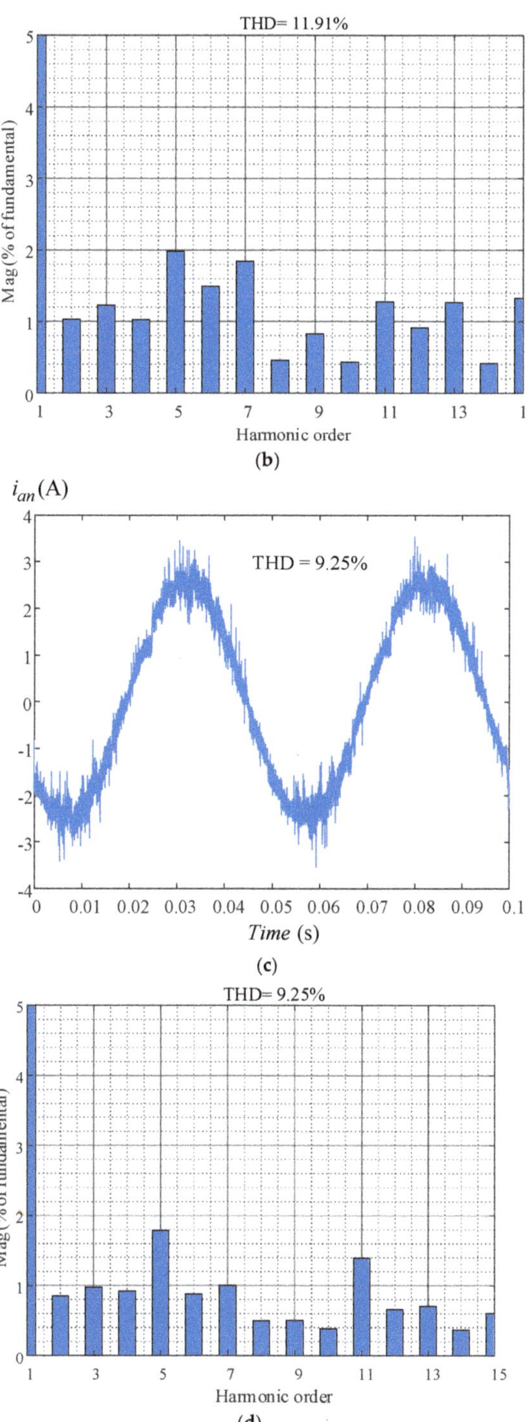

Figure 11. Measured a-phase output currents of matrix-converter using different current controllers. (**a**) PI, (**b**) THD, (**c**) predictive, (**d**) THD.

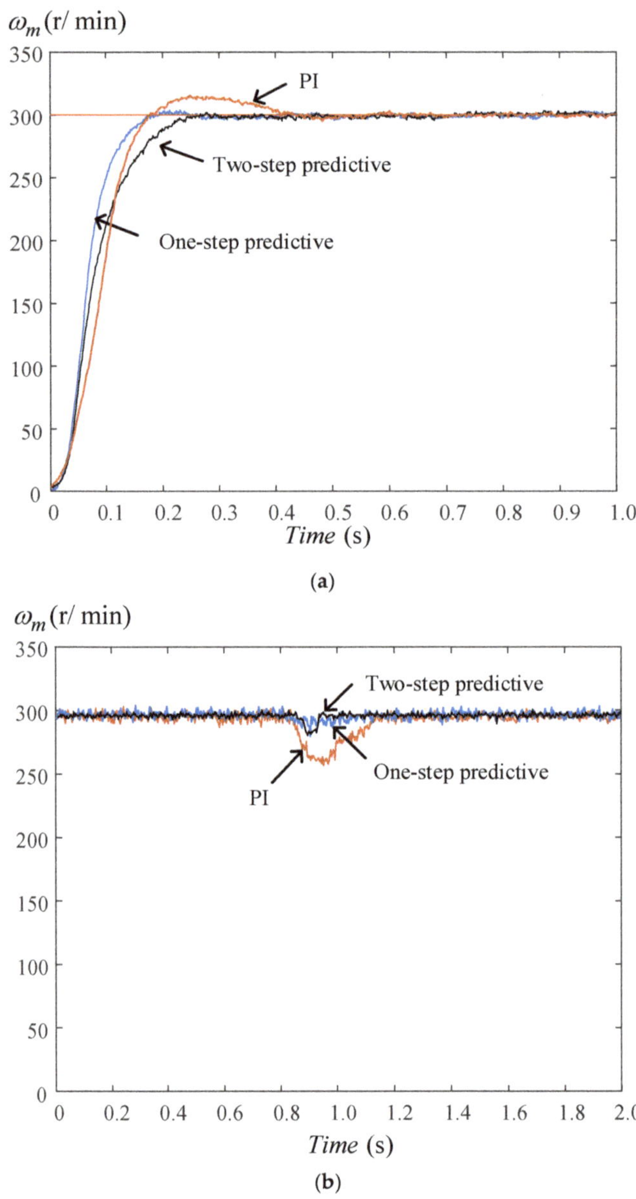

Figure 12. Measured speed responses. (**a**) Step-input responses, (**b**) load-disturbance responses.

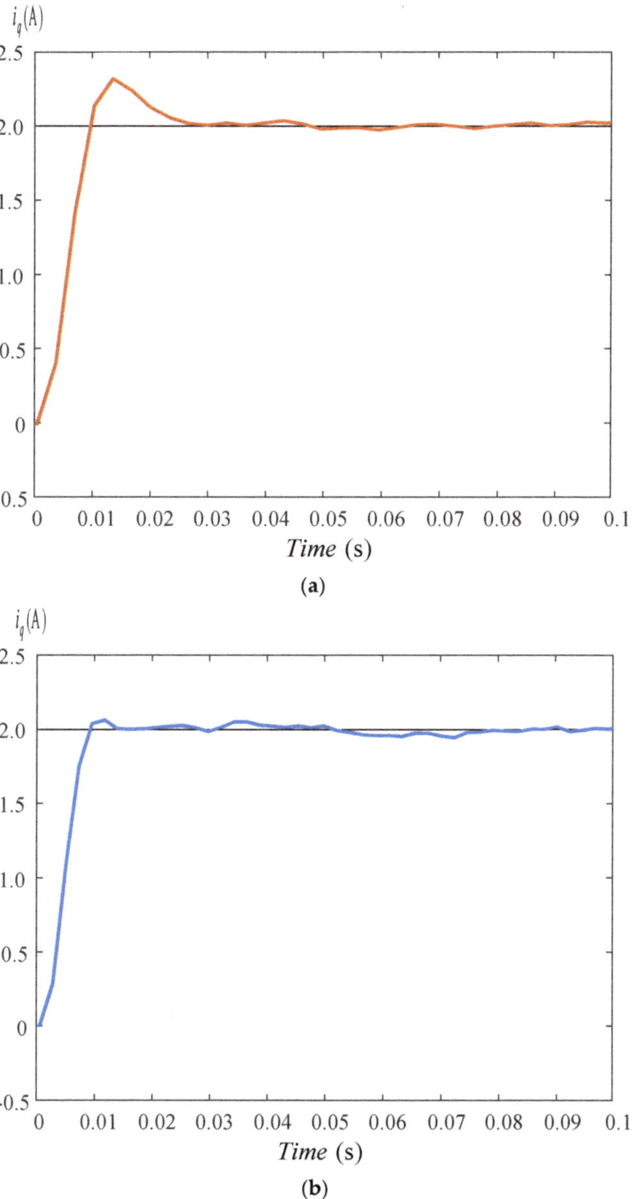

Figure 13. Measured current responses. (**a**) PI current controller, (**b**) predictive current controller.

Table 2. Comparison of input harmonic currents.

	5th Harmonic	7th Harmonic	THD
Without input filters	16.26%	11.61%	132.14%
Method 1	3.73%	1.87%	9.55%
Method 2	5.87%	2.09%	12.08%

Table 3. Comparison of speed responses.

Command	Controller	PI	One-Step Predictive	Two-Step Predictive
Step speed command 300 r/min	Rise time	0.15 s	0.13 s	0.17 s
	Settling time	0.4 s	0.24 s	0.26 s
	Overshoot	6.33%	2%	1%
	Steady state error	±2 r/min	±2 r/min	±1 r/min
Load-disturbance 2 N-m	Recovery time	0.28 s	0.17 s	0.2 s
	Speed drop	48 r/min	24 r/min	23 r/min

8. Conclusions

In this paper, two different design methods, which are simpler than the traditional numeric methods, using a computer for a three-phase input AC filter for matrix-converter PMSM-drive systems, are investigated and compared. The first method requires only analytic processes, which is simpler than the traditional numeric method using a computer. The second method uses frequency responses to determine the R-L-C parameters of the AC filter. In addition, a two-step predictive speed controller and a one-step predictive speed controller are investigated to improve the dynamic responses of speed-loop control systems. Moreover, a predictive current controller is designed to provide smaller harmonic currents than a PI current controller. Experimental results show that the proposed methods can effectively improve the performance of matrix converter-based PMSM drive systems, including obvious improvements in the input AC harmonic currents, output AC harmonic currents, and dynamic responses.

Author Contributions: Conceptualization, T.-H.L.; methodology, J.-H.L.; software, J.-H.L.; validation, T.-H.L.; formal analysis, T.-H.L. and J.-H.L.; investigation, T.-H.L. and J.-H.L.; resources, T.-H.L.; data curation, J.-H.L.; writing—original draft preparation, T.-H.L. and J.-H.L.; writing—review and editing, T.-H.L.; visualization, J.-H.L.; supervision, T.-H.L.; project administration, T.-H.L.; funding acquisition, T.-H.L. All authors have read and agreed to the published version of the manuscript.

Funding: This research was supported by Ministry of Science and Technology, under Grant MOST-110-2221-E-011-086.

Conflicts of Interest: The authors declare no conflict of interest.

References

1. Rodriguez, J.; Rivera, M.; Kolar, J.W.; Wheeler, P. A Review of Control and Modulation Methods for Matrix Converters. *IEEE Trans. Ind. Electron.* **2011**, *59*, 58–70. [CrossRef]
2. Lei, J.; Zhou, B.; Qin, X.; Wei, J.; Bian, J. Active damping control strategy of matrix converter via modifying input reference currents. *IEEE Trans. Power Electron.* **2014**, *30*, 5260–5271. [CrossRef]
3. Sahoo, A.K.; Basu, K.; Mohan, N. Systematic Input Filter Design of Matrix Converter by Analytical Estimation of RMS Current Ripple. *IEEE Trans. Ind. Electron.* **2014**, *62*, 132–143. [CrossRef]
4. Oser, D.; Mohan, N. A matrix converter ride-through configuration using input filter capacitors as an energy exchange mechanism. *IEEE Trans. Power Electron.* **2015**, *30*, 4377–4385. [CrossRef]
5. Dasgupta, A.; Sensarma, P. Filter Design of Direct Matrix Converter for Synchronous Applications. *IEEE Trans. Ind. Electron.* **2014**, *61*, 6483–6493. [CrossRef]
6. Kume, T.; Yamada, K.; Higuchi, T.; Yamamoto, E.; Hara, H.; Sawa, T.; Swamy, M.M. Integrated filters and their combined effects in matrix converter. *IEEE Trans. Ind. Appl.* **2007**, *43*, 571–581. [CrossRef]
7. Liu, S.; Ge, B.; Liu, Y.; Haitham, A.R.; Balog, R.S.; Sun, H. Modeling, analysis, and parameters design of LC-filter-integrated quasi-Z-source indirect matrix converter. *IEEE Trans. Power Electron.* **2016**, *31*, 7544–7555. [CrossRef]
8. Deng, W.; Li, S. Direct Torque Control of Matrix Converter-Fed PMSM Drives Using Multidimensional Switching Table for Common-Mode Voltage Minimization. *IEEE Trans. Power Electron.* **2020**, *36*, 683–690. [CrossRef]
9. Zhang, J.; Li, L.; Dorrell, D.G.; Guo, Y. Modified PI controller with improved steady-state performance and comparison with PR controller on direct matrix converters. *Chin. J. Electr. Eng.* **2019**, *5*, 53–66. [CrossRef]
10. Mubarok, M.S.; Liu, T.-H. Implementation of Predictive Controllers for Matrix-Converter-Based Interior Permanent Magnet Synchronous Motor Position Control Systems. *IEEE J. Emerg. Sel. Top. Power Electron.* **2018**, *7*, 261–273. [CrossRef]
11. Siami, M.; Khaburi, D.A.; Rodriguez, J. Simplified Finite Control Set-Model Predictive Control for Matrix Converter-Fed PMSM Drives. *IEEE Trans. Power Electron.* **2017**, *33*, 2438–2446. [CrossRef]

12. Xia, C.; Zhao, J.; Yan, Y.; Shi, T. A Novel Direct Torque Control of Matrix Converter-Fed PMSM Drives Using Duty Cycle Control for Torque Ripple Reduction. *IEEE Trans. Ind. Electron.* **2013**, *61*, 2700–2713. [CrossRef]
13. Nguyen-Duy, K.; Liu, T.-H.; Chen, D.-F.; Hung, J.Y. Improvement of Matrix Converter Drive Reliability by Online Fault Detection and a Fault-Tolerant Switching Strategy. *IEEE Trans. Ind. Electron.* **2011**, *59*, 244–256. [CrossRef]
14. Friedli, T.; Kolar, J.W.; Rodriguez, J.; Wheeler, P. Comparative Evaluation of Three-Phase AC–AC Matrix Converter and Voltage DC-Link Back-to-Back Converter Systems. *IEEE Trans. Ind. Electron.* **2011**, *59*, 4487–4510. [CrossRef]
15. Di, Z.; Xu, D.; Tarisciotti, L.; Wheeler, P. A Novel Predictive Control Method with Optimal Switching Sequence and Filter Resonance Suppression for Two-Stage Matrix Converter. *Energies* **2021**, *14*, 3652. [CrossRef]
16. Li, S.; Wang, Z.; Yan, Y.; Shi, T. Finite Set Model Predictive Control of a Dual-Motor Torque Synchronization System Fed by an Indirect Matrix Converter. *Energies* **2021**, *14*, 1325. [CrossRef]
17. Di, Z.; Xu, D.; Zhang, K. Continuous Control Set Model Predictive Control for an Indirect Matrix Converter. *Energies* **2021**, *14*, 4114. [CrossRef]
18. Dendouga, A. Conventional and Second Order Sliding Mode Control of Permanent Magnet Synchronous Motor Fed by Direct Matrix Converter: Comparative Study. *Energies* **2020**, *13*, 5093. [CrossRef]
19. Bu, H.; Cho, Y. GaN-based matrix converter design with output filters for motor friendly drive system. *Energies* **2020**, *13*, 971. [CrossRef]
20. Feng, S.; Wei, C.; Lei, J. Reduction of Prediction Errors for the Matrix Converter with an Improved Model Predictive Control. *Energies* **2019**, *12*, 3029. [CrossRef]
21. Orcioni, S.; Biagetti, G.; Crippa, P.; Falaschetti, L. A Driving Technique for AC-AC Direct Matrix Converters Based on Sigma-Delta Modulation. *Energies* **2019**, *12*, 1103. [CrossRef]
22. Tuyen, N.D.; Dzung, P.Q. Space Vector Modulation for an Indirect Matrix Converter with Improved Input Power Factor. *Energies* **2017**, *10*, 588. [CrossRef]
23. He, Q.; Liu, L.; Qiu, M.; Luo, Q. A Step-by-Step Design for Low-Pass Input Filter of the Single-Stage Converter. *Energies* **2021**, *14*, 7901. [CrossRef]
24. Trentin, A.; Zanchetta, P.; Clare, J.; Wheeler, P. Automated Optimal Design of Input Filters for Direct AC/AC Matrix Converters. *IEEE Trans. Ind. Electron.* **2011**, *59*, 2811–2823. [CrossRef]
25. Rodriguez, J.; Cortes, P. *Predictive Control of Power Converters and Electrical Drives*; John Wiley & Sons, Ltd.: West Sussex, UK, 2021.
26. Wang, L. *Model Predictive Control System Design and Implementation Using MATLAB*; Springer: London, UK, 2009.

Article

Investigation of Factors Affecting Partial Discharges on Epoxy Resin: Simulation, Experiments, and Reference on Electrical Machines

Dimosthenis Verginadis [1],*, Athanasios Karlis [1], Michael G. Danikas [1] and Jose A. Antonino-Daviu [2]

[1] Department of Electrical & Computer Engineering, Democritus University of Thrace (DUTh), 671 00 Xanthi, Greece; akarlis@ee.duth.gr (A.K.); mdanikas@ee.duth.gr (M.G.D.)
[2] Instituto Tecnologico de la Energia, Universitat Politècnica de València, 46022 Valencia, Spain; joanda@die.upv.es
* Correspondence: dvergina@ee.duth.gr

Abstract: In Power Systems, Synchronous Generators (SGs) are mostly used for generating electricity. Their insulation system, of which epoxy resin is a core component, plays a significant role in reliable operation. Epoxy resin has high mechanical strength, a characteristic that makes it a very good material for reliable SG insulation. Partial Discharges (PDs) are a constant threat to this insulation since they cause deterioration and consequential degradation of the aforementioned material. Therefore, it is very important to detect PDs, as they are both a symptom of insulation deterioration and a means to identify possible faults. Offline and Online PDs Tests are described, and a MATLAB/Simulink model, which simulates the capacitive model of PDs, is presented in this paper. Moreover, experiments are carried out in order to examine how the flashover voltage of epoxy resin samples is affected by different humidity levels. The main purpose of this manuscript is to investigate factors, such as the applied voltage, number, and volume of water droplets and water conductivity, which affect the condition of epoxy resin, and how these are related to PDs and flashover voltages, which may appear also in electrical machines' insulation. The aforementioned factors may affect the epoxy resin, resulting in an increase in PDs, which in turn increases the overall Electrical Rotating Machines (EMs) risk factor.

Keywords: electrical machines; insulation system; partial discharges; capacitive model; MATLAB/Simulink; flashover voltage; epoxy resin

Citation: Verginadis, D.; Karlis, A.; Danikas, M.G.; Antonino-Daviu, J.A. Investigation of Factors Affecting Partial Discharges on Epoxy Resin: Simulation, Experiments, and Reference on Electrical Machines. *Energies* **2021**, *14*, 6621. https://doi.org/10.3390/en14206621

Academic Editor: Lorand Szabo

Received: 7 September 2021
Accepted: 10 October 2021
Published: 14 October 2021

Publisher's Note: MDPI stays neutral with regard to jurisdictional claims in published maps and institutional affiliations.

1. Introduction

Electrical Rotating Machines (EMs) and especially Synchronous Generators (SGs) constitute the main source of electricity production. Therefore, SGs must be reliable, must have as long a lifespan as possible, and must operate reliably [1–5]. One of the main reasons for the failure of SGs is their stator insulation system, which has three main insulation levels [2–6]:

- Ground wall insulation;
- Insulation between turns/strands;
- Semi-conductive coating.
- The insulation system of the High Voltage (HV) EMs consists of [6–9]:
- mica, which is used in order to minimize the Partial Discharge (PD) activity;
- supporting material for proper mechanical strength;
- resin that fills the voids between mica and the supporting material.

The insulation system is subjected to different stresses acting alone or in combination with each other. The stator winding is one of the most stressed components of a generator. If a winding fault occurs during operation, the consequential damages to the machine can lead to significant outage times. These stresses are Thermal, Electrical, various Ambient factors, and Mechanical (TEAM), which create faults, such as delaminations and cracks and these can lead to the breakdown and overall damage of the insulation system. More specifically, PDs appear to be one of the major problems of the EMs' insulation system. If high PD magnitudes occur over a long period, deterioration of the EMs' insulation system may ensue. PDs are small electric spikes, which take place within air-filled cavities of the insulation or on its surface. PDs occur because the breakdown strength of air is much lower than the strength of the surrounding solid insulation. PDs cause small current and voltage pulses, while they are due to charges bombarding the cavity walls. These events may lead to insulation erosion. One could say that PDs are both a symptom of insulation deterioration and a means for the identification of possible faults. PDs may cause accelerated insulation ageing. Air cavities or other delaminations due to TEAM stresses increase the possibility of PD creation and can lead to insulation system deterioration due to chemical degradation and physical attack by nitrogen ions [10–15].

Epoxy resin is widely used and is proven to be a very important material achieving reliable and safe EM operation without undesirable shutdowns or additional maintenance. It helps EMs to perform adequately with reduced maintenance. Moreover, epoxy resin has high mechanical strength, good physical and electrical properties, high chemical resistance, adhesiveness, and low shrink ability, and it can be combined with other insulating materials, such as mica. Furthermore, epoxy resin is resistant to moisture and radiation; however, it may be affected by PDs and flashover voltages [2,16].

The present paper deals with the factors that may affect the proper operation and condition of epoxy resin and are related to PDs and flashover voltages. One significant goal of this manuscript is to investigate whether the presence of humidity leads to an increase or decrease in PD activity and to pinpoint the factors that lead to this result. Research works are ambiguous. Some mention that humidity leads to an increase in PD activity while others a decrease. Another important factor that this manuscript deals with is the investigation of which factors, such as applied voltage, water conductivity, and the geometry of enclosed voids, affect epoxy resin. Epoxy resin is one of the most significant insulation materials and is also widely employed in the SG. Technical data, PD measurements, and PD activity of the SG are described below. A combination of simulation and experimental results will lead to conclusions for all of the aforementioned factors and an overall view of the way that epoxy resin reacts to them.

2. Partial Discharges (PDs)

PDs are electrical discharges, which partially bridge the insulation between conductors [17]. They include internal discharges, slot discharges, and end-winding discharges.

2.1. PDs on EMs

SG are exposed to various defects in the insulation system. More specifically, TEAM [3–6] stresses are responsible for the deterioration of SGs' insulation. PDs usually occur as a result of the various defects, as described above. PDs in EMs cause typical patterns, the analysis of which can identify the origin of different PDs [18–21]. PDs can be categorized as [11]:

- Internal PDs, which occur on windings and can be caused due to poor manufacturing and design;
- Slot PDs, which occur on slot windings and are the most severe, and their cause can be poor manufacturing and improper maintenance;
- End-winding PDs, the cause of which can also be poor design and a lack of maintenance.

Different faults and problems of electrical erosion occurred from the presence of PDs in a SG, with a rated power of 202 MVA and a rated voltage of 15.75 kV at 50 Hz, located in a power plant in Greece. The stator insulation type is epoxy resin. Its degradation and aging due to electrical activity, and especially PDs, increasingly worsen with time.

Figure 1 shows a serious degradation of the insulation system of the stator bars due to mechanical and electrical stresses, both of which were caused by PDs that occurred during the EM operation time. The results of inspections and tests showed that the loosening of bars inside the slot is still at an initial stage except for some weak bars located in the bottom area of the stator core already at a critical stage. Given the above circumstances, preventative and/or corrective maintenance actions were performed in order to prevent the worsening of the EM condition. The standard conductive resin injection was chosen in order to tighten the bars inside their slots, to restore a proper contact between the insulation surface and core slot walls, and to inhibit further mechanical erosion and PD activity. These actions are shown in Figures 2–4. It must be noted that Figures 1–4 were captured during real on-site maintenance actions on the aforementioned SG.

Figure 1. Electrical Erosion (2012)—Stator Bars.

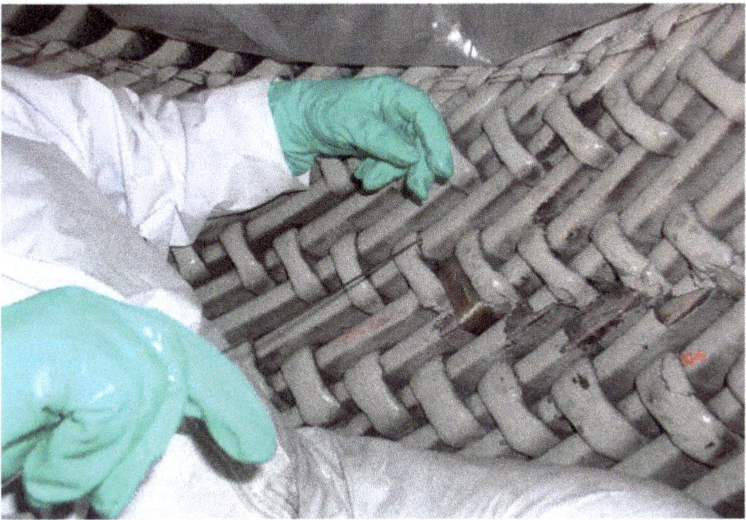

Figure 2. Restating ties with glass roving and resin.

Figure 3. Restored ties with glass roving and resin.

Figure 4. Painting of the restored area with resin.

The PDs were inner PDs arising from stress within the voids of the main insulation, surface PDs on the surfaces of the end-windings, and PDs between the insulation surfaces and the slot wall.

The factors that influence the characteristics of PDs are [22]:

- Applied Voltage: as the applied voltage increases, the PD magnitude increases;
- Type of Insulating Material: the PD amplitude depends on the relative permittivity of each insulating material used for EM insulation;
- Void Size and Shape: the bigger the void, the larger the discharge;
- Humidity: when there are more droplets on the surface of the insulating material, the PD amplitude tends to increase;
- The frequency of the PD;

- The applied voltage waveform.

To combat the aforementioned issues, proper maintenance plays a key role in detecting probable problems and, therefore, PD monitoring is an appropriate diagnostic method, as it is the most helpful in examining the insulation condition and identifying potential machine failures.

2.2. PD Measurement—General Information

Considering all of the above, it becomes clear that detecting PDs is a very important and useful diagnostic method in order to prevent possible failures in insulation systems. Moreover, PD measurement is a non-destructive diagnostic technique.

The principle of PD measurement is measuring the pulses, generated by charge displacements occurring within or on the stator winding insulation system. These pulses are captured using PD couplers temporary connected to the generator terminals.

The standard PD test (IEC 60034-27) consists of installing a set of coupling capacitors at the EM terminals and monitoring the high frequency currents that flow through them in the case of PDs in the insulation. This can be done in an on-line or an off-line mode [23,24].

In Offline PD tests, locations in the insulation system with weak dielectric conditions can be detected. The influence of noise is not as an important problem as for the Online PD test and it is easier to pinpoint the localization of the various fault areas. Moreover, Offline PD testing enables determination of PD inception and extinction voltage. On the other hand, since the EMs do not operate under real conditions, voltage stresses in the winding and mechanical stresses may not be representative. Furthermore, EMs must be taken out of operation, which may incur economic consequences. Finally, an external power supply is necessary.

Online PD testing [11,25] provides permanent monitoring of PDs as well as continuous information on stator winding conditions. It is possible to observe abnormal situations or sudden changes in the behaviour of the stator windings and to predict if the SG's insulation is close enough to degradation. Furthermore, Online PD measurements provide information about the real insulation condition and the aging progress of the generator. The machine operates under real conditions, and a separate power supply is not necessary. A serious problem is the difficulty in determining the origin of the failure since the electrical noise obscures the fault. Installation of sensors is required.

2.3. PD Measurements under Real Conditions

In this section, the PD measurements that were conducted on the aforementioned SG are analysed. Each phase of the generator was tested separately. The PD measurement was carried out on both sides of each generator phase: the phase was energized before the start point side, and the capacitive coupler was connected the high voltage side, Subsequently, this phase was energized on the HV side, and the capacitive coupler was connected on the star point side. The test voltage on each phase was increased in steps of 0.2 kV up to the nominal generator phase voltage. PD activity was measured and recorded in each step. The PD measurement results are presented in Phase-Resolved Partial Discharge (PRPD) diagrams, three-dimensional diagrams that show the PD activity related to the AC cycle.

The Online PD monitoring procedure dictates proper installation of PD couplers, a connection box, and the portable measurement device in order to perform manual or fully automatic periodic measurements. Figure 5 shows the Online PD measurement arrangement. Coupling capacitors are connected to each phase of the SG, while Phase A is connected to the PD sensor, which sends the data to the PD Monitoring System. An ethernet cable connects the PD monitoring system to a personal computer where data are processed.

The measurements of PDs were conducted with an acquisition time of 20 se the Dead time was 7 μs and there were 1002 periods. It is possible to identify the nature of the different PDs through appropriate data-handling techniques. Surface PDs arise on the surfaces of end-windings and they have a low risk factor. Slot PDs arise inside and between insulation surfaces and slot walls due to the presence of air gaps and a lack of slot corona protection. Inner PDs have a very low risk factor. Online PD measurements at different reactive load and winding temperatures could confirm the presence of slot PDs, which can cause serious problems.

Figures 6 and 7 show the PD activity at one of the phases (phase A) of the SG occurring at phase voltage (PRPD) [26]). The visualization of the graphs reveals a reduction in PD activity between 2010 and 2011 due to maintenance actions that took place on the insulation, as shown in Figures 2–4.

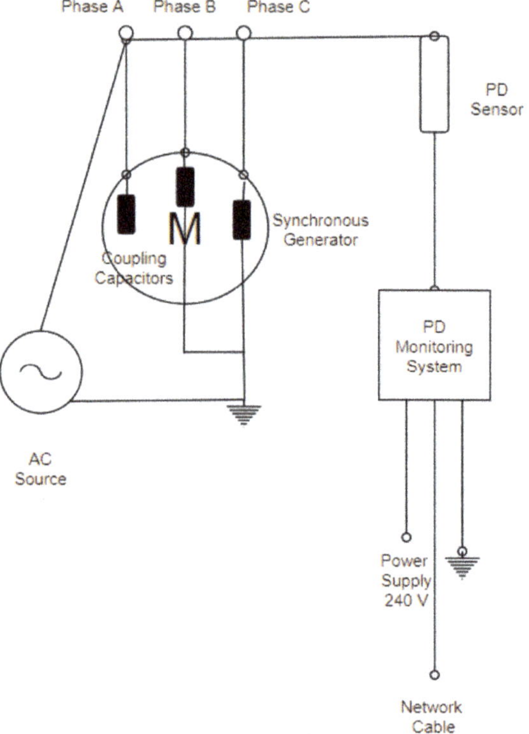

Figure 5. Online PD Measurement Arrangement.

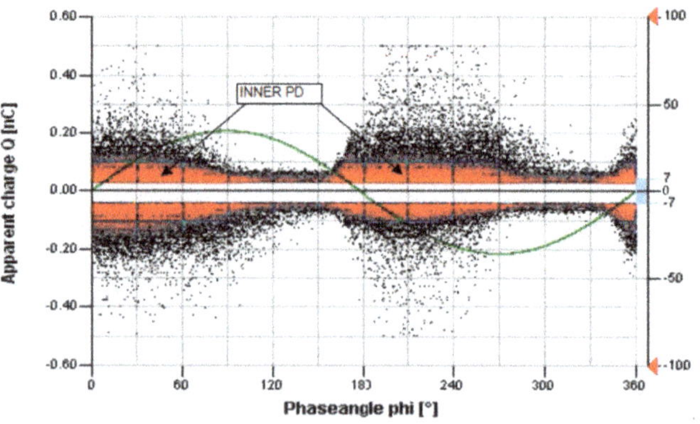

Figure 6. PD activity in Phase U (2010).

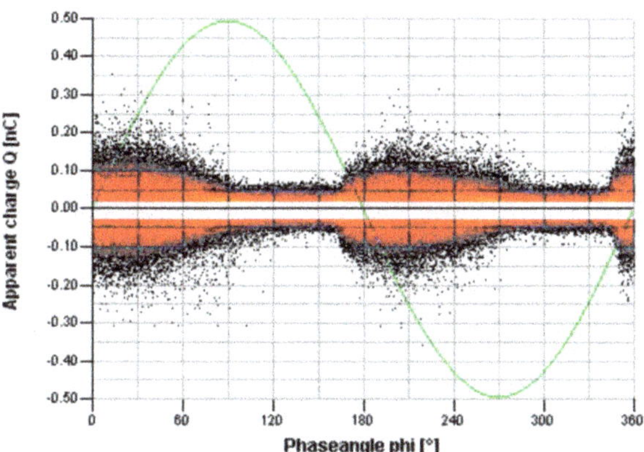

Figure 7. PD activity in Phase U (2011).

3. PD Model

3.1. Capacitive Model

Research would benefit from a PD model able to simulate the way epoxy resin reacts under varying conditions encountered in industrial applications. It is very important to investigate the PD activity on epoxy resin when there is a void inside the insulation, when this void becomes bigger or smaller, and when the voltage changes.

The PD model chosen is the capacitive model, a widely known and accepted model for PD studies in solid insulation materials. This model, referred to as the conventional method for detecting PD, is capable of describing mainly internal discharges, surface discharges as well as corona discharges. The capacitive model is an approximative representation of what happens inside an enclosed cavity. The classical capacitance model consists of three capacitors, which indicate the test object (C_c: Capacitance of the void in the solid material, C_b: Capacitance of the insulation material connected to the void, C_a: Capacitance of the remaining insulation material). The MATLAB/Simulink model (IEC 60270 Standard), which was created in order to simulate the PD, is shown in Figure 8. The circuit consists of [27–31]:

- One AC High Voltage (HV) Source;
- One resistor (R_1), which plays the role of a HV filter and it is used in order to reduce the noise of the AC HV Source;
- Two capacitors (C_m: HV measuring capacitor,C_k: Coupling capacitor, which must have low inductance in order to keep PD pulses low as well as act as a filter. These two capacitors consist of the most common sensor for measuring PDs in EM, as when a PD occurs, coupling capacitors provide the test object with a displacement current, which is measurable by the measuring capacitor and the MI);
- Three capacitors, which indicate the test object;

One Measuring Impedance (MI) RLC, which is used in order to collect the PD signals. The capacities C_a, C_b, C_c are calculated, considering the insulation material, by:

$$C_a = \frac{\varepsilon_0 \varepsilon_r (a - 2r) b}{d},$$ (1)

$$C_b = \frac{\varepsilon_0 \varepsilon_r r^2 \pi}{d - h},$$ (2)

$$C_c = \frac{\varepsilon_0 r^2 \pi}{h},$$ (3)

where ε_0 is the dielectric constant in vacuum, ε_r is the relative permittivity (dielectric constant) of the insulating material, a is the length, b is the weight and d is the height of the test object, r is the radius, and h is the height of the void.

When AC voltage V is applied to the insulation material and it reaches a voltage higher than the insulation is able to withstand, a discharge occurs [32–34]. The voltage V_c appears in the void according to [33,34]:

$$V_c = \frac{C_b}{C_b + C_c} V, \tag{4}$$

$$C_c = \frac{C_a C_b}{C_a + C_b}, \tag{5}$$

When a partial discharge takes place, a charge will be transferred from one side of the void to the other, and the voltage due to the PD (ΔV_c) will appear on C_c. The charge is too small to be measured; therefore, instead of the charge, the apparent charge is measured, and it is calculated by:

$$q_a \cong \Delta V_c C_b, \tag{6}$$

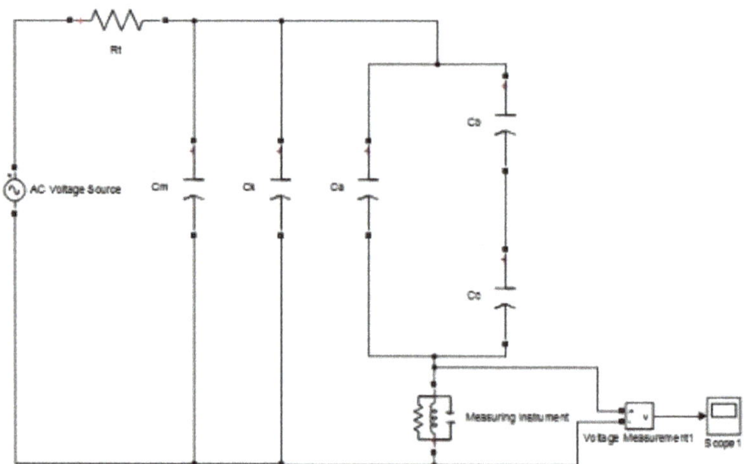

Figure 8. Capacitive PD Model—Simulink.

3.2. Simulation Results

Table 1 shows the simulation parameters for the epoxy resin. Tests specifying the values for measuring the capacitor, coupling capacitor, R_m, L_m, and C_m were conducted in addition to consulting other research work [27–30]. Epoxy resin is the insulating material used in the aforementioned SG and thus used in the capacitive model.

The applied voltages, used for the simulations, were the same as the voltages used in the PD measurements of the SG, described below. PD characteristics do not appear for a long time period and that is why the simulation time was chosen to be 0.02 s. Some results of the simulations are shown in the next figures. Figures 9–11 show the resulting PDs at 5 kV, 10 kV, and 15 kV respectively. PDs are shown as spikes in these figures. More specifically, in Figure 9, the most important spikes are at 0.015 s, 0.017 s, and 0.019 s, in Figure 10, the spikes are at 0.0005 s and 0.017 s, and in Figure 11, the spikes are at 0.005 s, 0.011 s, 0.0125 s, and 0.0165 s. It is obvious, that while increasing the applied voltage, both the number and the amplitude of the PD increase.

Table 1. Simulation Parameters (Epoxy Resin).

Parameter	Value
Measuring Capacitor	1000 pF
Coupling Capacitor	1000 µF
Dielectric Constant	8.85×10^{-12} F/m
Relative Permittivity	3.6
R_m	50 Ω
L_m	0.60 mH
C_m	0.45 µF
C_a	4.460×10^{-12} F
C_b	5.56062×10^{-14} F
C_c	1.39015×10^{-13} F

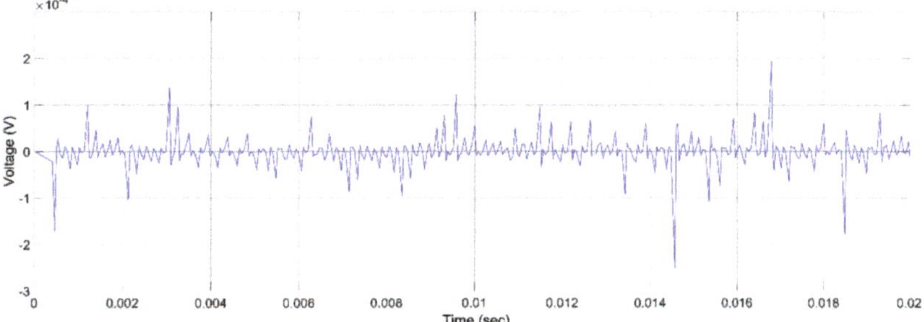

Figure 9. PD at 5 kV.

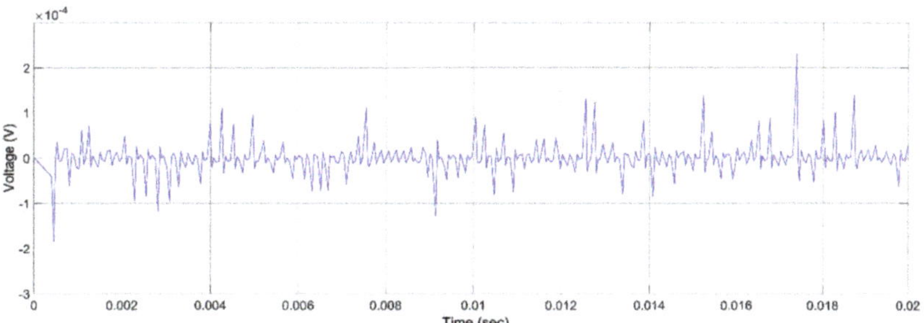

Figure 10. PD at 10 kV.

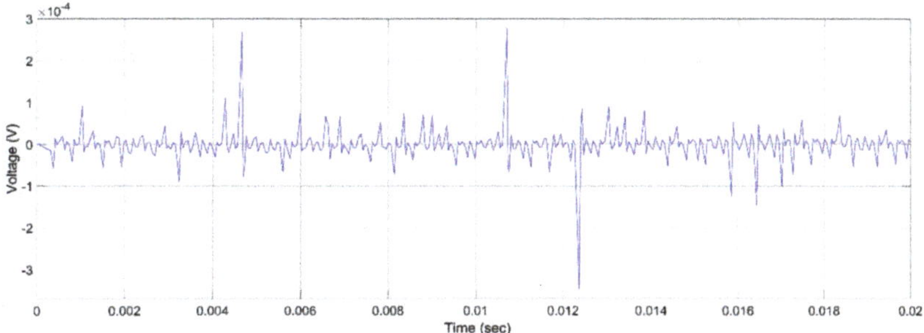

Figure 11. PD at 15 kV.

3.3. Simulation Results—Half Height of the Void

The same simulations were performed with some changes in the volume of the void. The height of the void varied by half the value and the radius was kept stable. Table 2 shows how the capacities change according to the variable volume of the void. As mentioned above, in Figure 12, the most important spikes (PDs) are at 0.0005 s, 0.0037 s, and 0.018 s, in Figure 13, the spikes are at 0.0004 s and 0.0022 s, and in Figure 14, the spikes are at 0.0003 s.

Table 2. Simulation Parameters (Epoxy Resin).

Parameter	Value
C_a	4.460×10^{-12} F
C_b	5.26795×10^{-14} F
C_c	2.78031×10^{-13} F

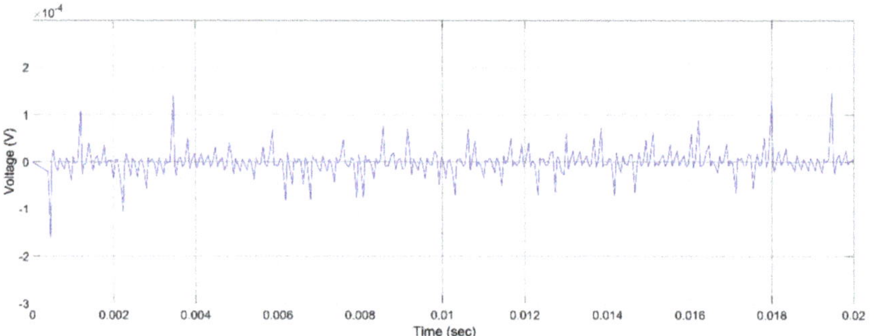

Figure 12. PD at 5 kV-Half Height.

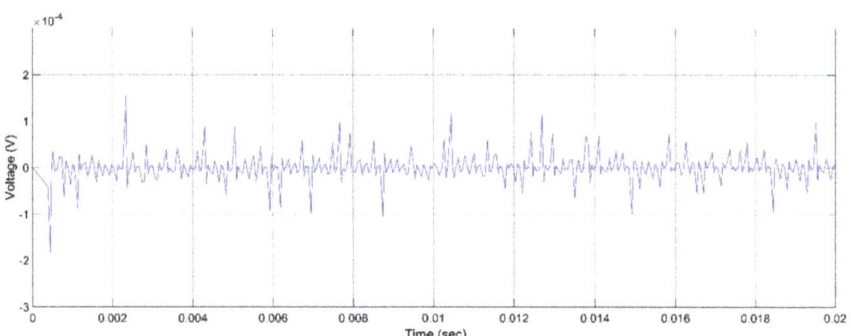

Figure 13. PD at 10 kV-Half Height.

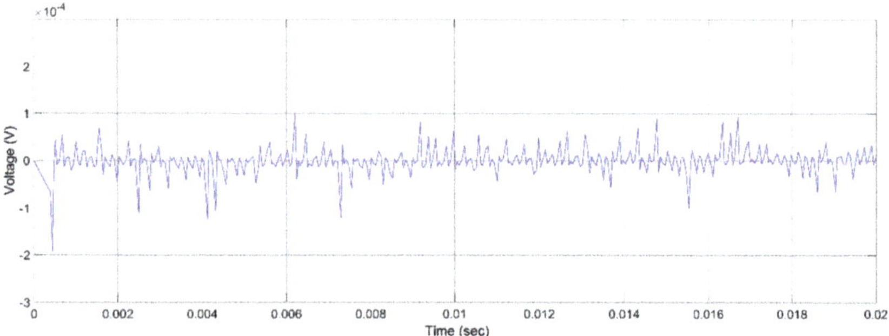

Figure 14. PD at 15 kV-Half Height.

3.4. Simulation Results—Half Radius of the Void

The same simulations were performed with the height of the radius changed by half and the height kept stable. Table 3 shows how the capacities change according to the above changes. As mentioned above, in Figure 15, the most important spikes (PDs) are at 0.0005 s, 0.001 s, and 0.018 s, in Figure 16, the spikes are at 0.0004 s, 0.008 s, and 0.009 s, and in Figure 17, the spikes are at 0.009 s and 0.0013 s.

Table 3. Simulation Parameters (Epoxy Resin).

Parameter	Value
C_a	4.46197×10^{-12} F
C_b	1.93077×10^{-14} F
C_c	3.47539×10^{-14} F

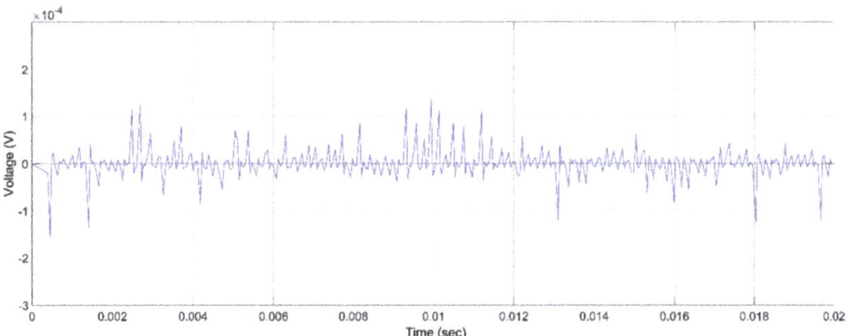

Figure 15. PD at 5 kV-Half Radius.

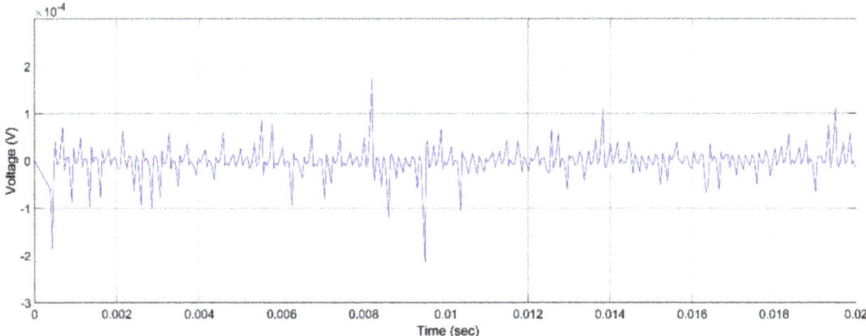

Figure 16. PD at 10 kV-Half Radius.

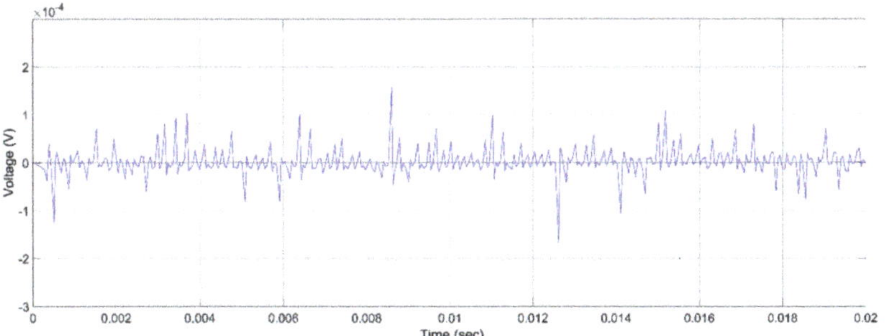

Figure 17. PD at 15 kV-Half Radius.

Comparing the above figures, it becomes clear that when the volume of the void was reduced, either by reducing the radius or the height of the void, the PDs were reduced. The aforementioned results are similar to results of other research, which indicate that PD activity is related to void geometry and applied voltage [34,35].

4. Effect of Humidity on PDs

The presence of humidity leads to an increase in electrical stresses, resulting in an increase in PD activity and thus affecting the state of EM insulation (possible deterioration). There are some papers and research works which claim that the presence of humidity may decrease the number of PDs and PD activity. There are no clear results from previous research whether humidity leads to an increase or a decrease in PD activity [36]. Moreover, some test results showed that if the relative humidity is increased, the magnitude of the PDs will be increased as well. The absolute humidity should be taken into consideration [37,38] as it affects the Partial Discharge Inception Voltage (PDIV). References [39,40] show that an increase in absolute humidity leads to a pertinent increase in PDIV, while other research work [41] disagrees with the aforementioned results.

Experiments were carried out to investigate the effect of humidity on flashover voltages. The sample was epoxy resin, which is the same insulation material used for the previous simulations and the aforementioned SG. The studied factors were water droplet volume, water conductivity, water droplet distance from the electrodes, and the number of water droplets on the epoxy resin surface.

Eight different water droplet setups were chosen (Figures 18–21), which one by one, were positioned on the sample surface of the epoxy resin. Eight different water conductivies were used, from 1.4 µS/cm to 10,000 µS/cm, in order to investigate how they affect the PD. The experiment procedure was analyzed in-depth in previous work [42]:

After the flashover occurred, the voltage was recorded and the same water droplet arrangement was positioned on the epoxy resin surface. The voltage was raised up to the previous voltage minus 1.2 kV, and the arrangement stayed for 5 min. If flashover did not happen, the voltage was raised by 0.4 kV and so on. When the flashover occurred, the final voltage was recorded and used in order to create Figure 22, containing the results of the different water droplet arrangements. The reason for allowing the voltage value for 5 min at each voltage level was because a certain time was required for the droplets to deform and for the surface discharges to start.

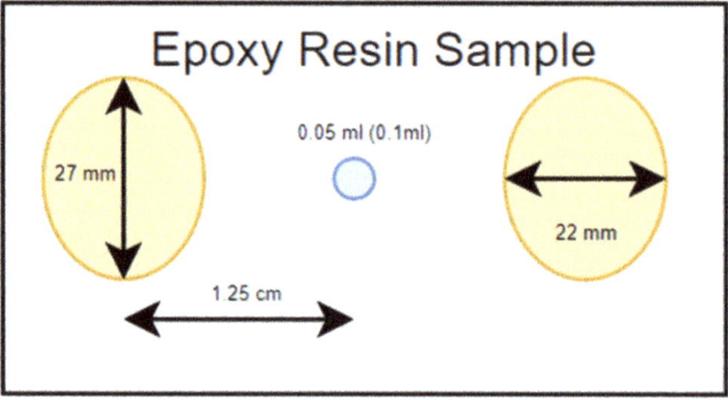

Figure 18. One-Droplet Arrangement (1D).

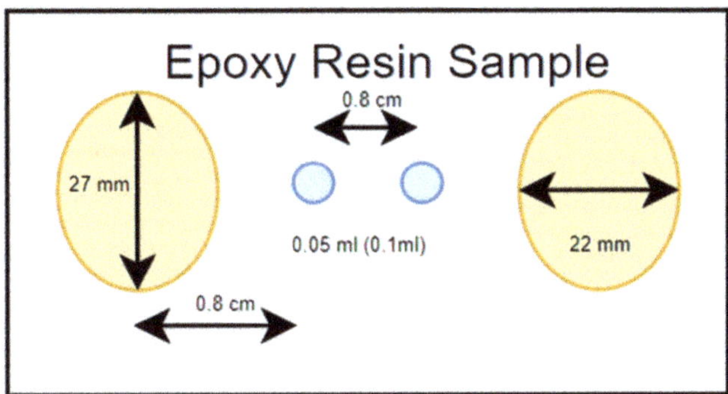

Figure 19. Two-Droplets Arrangement with one next to the other (2.1D).

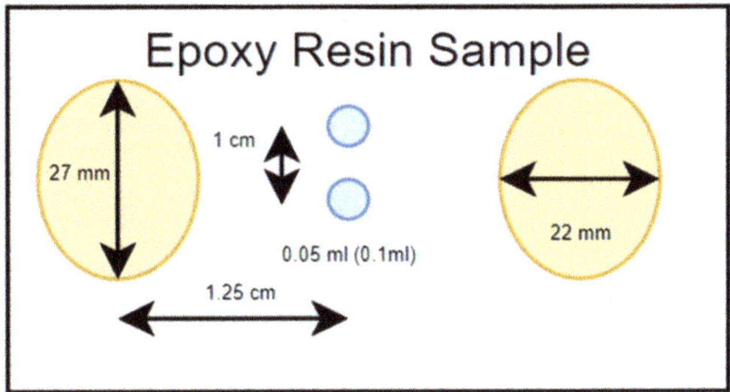

Figure 20. Two-Droplets Arrangement with one on top of the other (2.2D).

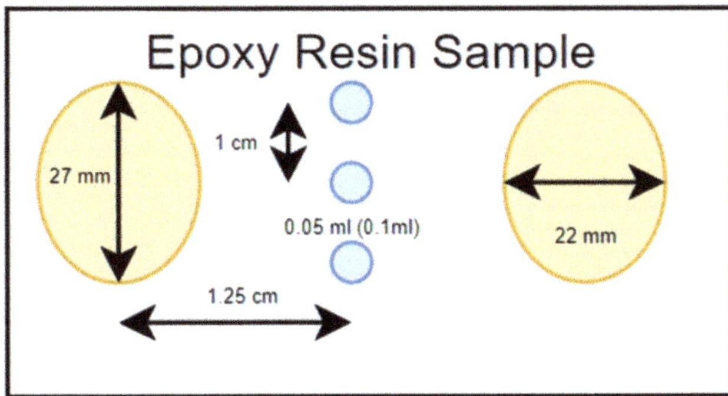

Figure 21. Three-Droplet Arrangement (3D).

Figure 22 shows the variation of the flashover voltage with water conductivity for the epoxy resin sample. The colours of the different lines are explained next to the diagram.

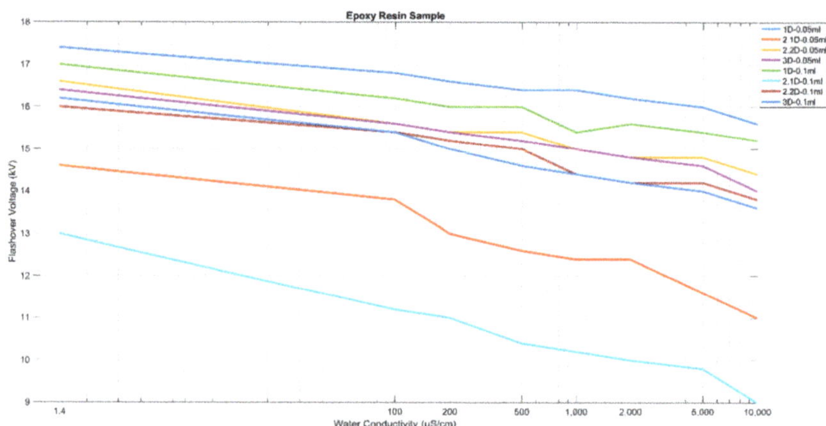

Figure 22. Comparison of different flashover voltages for the epoxy resin sample.

First, the water conductivity affects the flashover voltage, e.g., when the water conductivity increases, the flashover voltage decreases. Furthermore, in most experiments, the flashover voltage of the arrangements of 0.1 mL was less compared to the same arrangement of 0.05 mL. Moreover, the more droplets on the sample surface, the lower the flashover voltage. Last but not least, the two-droplet arrangement one next to the other presents the minimum values of the flashover voltage, and this happens because the droplets are closer to the electrodes.

5. Discussion

EMs are one of the most significant elements in industry and in generating electric-ity. Long lifespan and proper operation are of paramount importance. The EM insulation system condition is one of the defining factors of its overall health condition [43,44]. One of the most widely used insulation materials is epoxy resin, as it has many advantages, such as tensile strength [45], and this paper aimed to study the influence of various factors on epoxy resin. Delaminations and voids in a solid dielectric could lead to the deterioration and the degradation of the insulation system. If a void of a certain volume is present in a solid insulation material, this could lead to more PDs as well as a higher amplitude of PDs as the applied voltage increases. In addition to this, as the volume of the void increases, PDs increase as well.

The MATLAB/Simulink model proposed in this paper was created in order to simulate the way that a common insulation material that is used in various EMs, epoxy resin, reacts to different situations. The combination of these results with the experimental tests, which give significant information about the humidity and how it affects epoxy resin, leads to a more general point of view of this insulation material. According to simulation results, it is obvious that when increasing the applied voltage, both the number and the amplitude of the PDs increase. In addition to this, when the volume of the void is increased, the PDs increase.

Another important factor in the epoxy resin condition seems to be humidity. Other research efforts present different approaches on how humidity affects PD activity and flashover voltages, as mentioned above. According to experiments, humidity affects the deterioration of the insulation because it is related to the increase in flashover voltage as well as the PD activity. The factors that affect the PD activity are water conductivity, the distance of the water droplets from the electrodes, and the volume and number of droplets on the sample surface. As the number of droplets or the volume increases, the flashover voltage decreases and facilitates PD occurrence. In addition, when the water droplets are closer to electrodes, the flashover voltage decreases. Finally, the water conductivity affects flashover voltages and PD activity, as the smallest values for flashover voltages occur in experiments with the highest value of water conductivity.

During the experiments, one could notice the following: before the breakup of the water droplets there was an oscillation and a characteristic noise. More specifically, in the two-droplet arrangement with one droplet next to the other, the two droplets were approaching each other before the PD occurred. On the other hand, in the three-droplet arrangement, most of the time, the middle droplet was the one that deformed and broke up. Further experiments will be developed in the future in order to verify the conclusions across a wider range of machines and cases. Both the simulations and the experiments can be used in order to examine the effect of the PDs on EMs, as PD activity is both a cause and a symptom of the insulation aging. As mentioned above, one of the main problems that occurs in EMs is PDs, which occur due to various defects on their insulation system. The presence of PDs can lead to insulation degradation and failure. As such, it is very important to detect the PDs immediately either via Offline or Online PD tests, as described above. First, the installation of an Online PD monitoring system is recommended in order to prevent an abnormal situation or sudden changes in the behaviour of stator winding. An Online PD monitoring system can give important information about future insulation conditions, possible faults, and problems. Online PD tests are more expensive but they offer better results in EM monitoring. In any case, the systematic cleaning of the EMs, aimed to reduce humidity and contamination to minimize the presence of PDs, should always be considered.

6. Conclusions

The purpose of this manuscript is to discuss the factors affecting the PD and their detection in EMs and especially epoxy resin, one of the most significant components of EMs' insulation system. Epoxy resin was chosen, as it is used not only on the insulation system of the SG described above but also in many other similar systems. It was therefore essential for this research effort to simulate the insulation of this SG in order to investigate how the above (and many other) factors affect its "health" condition and prognose possible faults. PD tests, Offline or Online, are the most suitable tests for evaluating the insulation condition. Offline tests are mainly used for determining the point where a potential fault starts, while Online PD tests are able to monitor the insulation condition continuously during operation.

The simulation results show that, when increasing the applied voltage, the number and the amplitude of PDs increase. Furthermore, if the volume of the void decreases, the PDs decrease. Other factors that play a significant role in PDs are humidity and contamination. Experiments with epoxy resin samples and different droplet arrangements were carried out in order to investigate which factors can affect the PDs and provided interesting results. Water droplet conductivity, the distance of the water droplets from the electrodes, and the volume and number of the water droplets on the surface of the epoxy resin affect the flashover performance and the PDs.

As for future work, experiments with different water conductivities and water droplet arrangements would be useful in confirming the results and providing statistical significance. As for the simulations, more iterations with different geometries and dimensions of voids would be useful. The simulations and the experiments could be combined with the onsite measurements of PDs across a wider range of existing SGs in power plants in order to determine the situation of their insulation system for planning maintenance actions more efficiently.

Author Contributions: Conceptualization, D.V., A.K., M.G.D. and J.A.A.-D.; methodology, D.V., A.K., M.G.D. and J.A.A.-D.; software, D.V.; validation, D.V., A.K., M.G.D. and J.A.A.-D.; formal analysis, D.V., A.K., M.G.D. and J.A.A.-D.; investigation, D.V., A.K., M.G.D. and J.A.A.-D.; resources, D.V., A.K. and M.G.D.; data curation, D.V.; writing—original draft preparation, D.V., A.K., M.G.D. and J.A.A.-D.; writing—review and editing, D.V., A.K., M.G.D. and J.A.A.-D.; visualization, D.V., A.K., M.G.D. and J.A.A.-D.; supervision, A.K., M.G.D. and J.A.A.-D.; project administration, D.V., A.K.,

M.G.D. and J.A.A.-D.; funding acquisition, A.K. All authors have read and agreed to the published version of the manuscript.

Funding: This research received no external funding.

Institutional Review Board Statement: Not applicable.

Informed Consent Statement: Not applicable.

Data Availability Statement: Not applicable.

Acknowledgments: The authors of this paper would like to thank Pericles Stratigopoulos, Director of Komotini Power Plant, Public Power Corporation S.A.—Hellas, for providing the pictures and the results of the PD measurements for this SG.

Conflicts of Interest: The authors declare no conflict of interest.

Abbreviations

The following abbreviations are used in this manuscript:

EMs	Electrical Rotating Machines
HV	High Voltage
MI	Measuring Impedance
PDs	Partial Discharges
PDIV	Partial Discharge Inception Voltage
PRPD	Phase Resolved Partial Discharge
SGs	Synchronous Generators
TEAM	Thermal, Electrical, Ambient, and Mechanical

References

1. Afrandideh, S.; Milasi, M.E.; Haghjoo, F.; Cruz, S.M.A. Turn to Turn Fault Detection, Discrimination, and Faulty Region Identification in the Stator and Rotor Windings of Synchronous Machines Based on the Rotational Magnetic Field Distortion. *IEEE Trans. Energy Convers.* **2020**, *35*, 292–301. [CrossRef]
2. Stone, G.C.; Boulter, E.A.; Culbert, I.; Dhirani, H. *Electrical Insulation for Rotating Machines: Design, Evaluation, Aging, Testing, and Repair*; A John Wiley & Sons, Inc.: Toronto, ON, Canada, 2004.
3. Shankar, R.; Dave, S. Degradation and Diagnostics of Electrical Machines. Master's Thesis, Department of Electrical Engineering, Chalmers University of Technology, Gothenburg, Sweden, 2021.
4. Seri, P.; Ghosh, R.; Montanari, G.C. An Unsupervised Approach to Partial Discharge Monitoring in Rotating Machines: Detection to Diagnosis with Reduced Need of Expert Support. *IEEE Trans. Energy Convers.* **2021**, *36*, 2485–2492. [CrossRef]
5. Madonna, V.; Giangrande, P.; Zhao, W.; Zhang, H.; Gerada, C.; Galea, M. On the Design of Partial Discharge-Free Low Voltage Electrical Machines. In Proceedings of the 2019 IEEE International Electric Machines & Drives Conference (IEMDC), San Diego, CA, USA, 12–15 May 2019; pp. 1837–1842.
6. Hemmati, R.; Wu, F.; El-Refaie, A. Survey of Insulation Systems in Electrical Machines. In Proceedings of the 2019 IEEE International Electric Machines & Drives Conference (IEMDC), San Diego, CA, USA, 12–15 May 2019; pp. 2069–2076.
7. Muto, H. Dielectric Breakdown and Partial Discharge Properties of Nanocomposites for Epoxy/Mica Insulation system in Large-size Rotating Machines. In Proceedings of the 2020 International Symposium on Electrical Insulating Materials (ISEIM), Tokyo, Japan, 13–17 September 2020.
8. Brütsch, R.; Chapman, M. Insulating systems for high voltage rotating machines and reliability considerations. In Proceedings of the 2010 IEEE International Symposium on Electrical Insulation, San Diego, CA, USA, 6–9 June 2010; pp. 1–5.
9. Nathaniel, T. Diagnostics of Stator Insulation by Dielectric Response and Variable Frequency Partial Discharge Measurements. Ph.D. Thesis, KTH Royal Institute of Technology Stockholm, Stockholm, Sweden, November 2006.
10. Madonna, V.; Giangrande, P.; Zhao, W.; Buticchi, G.; Zhang, H.; Gerada, C.; Galea, M. Reliability vs. Performances of Electrical Machines: Partial Discharges Issue. In Proceedings of the 2019 IEEE Workshop on Electrical Machines Design, Control and Diagnosis (WEMDCD), Athens, Greece, 22–23 April 2019; pp. 77–82.
11. Luo, Y.; Li, Z.; Wang, H. A Review of Online Partial Discharge Measurement of Large Generators. *Energies* **2017**, *10*, 1694. [CrossRef]
12. Montanari, G.C. Insulation Diagnosis of High Voltage Apparatus by Partial Discharge Investigation. In Proceedings of the 2006 IEEE 8th International Conference on Properties & Applications of Dielectric Materials, Bali, Indonesia, 26–30 June 2006; pp. 1–11.
13. Pan, C.; Chen, G.; Tang, J.; Wu, K. Numerical modeling of partial discharges in a solid dielectric-bounded cavity: A review. *IEEE Trans. Dielectr. Electr. Insul.* **2019**, *26*, 981–1000. [CrossRef]

14. Moghadam, D.E.; Herold, C.; Zbinden, R. Effects of Resins on Partial Discharge Activity and Lifetime of Insulation Systems Used in eDrive Motors and Automotive Industries. In Proceedings of the 2020 IEEE Electrical Insulation Conference (EIC), Knoxville, TN, USA, 22 June–3 July 2020; pp. 221–224.
15. Höpner, V.N.; Volmir, E.W. Insulation Life Span of Low-Voltage Electric Motors—A Survey. *Energies* **2021**, *14*, 1738. [CrossRef]
16. Wang, P.; Hui, S.; Akram, S.; Zhou, K.; Nazir, M.T.; Chen, Y.; Dong, H.; Javed, M.S.; Ul Haq, I. Influence of Repetitive Square Voltage Duty Cycle on the Electrical Tree Characteristics of Epoxy Resin. *Polymers* **2020**, *12*, 2215. [CrossRef] [PubMed]
17. Stone, G.C. Partial discharge diagnostics and electrical equipment insulation condition assessment. *IEEE Trans. Dielectr. Electr. Insul.* **2005**, *12*, 891–904. [CrossRef]
18. Sabat, A.; Karmakar, S. Simulation of Partial Discharge in High Voltage Power Equipment. *Int. J. Electr. Eng. Inform.* **2011**, *3*, 234–247. [CrossRef]
19. Danikas, M.G.; Adamidis, G. Partial discharges in epoxy resin voids and the interpretational possibilities and limitations of Pedersen's model. *Archiv fuer Elektrotech.* **1997**, *80*, 105–110. [CrossRef]
20. Lv, Z.; Rowland, S.M.; Chen, S.; Zheng, H.; Iddrissu, I. Evolution of partial discharges during early tree propagation in epoxy resin. *IEEE Trans. Dielectr. Electr. Insul.* **2017**, *24*, 2995–3003. [CrossRef]
21. Fenger, M.; Stone, G. Investigations into the Effect of Humidity on Stator Winding Partial Discharges. *IEEE Trans. Dielectr. Electr. Insul.* **2005**, *12*, 341–365. [CrossRef]
22. Pandit, I.; Sinha, A.K.; Kumar, P.; Hati, A.S. Partial Discharge in Solid Insulating Materials, Causes, Effects and Factors of Dependence–A Comparative Investigation. *IJEREEE* **2018**, *4*, 190–196.
23. Waldi, E.P.; Murakami, Y.; Nagao, M. Effect of humidity on breakdown of low density polyethylene film due to partial discharge. In Proceedings of the Condition Monitoring and Diagnosis International Conference on, Beijing, China, 21–24 April 2008; pp. 655–658.
24. Nawawi, Z.; Murakami, Y.; Hozumi, N.; Nagao, M. Effect of Humidity on Time Lag of Partial Discharge in Insulation-Gap-Insulation System. In Proceedings of the Properties and Applications of Dielectric Materials, 8th International Conference on Bali, Bali, Indonesia, 26–30 June 2006; pp. 199–203.
25. Danikas, M.G.; Sarathi, R. Electrical machine insulation: Traditional insulating materials, Nanocomposite polymers and the question of electrical trees. *Funktechnikplus J.* **2014**, *1*, 7–32.
26. Illias, H.; Yuan, T.S.; Bakar, A.H.A.; Mokhlis, H.; Chen, G.; Lewin, P.L. Partial discharge patterns in high voltage insulation. In Proceedings of the 2012 IEEE International Conference on Power and Energy (PECon), Kota Kinabalu, Malaysia, 2–5 December 2012; pp. 750–755.
27. Sharma, P.; Bhanddakkar, A. Simulation Model of Partial Discharge in Power Equipment. *IJEER* **2015**, *3*, 149–155.
28. di Silvestre, M.L.; Miceli, R.; Romano, P.; Viola, F. Simplified Hybrid PD Model in Voids: Pattern Validation. In Proceedings of the 4th International Conference on Power Engineering, Energy and Electrical Drives, Istanbul, Turkey, 13–17 May 2013.
29. Gunawardana, S.D.M.S.; Kanchana, A.A.T.; Wijesingha, P.M.; Perera, H.A.P.B.; Samarasinghe, R.; Lucas, J.R. A Matlab Simulink Model for a Partial Discharge Measuring System. In *Electrical Engineering Conference*; University of Moratuwa: Sri Jayawardana-pura Kotte, Sri Lanka, 2015.
30. Srinivasa, D.; Harish, B.; Harisha, K. Analysis Study on Partial Discharge Magnitudes to the Parallel and Perpendicular Axis of a Cylindrical Cavity. *Int. J. Eng. Trends Technol. (IJETT)* **2017**, *45*, 334–337.
31. Aakre, T.G.; Ildstad, E.; Hvidsten, S. Partial discharge inception voltage of voids enclosed in epoxy/mica versus voltage frequency and temperature. *IEEE Trans. Dielectr. Electr. Insul.* **2020**, *27*, 214–221. [CrossRef]
32. Girish, S.K.; Sumangala, D.B.V. Analysis of Partial Discharge Activity in Pressboards Using MATLAB SIMULINK. *Int. J. Adv. Eng. Res. Dev.* **2017**, *4*, 255–260.
33. Singh, N.; Debdas, S.; Chauhan, R. Simulation and experimental study of Partial Dischage in insulating material for High Voltage Power Equipements. *Int. J. Sci. Eng. Res.* **2013**, *4*, 1677–1684.
34. Gaurkar, B.P.; Bonde, U.G. Partial Discharge Testing of Insulator using MATLAB Simulink. *Int. J. Sci. Res. Dev.* **2017**, *5*.
35. Kumar, D.; Singh, R. Simulation of Partial Discharge for Different Insulation Material Using MATLAB. *Int. J. Sci. Res. Dev. (IJSRD)* **2015**, *3*, 660–663.
36. Warren, V. Partial discharge testing: A progress report. In Proceedings of the Iris Rotating Machinery Conference, Santa Monica, CA, USA, 20–23 June 2011; pp. 1–17.
37. Driendl, N.; Pauli, F.; Hameyer, K. Influence of Ambient Conditions on the Qualification Tests of the Interturn Insulation in Low-Voltage Electrical Machines. In *IEEE Transactions on Industrial Electronics*; IEEE: New York, NY, USA, 2021.
38. Nawawi, Z.; Hozumi, Y.M.N.; Nagao, M. Effect of Humidity on Partial Discharge Characteristics. In Proceedings of the 7th International Conference on Properties and Applications of Dielectric Materials, Nagoya, Japan, 1–5 June 2003; Volume 30.
39. Moonesan, M.; Jayaram, S.; Cherney, E. Effect of Temperature and Humidity on Surface Discharge Activities under High Voltage Unipolar Pulses. In Proceedings of the ESA Annual Meeting on Electrostatics, Pomona, CA, USA, 16–18 June 2015; Volume 31.
40. Dang, H. Investigation of the Effects of Humidity and Surface Charge on Partial Discharge in Insulator Posts under DC Voltage. Master's Thesis, University of Manitoba, Winnipeg, MB, Canada, 2020.
41. Rahman, M.F.; Rao, B.N.; Nirgude, P.M. Influence of moisture on partial discharge characteristics of oil impregnated pressboard under non-uniform field. In Proceedings of the International Symposium on High Voltage Engineering, Buenos Aires, Argentina, 28 August–1 September 2017; Volume 34, pp. 1–6.

42. Verginadis, D.; Danikas, M.G.; Sarathi, R. Study of the Phenomena of Surface Discharges and Flashover in Nanocomposite Epoxy Resin Under the Influence of Homogeneous Electric Fields. *Eng. Technol. Appl. Res. (ETASR)* **2019**, *9*, 4315–4321.
43. Verginadis, D.; Antonino-Daviu, J.; Karlis, A.; Danikas, M.G. Diagnosis of Stator Faults in Synchronous Generators: Short Review and Practical Case. In Proceedings of the 2020 International Conference on Electrical Machines (ICEM), Gothenburg, Sweden, 23–26 August 2020; pp. 1328–1334.
44. Hassan, W.; Hussain, G.A.; Mahmood, F.; Amin, S.; Lehtonen, M. Effects of Environmental Factors on Partial Discharge Activity and Estimation of Insulation Lifetime in Electrical Machines. *IEEE Access* **2020**, *8*, 108491–108502. [CrossRef]
45. Credo, A.; Villani, M.; Popescu, M.; Riviere, N. Application of Epoxy Resin in Synchronous Reluctance motors with fluid-shaped barriers for e mobility. In *IEEE Transactions on Industry Applications*; IEEE: New York, NY, USA, 2021.

Article

Analysis of Modular Stator PMSM Manufactured Using Oriented Steel

Anmol Aggarwal [1,*], **Matthew Meier** [1], **Elias Strangas** [1] **and John Agapiou** [2]

[1] Department of Electrical Engineering, Michigan State University, East Lansing, MI 48824, USA;
meiermat@msu.edu (M.M.); strangas@egr.msu.edu (E.S.)
[2] Manufacturing Systems Research Laboratory, General Motors, Warren, MI 48092, USA;
john.agapiou@gm.com
* Correspondence: aggarw43@msu.edu

Abstract: Oriented steel has higher permeability and lower losses in the direction of orientation (the rolling direction) than non-oriented steel. However, in the transverse direction, oriented steel typically has lower permeability and higher losses. The strategic use of oriented steel in a modular Permanent Magnet Synchronous Machine (PMSM) stator can improve machine performance, particularly when compared to a machine designed with non-oriented steel, by increasing both torque and efficiency. Typically, steel manufacturers provide magnetic properties only in the rolling and transverse directions. Furthermore, in modern Finite Element Analysis (FEA) software, the magnetic properties between the rolling and transverse directions are interpolated using an intrinsic mathematical model. However, this interpolation method has proven to be inaccurate; to resolve this issue, an improved model was proposed in the literature. This model requires the magnetic properties of the oriented steel in between the rolling and transverse directions. Therefore, a procedure for extracting the magnetic properties of oriented steel is required. The objective of this work is to propose a method of determining the magnetic properties of oriented steel beyond just the oriented and transverse directions. In this method, flux-injecting probes, also known as sensors, are used to inject and control the flux density in an oriented steel segmented stator in order to extract the properties of the oriented steel. These extracted properties are then used to model an oriented steel modular stator PMSM. The machine's average torque and core losses are compared with conventional, non-modular, non-oriented steel stator PMSM, and modular, non-oriented steel stator PMSM. It is shown that both the average torque and the core loss of the oriented steel modular stator PMSM have better performance at the selected number of segments than either of the two non-oriented steel stators.

Keywords: electric machines; electromagnetic analysis; electromagnetic measurements; core losses; rotor flux linkage; modular stator; oriented steel; finite element analysis; flux-injecting probes

Citation: Aggarwal, A.; Meier, M.; Strangas, E.; Agapiou, J. Analysis of Modular Stator PMSM Manufactured Using Oriented Steel. *Energies* **2021**, *14*, 6583. https://doi.org/10.3390/en14206583

Academic Editor: Athanasios Karlis

Received: 21 September 2021
Accepted: 8 October 2021
Published: 13 October 2021

1. Introduction

Improved manufacturing techniques reduce the cost of electrical machines while improving performance. Segmented stators simplify the winding process, thus increasing the slot fill factor and the ease of handling and assembly [1,2]. Segmented stator design also enables the use of different materials for the stator and rotor while reducing waste. As a result, the stator can be built with lower-loss magnetic steel while the rotor is built with materials of higher tensile strength, as required by high-speed operation. Due to the separate production of the rotor and stator laminations, the width of the air gap is no longer determined by the limitations of the punching tool. Hence, the air gap can be further reduced. Additionally, this can increase the torque density of the machine. Faults may lead to adverse effects on the operation of the machine and are dangerous to the system and to human safety [3–7]. High fault tolerance is also achieved in segmented stator construction due to the physical separation of the segments [8]. The construction of the segmented stator also allows for the use of oriented steel that shows superior magnetic properties, higher permeability, and lower core losses in the rolling direction [9–11]. Therefore, the proper design of oriented steel segments for the stator construction may lead to the enhanced performance of the machine, along with the other advantages of modular stator construction.

In order to design a machine with the use of oriented steel segments, an accurate modelling of the magnetic properties of the steel is required for numerical simulations. The magnetic properties of oriented steel are distinct in different directions, and knowledge of the magnetic properties of all the directions is required in order to accurately design the machine. However, the magnetic properties of only the rolling and transverse directions are supplied by steel manufacturers. Moreover, in modern FEA software, the magnetic properties of the rolling (0° direction) and transverse directions (90° direction) of the oriented steel are utilized to interpolate the magnetic properties in between these directions with the use of an internal algorithm. Several techniques have been proposed in the literature to model oriented steel in FEA. In [12], an elliptical model was used to simulate the oriented steel. This method assumes that the permeability of the oriented steel follows an elliptical curve at a given flux density, in the rolling and transverse directions. Therefore, from the BH curves in the rolling and transverse directions, the BH curves for any direction can be interpolated. However, it is shown in [13] that the permeability of the oriented steel is the lowest at around the 50–60° direction, hence the interpolation method using an ellipse is inaccurate. In [14], the magnetic field cross effects using a non-linear method is proposed. The cross field exists when the magnetic field is applied in a direction other than the principal direction, which leads to the non-parallel magnetic field intensity (H) and the flux density (B). However, the cross effects are negligible when the field is applied in the principal direction. In the proposed method, the equivalent magnetic field components are substituted when the cross effects present are derived from the co-energy densities. Using this technique, the equivalent behavior of anisotropic materials can be simulated in any direction. This method has shown improvement in the modelling of the oriented steel over the elliptical. However, in [13], it is shown that variation of the magnetic properties of the oriented steel in between the rolling and transverse directions is unique for every material. Therefore, a mathematical model based on the interpolation of the magnetic properties of the rolling and transverse directions cannot ensure the accuracy of the modelling of the oriented steel. In [15], the reluctivity is modelled as the tensor for the oriented steel. The reluctivity matrix has non-zero, non-diagonal elements that have resulted from the consideration of reciprocal parameters along the principal direction. By choosing the appropriate axis, the tensor matrix is reduced to a diagonal matrix, which predicts the properties of the oriented steel in all the directions between the rolling and the transverse. This method is experimentally validated on a transformer core; however, this method is very complex to apply. In [16], the oriented steel is used to improve the performance of the high-speed traction motor. However, the details of the numerical simulations using FEA is not provided.

In [17], a relatively simple method of modelling a PMSM with an oriented steel modular stator is proposed. The stator is created as a piecewise isotropic model where each isotropic section utilizes the properties of the oriented steel based on the flux flow direction in the stator. This proposed method is based on the fact that flux generally flows radially in the teeth, and then flows circumferentially in the back iron between the adjacent teeth. This modelling method requires the properties of the oriented steel in the direction of each tooth, and the portion of the back iron between two adjacent teeth with respect to the orientation direction. The higher the number of teeth in a stator, the smaller the angular displacement between adjacent teeth. Consequently, a larger number of teeth requires BH and loss curves for more distinct angular orientations with respect to the rolling direction. In such situations, the use of a standard Epstein frame test is expensive and tedious as a lot of the oriented steel sheets that cut along the different directions of orientation are required for testing. Hence, a method that is more practical and comparatively less expensive is required to extract the properties of the oriented steel.

The purpose of this work is to develop a relatively inexpensive method for extracting the properties of the oriented steel relative to the rolling and transverse directions. The idea of the proposed methodology is to develop and use sensors to inject flux of a desired magnitude and frequency into the oriented steel segmented stator. The first contribution provided systematic steps for the acquisition of the properties of a particular oriented steel, which could then be used to design a machine. The obtained magnetic characteristics were used to model an existing PMSM with a modular stator, and the performance of this stator was compared with the same stator using modular and non-modular non-oriented steel. It was observed that the core losses and average torque in the PMSM with an oriented steel modular stator were decreased and increased, respectively, for some specific number of segments as compared to those in the PMSM using both modular and non-modular non-oriented steel stator.

This paper is organized as follows: Section 2 discusses the modelling of anisotropic steel in segmented stators. Section 3 discusses the proposed method for the extraction of magnetic properties of segments. Section 4 describes the experimental set-up, the procedure for the estimation of BH and loss curves of the oriented steel from experiments, and the experimental results. Section 5 makes a comparison of the modular stator PMSM and the PMSM using both modular and non-modular non-oriented steel stators. Section 6 discusses the selection of the number of segments for best performance. Finally, Section 7 draws the conclusions of the paper.

2. Modelling of Anisotropic Steel in the Segmented Stators

Oriented steel (anisotropic steel) has crystal grains oriented in one direction, known as the rolling direction. Due to this internal morphology, the magnetic characteristics of oriented steel depend on the flux direction. Such behavior is different in non-oriented or quasi-isotropic steel, which has almost the same characteristics regardless of the direction. Moreover, anisotropic steel has different BH and loss curves in every direction, unique for every steel, which makes it difficult to model in finite-element simulations. In today's FEA software, the BH and loss curves of the rolling and transverse directions are used to interpolate the magnetic properties in between these two directions. Moreover, the method of interpolation is not controlled by the user, which may lead to inaccurate results.

In [17], a method that could be used in FEA to accurately analyze the performance of a PMSM with oriented steel stator laminations was developed. In this method, the stator was divided into sections, and based on the direction of the flux, the BH and loss curves were assigned to individual sections. Thus, the simulation was reduced to a connected structure of isotropic materials. Because of the structure of the proposed model, it is termed here as a piecewise isotropic model. The machine used for analysis in [17] was a 12-slot/14-pole PMSM consisting of six segments where each segment was generally oriented in the direction of the teeth. The direction of orientation of one segment is shown in Figure 1.

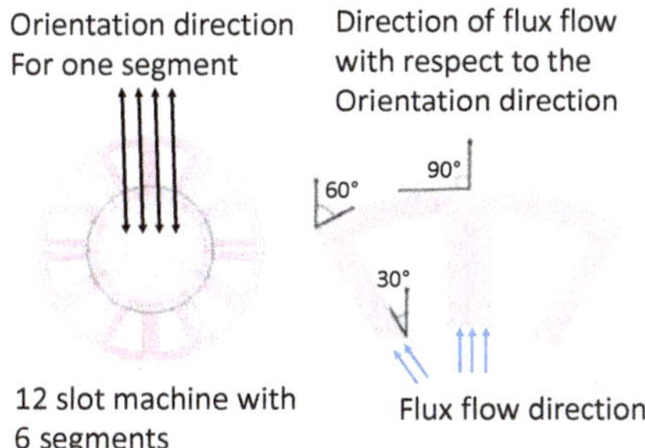

Figure 1. Direction of orientation and flux flow in the whole machine and in one segment [17].

For one segment, the direction of flux in the middle tooth is in the direction of alignment, i.e., $0°$ away from the rolling direction; for the two adjacent teeth, the direction of flux is 30° away from the rolling direction. For the back iron between the middle tooth and the adjacent tooth, the flux changes the direction from 90° to 60° as it flows through the back iron. The proposed model in [17] was obtained by connecting the segments with the BH and loss curves of the 0° direction on the central tooth, and with the BH and loss curves of the 30° direction on the lateral teeth; the back iron was modelled using the BH and loss curves of the 75° direction, which is the average value of the 60° and 90° angles. The model of one segment is given in Figure 2, and six segments connected together along the circumference are used to model the complete machine.

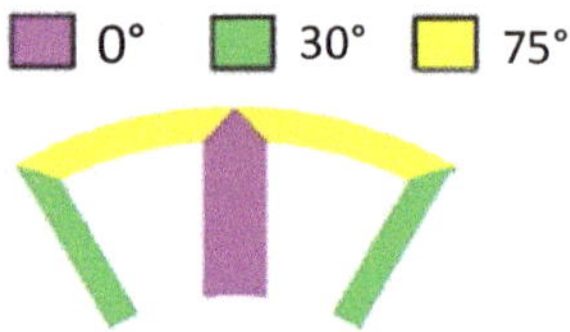

Figure 2. One of the segments of the piecewise isotropic model used for analysis of the 12-slot machine consisting of six segments. Each colour corresponds with the angle away from the rolling direction that uses the magnetic properties of the oriented steel for that angle [17].

While the properties of the 0° segment are known from the manufacturer's data sheet, the other two are generally not and must be measured. Moreover, the number of teeth and the segmentation of the stator determine how many different isotropic segments need to be modelled. It is clear that in order to overcome the shortcomings of the current FEA software and accurately model the segmented oriented steel stator, the magnetic properties of the oriented steel in the desired directions are required. A method of extracting these properties is presented in the next section.

3. Proposed Method for the Extraction of the Magnetic Properties of Oriented Steel

The proposed method is applicable to modular stators consisting of circumferential segmentation, as discussed in [2]. This study focuses on machines where the orientation of each segment is in the direction of the teeth, as shown Figure 1, although the proposed method can be expanded to machines in which the orientation is in the direction of the back iron. The core idea of the proposed method is to calculate the core loss and MMF drop at different levels of flux densities and frequencies of the teeth and the back iron segments of the segmented stator, as shown in Figure 2. These values are subsequently used to estimate the BH and Loss curves in between the rolling and transverse directions. A sensor measures the core loss or MMF drop of a portion of the stator, which is the combination of the core loss of the teeth and back iron. Therefore, to calculate the core loss and MMF drop of each tooth and back iron segment, sensors of different spans are used. In this section, the use of oriented steel segmented stators and sensors is first presented, along with the basic idea of the proposed method. Then, the mathematical model is presented in order to determine the core loss and MMF drop in each tooth and back iron segment.

3.1. Basic Idea of the Proposed Method

Consider a machine consisting of an oriented steel modular stator with N_t teeth. The adjacent teeth of this machine are located $360/N_t$ degrees apart, and the steel orientation of these two teeth are also $360/N_t$ degrees apart. For example, for a 12-slot machine, the two teeth are 30° apart, as shown in Figure 2. As the number of teeth increases, the change in their orientation decreases; hence, data at more discrete orientations are needed to accurately model the machine. This implies that the Epstein frame tests become more impractical as N_t increases. In the proposed method, the stator used to measure the magnetic properties should be the same as the final desired stator design, and the number of segments should be chosen to provide enough different orientation angles for the final analysis. This is demonstrated in Figure 3, where the magnetic properties of the teeth and back iron in one segment are shown for a machine consisting of 72 teeth and N_s number of segments. Clearly, the machine with $N_s = 4$ can provide magnetic data for the oriented steel at more orientation angles compared to the machine with $N_s = 6$. The components with similar magnetic properties are similarly coloured.

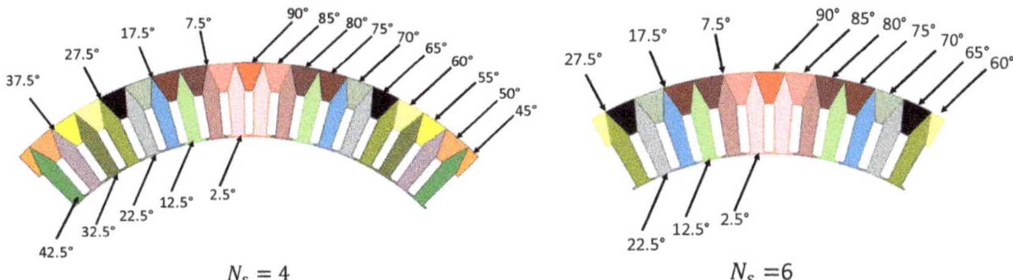

Figure 3. An example of a 72-tooth machine with 4 and 6 segments showing which teeth and back iron sections have similar magnetic properties.

A common starting point in designing the modular stator is the stator used in its non-modular counterpart [2]. In this work, the geometric design of the stator already used for testing is the initial design of the non-modular stator.

The second requirement of the proposed method is the sensor. The sensor imposes a time-dependent magnetic flux of controlled amplitude and frequency in the stator teeth and back iron. It consists of an H-shaped core of high-quality magnetic steel laminations and two coils: a drive coil to impose a current and the resulting magneto motive force (MMF) onto the circuit, and a pickup coil to measure the induced voltage. Figure 4 shows the experimental set-up consisting of oriented steel stator and sensor, indicating the drive coil or primary coil and the pick-up coil or secondary coil.

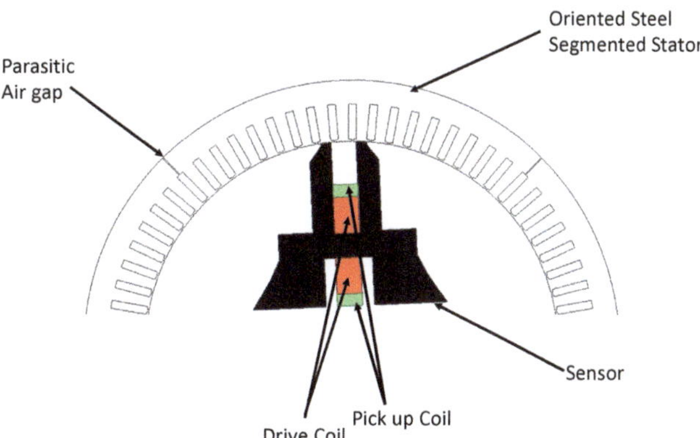

Figure 4. Experimental set-up consisting of oriented steel segments, a sensor, a drive coil, and a pick-up coil.

The motivation to use sensors such as these to inject the flux into the stator is to emulate the direction of flux in the stator teeth and back iron, similar to that during machine operation. The applied flux passes through the teeth, follows the path from the back iron spanned by the sensor, and then returns through the second tooth opposite to the second sensor limb. The teeth directly in line with the sensor limbs provide the properties of the oriented steel in the orientation of the said teeth. The same is true for the portion of the back iron in between the adjacent teeth. A third component is the flux in the outer portions of the teeth between the sensor limbs, but this is not used for analysis due to the non-uniformity of the flux flow in these regions. The flux lines in the portion of the segment injected by the sensor that spans over one tooth with the orientation angle is shown in Figure 5. The relative placement of the teeth for this example is 5°.

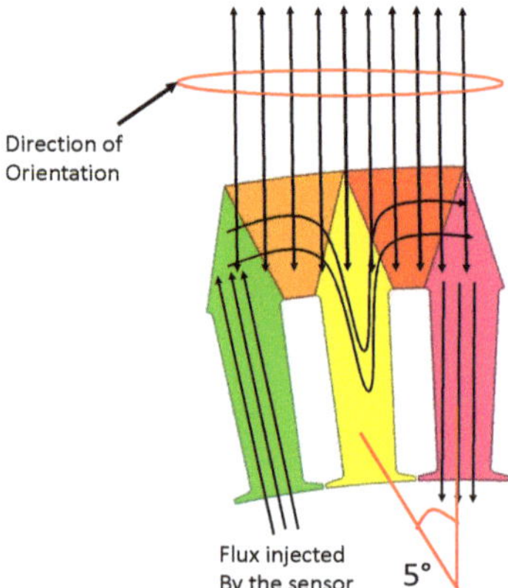

Figure 5. Conceptual drawing of the flux lines in the portion of the segment injected by the sensor spanning over one tooth.

The flux in the pink and green teeth emulates the flux of the machine teeth under operation. Similarly, the red and orange components of the back iron emulates the flux of the machine back under operation. As discussed in Section 2, the pink teeth, green teeth, red back iron, and orange back iron represent the magnetic properties of 0°, 10°, 87.5°, and 82.5° away from the rolling direction. However, due to the non-uniformity of the flux in the yellow portion, the magnetic properties in those teeth cannot be determined with confidence. Therefore, in this work, the magnetic properties of these components are not used to estimate the magnetic properties of the oriented steel. The core losses or MMF drops measured by the sensors are denoted by "Y". Hence, the core losses or MMF drops measured by the sensor from Figure 5 is calculated as follows:

$$Y_{sensor} = Y_{\text{Pink Teeth}} + Y_{\text{Green Teeth}} + Y_{\text{Red Back Iron}} + Y_{\text{Orange Back Iron}} + Y_{\text{Yellow Portion}} \quad (1)$$

In this work, the back iron, teeth, and the teeth with non-uniform flux are denoted by X, T, and $\hat{T}$, respectively. Thus, Equation (1) is reduced to:

$$Y_{sensor} = Y_{T_{0°}} + Y_{T_{10°}} + Y_{X_{87.5°}} + Y_{X_{82.5°}} + Y_{\hat{T}_{5°}} \quad (2)$$

where the angles in the suffix show the magnetic properties X or T. Notice that for $\hat{T}$, the suffix is denoted by 5° just to indicate the location of $\hat{T}$ with respect to the orientation direction. The sensors are used to measure the core losses and MMF drops at different values of flux densities and frequencies at different positions in the stator. For the values of the flux density and frequency, the values of the core loss and MMF drop for each X and T are calculated and then finally used to estimate the core loss and BH curves.

In order to calculate the core loss or MMF drop of X and T, the sensors of different spans are used. The measured values of core losses and MMF drops at different positions, from different sensors, within the oriented steel stator are represented by a set of linear equations. These equations are solved, and the values of the MMF drops or the core losses of individual teeth and back iron are obtained. The number and span of sensors required to achieve this depend on the number of segments and the number of teeth in the oriented steel stator used for testing. The selection of span and number of sensor is illustrated by an example.

Suppose one of the segments consists of an even number of teeth, denoted by "n". Each segment of the stator is symmetric along the axis that divides the segment into two identical parts. Figure 6 shows the line of symmetry, and all the components are coloured the same to represent identical magnetic properties.

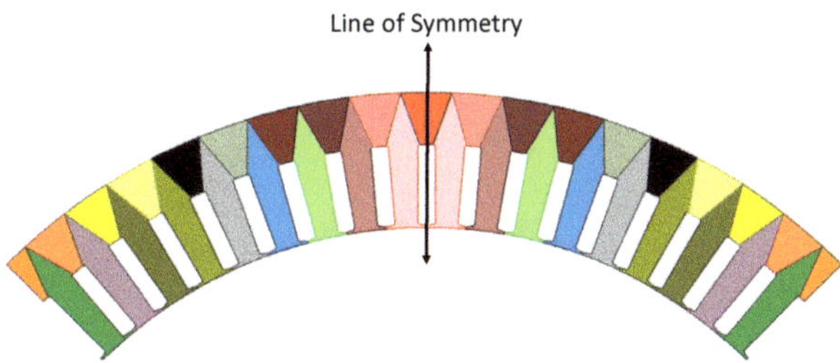

Figure 6. An example showing the line of symmetry that divides the segment into two identical halves.

This means there are n/2 different T and $\hat{T}$, and n/2 + 1 different X that may show different values of core losses or MMF drops for the same values of flux density and frequency. This results in 3n/2 + 1 variables to be solved with different spans and numbers of sensors. The sensor spanning over an odd number of teeth provides n/2 unique equations of the combinations of all three components, while the sensor spanning over an even number of teeth provides n/2 + 1 unique equations. These are obtained when the sensor measures the core losses or MMF drops in the stator at different positions. These equations, when combined, may not yield a total of 3n/2 + 1 unique equations. However, it can be shown that these equations can be solved with the use of simple assumptions. This implies that a combination of two sensors spanning an odd number of teeth and one sensor spanning an even number of teeth can be used to obtain the values of all the components. For the sake of simplicity, the measured core losses or MMF drops, which are the linear combinations of the three components, are arranged in the form of the matrices:

$$Y_{oddsen1}_{\frac{n}{2} \times 1} = A_{oddsen1}_{\frac{n}{2} \times \frac{3n+2}{2}} \times K_{\frac{3n+2}{2} \times 1} \tag{3}$$

$$Y_{oddsen2}_{\frac{n}{2} \times 1} = A_{oddsen2}_{\frac{n}{2} \times \frac{3n+2}{2}} \times K_{\frac{3n+2}{2} \times 1} \tag{4}$$

$$Y_{evensen}_{\frac{n+2}{2} \times 1} = A_{evensen}_{\frac{n+2}{2} \times \frac{3n+2}{2}} \times K_{\frac{3n+2}{2} \times 1} \tag{5}$$

where vector Y consists of the measured values at different sensor positions, matrix A accounts for the number of times each discrete steel orientation angle appears in the sensor span, and vector K represents the core losses or MMF drops of all three components. The same discussion can be extended to measurements with a sensor spanning an odd number of teeth. The application of the proposed method is discussed in the next section.

3.2. Application of the Proposed Method

The stator used for testing had 72 teeth, and the selected number of segments was four. Four segments were selected to extract data at all angles away from the rolling direction in 5° steps, as shown in Figure 3. Therefore, the information extracted from the four segments could still be used to model a machine consisting of two or three segments. The three designed sensors were named Sensor A, B, and C. The spans for the three different sensors are as follows:

1. Sensor spans one tooth in stator (Sensor A);
2. Sensor spans six teeth in stator (Sensor B);
3. Sensor spans thirteen teeth in stator (Sensor C).

The back iron division with respect to the line of symmetry is shown in Figure 7. The pairs of X_i, $\forall\, i \in [2, 10]$, are aligned identically with respect to the rolling direction, and hence they have similar magnetic properties. Moreover, X_1 represents the properties of $90°$ away from the rolling direction, X_2 represents properties of $85°$ away from the rolling direction, and so on.

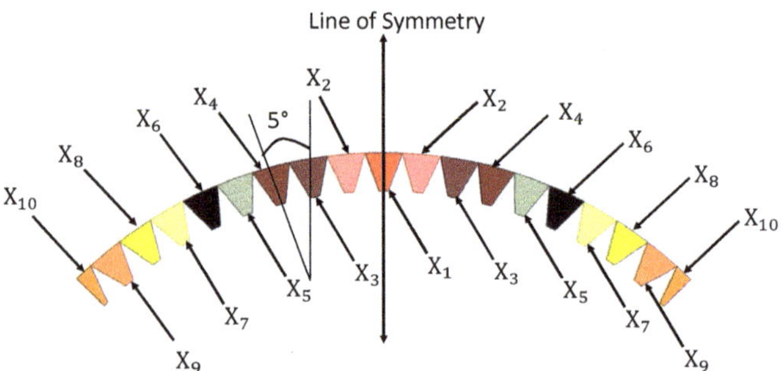

Figure 7. Symmetry of the back iron sections of one segment for a 72-tooth machine consisting of four segments.

The placement of the teeth with respect to the line of symmetry is shown in Figure 8. The pairs of T_j, $\forall\, j \in [1,9]$, have identical magnetic properties due to symmetry. Moreover, T_1 represents the property of $2.5°$ away from the rolling direction, T_2 represents the property of $7.5°$ away from the rolling direction, and so on. Similarly, symmetry holds for $\hat{T}_j$.

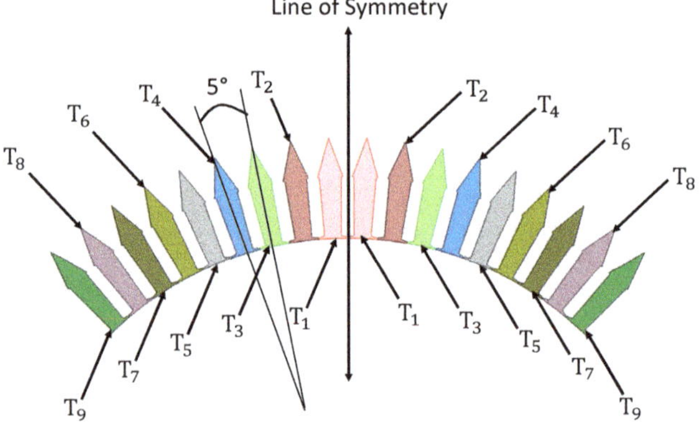

Figure 8. The division of one segment showing the relative placement of the teeth for a machine consisting of 72 teeth and four segments based on symmetry.

The lower values of i and j correspond to the positions of X_i, T_j, or $\hat{T}_j$ on the symmetrical axis, as shown in Figures 7 and 8. Starting from 1, the value of i increases by 1 for every $5°$ shift. Each sensor is used to measure the core losses and the MMF drops of different pieces of the segmented stator at different positions in the stator. To denote the position of the sensor, the following nomenclature is used:

1. The symmetric placement of the sensor is denoted by "Sensor name$_1$". For example, for sensors A, B, and C the symmetric placement of sensors with respect to the symmetrical line are denoted by A_1, B_1, and C_1, respectively;
2. The second position of the sensor is its $5°$ anti-clockwise shift from position 1. For sensors A, B, and C, these positions are denoted by A_2, B_2, and C_2, respectively;
3. Similarly, the third position of the sensor is its $10°$ anti-clockwise shift from position 1.

Positions A_1 and A_2 in Figure 9 show how the position of the sensor with respect to the stator is denoted in this method.

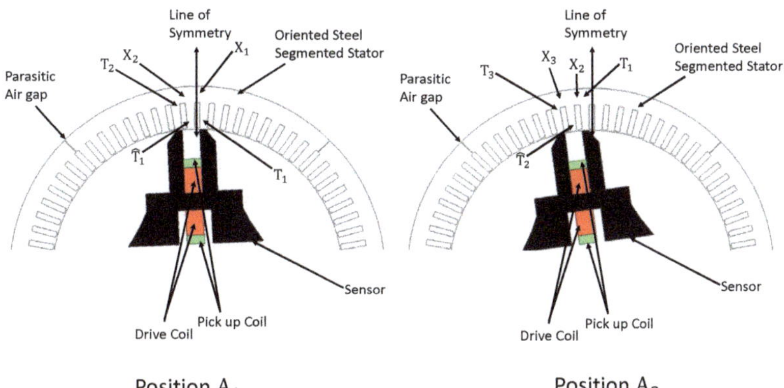

Figure 9. Positioning of Sensor A in positions A_1 and A_2.

The core losses or MMF drops measured by the sensors are denoted by "Y". Therefore, the core losses or MMF drops measured in a symmetrical position using sensor A are denoted by Y_{A_1}. Similarly, the core losses or MMF drops of element X_1 are denoted by Y_{X_1}, and so on. Therefore, the loss measured by sensor A in position A_1, as shown in Figure 9, is written as the sum of five components:

$$Y_{A_1} = Y_{X_1} + Y_{X_2} + Y_{\hat{T}_1} + Y_{T_1} + Y_{T_2} \tag{6}$$

This method divides the core losses or MMF drops in one portion of the segmented stator into the three components, as presented in Equation (6). Using these equations, the core losses and MMF drops for all values of X_i, T_i and $\hat{T}_i$ are determined. For convenience, equations are arranged in the form of matrices. The equations in matrix form for sensors A, B, and C are given as follows:

$$
\begin{bmatrix} Y_{A_1} \\ Y_{A_2} \\ Y_{A_3} \\ Y_{A_4} \\ Y_{A_5} \\ Y_{A_6} \\ Y_{A_7} \\ Y_{A_8} \\ Y_{A_9} \end{bmatrix} =
\begin{bmatrix}
1 & 1 & 0 & 0 & 0 & 0 & 0 & 0 & 0 & 0 & 1 & 1 & 0 & 0 & 0 & 0 & 0 & 0 & 0 & 1 & 0 & 0 & 0 & 0 & 0 & 0 & 0 & 0 \\
0 & 1 & 1 & 0 & 0 & 0 & 0 & 0 & 0 & 0 & 1 & 0 & 1 & 0 & 0 & 0 & 0 & 0 & 0 & 0 & 1 & 0 & 0 & 0 & 0 & 0 & 0 & 0 \\
0 & 0 & 1 & 1 & 0 & 0 & 0 & 0 & 0 & 0 & 0 & 1 & 0 & 1 & 0 & 0 & 0 & 0 & 0 & 0 & 0 & 1 & 0 & 0 & 0 & 0 & 0 & 0 \\
0 & 0 & 0 & 1 & 1 & 0 & 0 & 0 & 0 & 0 & 0 & 0 & 1 & 0 & 1 & 0 & 0 & 0 & 0 & 0 & 0 & 0 & 1 & 0 & 0 & 0 & 0 & 0 \\
0 & 0 & 0 & 0 & 1 & 1 & 0 & 0 & 0 & 0 & 0 & 0 & 0 & 1 & 0 & 1 & 0 & 0 & 0 & 0 & 0 & 0 & 0 & 1 & 0 & 0 & 0 & 0 \\
0 & 0 & 0 & 0 & 0 & 1 & 1 & 0 & 0 & 0 & 0 & 0 & 0 & 0 & 1 & 0 & 1 & 0 & 0 & 0 & 0 & 0 & 0 & 0 & 1 & 0 & 0 & 0 \\
0 & 0 & 0 & 0 & 0 & 0 & 1 & 1 & 0 & 0 & 0 & 0 & 0 & 0 & 0 & 1 & 0 & 1 & 0 & 0 & 0 & 0 & 0 & 0 & 0 & 1 & 0 & 0 \\
0 & 0 & 0 & 0 & 0 & 0 & 0 & 1 & 1 & 0 & 0 & 0 & 0 & 0 & 0 & 0 & 1 & 0 & 1 & 0 & 0 & 0 & 0 & 0 & 0 & 0 & 1 & 0 \\
0 & 0 & 0 & 0 & 0 & 0 & 0 & 0 & 1 & 1 & 0 & 0 & 0 & 0 & 0 & 0 & 0 & 1 & 1 & 0 & 0 & 0 & 0 & 0 & 0 & 0 & 0 & 1
\end{bmatrix} \cdot K \tag{7}
$$

where K is a column vector given by:

$$K^T = [Y_{X_1}\ Y_{X_2}\ Y_{X_3}\ Y_{X_4}\ Y_{X_5}\ Y_{X_6}\ Y_{X_7}\ Y_{X_8}\ Y_{X_9}\ Y_{X_{10}}\ Y_{T_1}\ Y_{T_2}\ Y_{T_3}\ Y_{T_4}\ Y_{T_5}\ Y_{T_6}\ Y_{T_7}\ Y_{T_8}\ Y_{T_9}\ Y_{\hat{T}_1}\ Y_{\hat{T}_2}\ Y_{\hat{T}_3}\ Y_{\hat{T}_4}\ Y_{\hat{T}_5}\ Y_{\hat{T}_6}\ Y_{\hat{T}_7}\ Y_{\hat{T}_8}\ Y_{\hat{T}_9}]$$

Similarly for sensor B and sensor C, the two matrices are given as follows:

$$
\begin{bmatrix} Y_{B_1} \\ Y_{B_2} \\ Y_{B_3} \\ Y_{B_4} \\ Y_{B_5} \\ Y_{B_6} \\ Y_{B_7} \\ Y_{B_8} \\ Y_{B_9} \\ Y_{B_{10}} \end{bmatrix} =
\begin{bmatrix}
1 & 2 & 2 & 2 & 0 & 0 & 0 & 0 & 0 & 0 & 0 & 0 & 0 & 2 & 0 & 0 & 0 & 0 & 0 & 2 & 2 & 2 & 0 & 0 & 0 & 0 & 0 & 0 \\
1 & 2 & 2 & 1 & 1 & 0 & 0 & 0 & 0 & 0 & 0 & 1 & 0 & 1 & 0 & 0 & 0 & 0 & 0 & 2 & 2 & 1 & 1 & 0 & 0 & 0 & 0 & 0 \\
1 & 2 & 1 & 1 & 1 & 1 & 0 & 0 & 0 & 0 & 0 & 1 & 0 & 0 & 0 & 1 & 0 & 0 & 0 & 2 & 1 & 1 & 1 & 1 & 0 & 0 & 0 & 0 \\
1 & 1 & 1 & 1 & 1 & 1 & 1 & 0 & 0 & 0 & 1 & 0 & 0 & 0 & 0 & 0 & 1 & 0 & 0 & 1 & 1 & 1 & 1 & 1 & 1 & 0 & 0 & 0 \\
0 & 1 & 1 & 1 & 1 & 1 & 1 & 1 & 0 & 0 & 1 & 0 & 0 & 0 & 0 & 0 & 0 & 1 & 0 & 0 & 1 & 1 & 1 & 1 & 1 & 1 & 0 & 0 \\
0 & 0 & 1 & 1 & 1 & 1 & 1 & 1 & 1 & 0 & 0 & 1 & 0 & 0 & 0 & 0 & 0 & 0 & 1 & 0 & 0 & 1 & 1 & 1 & 1 & 1 & 1 & 0 \\
0 & 0 & 0 & 1 & 1 & 1 & 1 & 1 & 1 & 1 & 0 & 0 & 1 & 0 & 0 & 0 & 0 & 0 & 1 & 0 & 0 & 0 & 1 & 1 & 1 & 1 & 1 & 1 \\
0 & 0 & 0 & 0 & 1 & 1 & 1 & 1 & 1 & 2 & 0 & 0 & 0 & 1 & 0 & 0 & 0 & 1 & 0 & 0 & 0 & 0 & 1 & 1 & 1 & 1 & 1 & 2 \\
0 & 0 & 0 & 0 & 0 & 1 & 1 & 1 & 2 & 2 & 0 & 0 & 0 & 0 & 1 & 0 & 1 & 0 & 0 & 0 & 0 & 0 & 0 & 1 & 1 & 2 & 2 & 0 \\
0 & 0 & 0 & 0 & 0 & 0 & 1 & 1 & 2 & 2 & 0 & 0 & 0 & 0 & 0 & 2 & 0 & 0 & 0 & 0 & 0 & 0 & 0 & 0 & 2 & 2 & 2 & 0
\end{bmatrix} \cdot K \tag{8}
$$

$$
\begin{bmatrix} Y_{C_1} \\ Y_{C_2} \\ Y_{C_3} \\ Y_{C_4} \\ Y_{C_5} \\ Y_{C_6} \\ Y_{C_7} \\ Y_{C_8} \\ Y_{C_9} \end{bmatrix} = \begin{bmatrix}
1 & 2 & 2 & 2 & 2 & 2 & 2 & 1 & 0 & 0 & 0 & 0 & 0 & 0 & 0 & 0 & 1 & 1 & 0 & 2 & 2 & 2 & 2 & 2 & 1 & 0 & 0 \\
1 & 2 & 2 & 2 & 2 & 2 & 1 & 1 & 1 & 0 & 0 & 0 & 0 & 0 & 0 & 1 & 0 & 0 & 1 & 2 & 2 & 2 & 2 & 2 & 1 & 1 & 1 & 0 \\
1 & 2 & 2 & 2 & 2 & 1 & 1 & 1 & 1 & 1 & 0 & 0 & 0 & 0 & 1 & 0 & 0 & 0 & 1 & 2 & 2 & 2 & 2 & 1 & 1 & 1 & 1 & 1 \\
1 & 2 & 2 & 2 & 1 & 1 & 1 & 1 & 2 & 1 & 0 & 0 & 0 & 1 & 0 & 0 & 0 & 1 & 0 & 2 & 2 & 2 & 1 & 1 & 1 & 1 & 1 & 2 \\
1 & 2 & 2 & 1 & 1 & 1 & 1 & 2 & 2 & 1 & 0 & 0 & 1 & 0 & 0 & 0 & 1 & 0 & 0 & 2 & 2 & 1 & 1 & 1 & 1 & 1 & 2 & 2 \\
1 & 2 & 1 & 1 & 1 & 1 & 2 & 2 & 2 & 1 & 0 & 1 & 0 & 0 & 0 & 1 & 0 & 0 & 0 & 2 & 1 & 1 & 1 & 1 & 1 & 2 & 2 & 2 \\
1 & 1 & 1 & 1 & 1 & 2 & 2 & 2 & 2 & 1 & 1 & 0 & 0 & 0 & 1 & 0 & 0 & 0 & 0 & 1 & 1 & 1 & 1 & 1 & 2 & 2 & 2 & 2 \\
0 & 1 & 1 & 1 & 2 & 2 & 2 & 2 & 2 & 1 & 1 & 0 & 0 & 1 & 0 & 0 & 0 & 0 & 0 & 0 & 1 & 1 & 1 & 2 & 2 & 2 & 2 & 2 \\
0 & 0 & 1 & 2 & 2 & 2 & 2 & 2 & 2 & 1 & 0 & 1 & 1 & 0 & 0 & 0 & 0 & 0 & 0 & 0 & 0 & 1 & 2 & 2 & 2 & 2 & 2 & 2
\end{bmatrix} \cdot K \qquad (9)
$$

All the equations given in (7)–(9) are used to calculate the values of Y_{X_i} and Y_{T_i}.

4. Experimental Set-Up, Estimation of BH and Loss Curves, and Experimental Results

4.1. Experimental Set-Up

All the sensor limbs are identical, while the control coil voltage, a wound on the sensor limb, is maintained as a sinusoidal signal. The amplitude of the control coil voltage is adjusted by modifying the primary supply voltage, as discussed in [18], in order to obtain the desired flux density in the sensor limb ($\hat{B}_{limb}$). The sensor measures the total MMF drop of the sensor stator arrangement consisting of a portion of the segment, the sensor itself, and the air gaps between the sensor limb and the stator tooth. It also measures the total core loss of the sensor and the corresponding portion of the segment. To calculate the MMF drop/core loss of the portion of the stator segment, the MMF drop of the sensor and the air gaps/core loss of the sensor is subtracted. The MMF drop/core loss of the sensor and the two air gaps/sensor are calculated by measuring the MMF drop and core loss of two identical sensors placed head, on as shown in Figure 10. Both the primary windings are excited with the same primary current to obtain the desired flux density ($\hat{B}_{limb}$) in the sensor limb. The total MMF drop of the set-up consists of two sensors and four air gaps, each with a piece of steel and a sensor limb, and the total core loss of two sensors. Hence, the MMF drop/core loss of the sensor and two air gaps are half of the total. The error due to the presence of a piece of steel in the estimation of core losses and MMF drops is reduced to an acceptable limit by making the area of the piece of steel large enough such that the flux density is small.

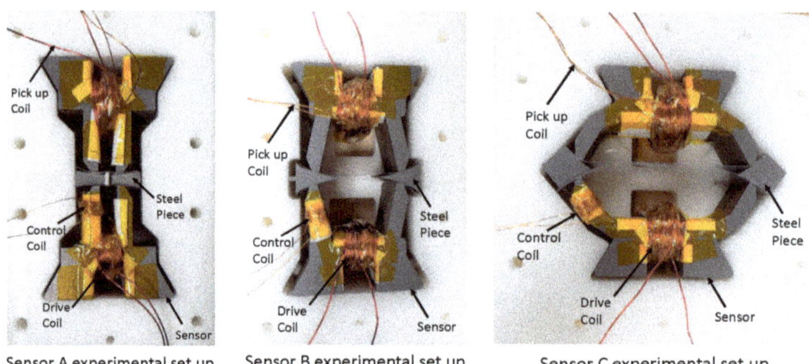

Figure 10. Sensor characterization experimental set-up with small steel pieces for all sensors.

The experimental set-up consisting of a sensor and an oriented steel segmented stator is shown in Figure 11. A two-sensor arrangement, shown in [19], is used for testing to avoid flux leakage through the back iron.

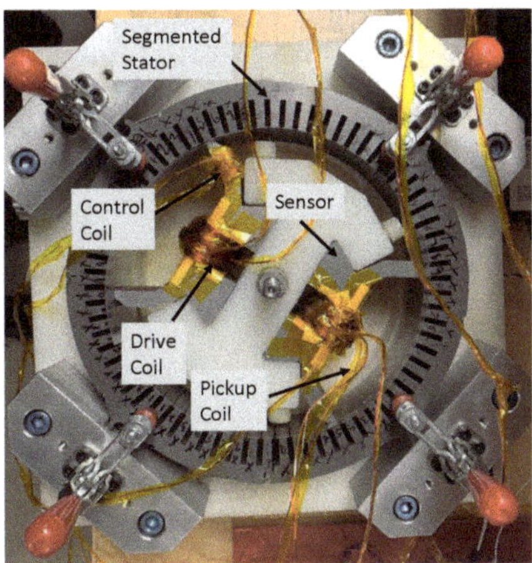

Figure 11. Experimental set-up consisting of a sensor and a stator with the drive, pick-up, and control coils.

The drive coil voltage is applied through the amplifier, while the control coil voltage, pick-up voltage, and primary current are recorded using LabVIEW through an SCB box from National Instruments. The overall experimental set-up consisting of a computer that controls LabVIEW, which is used to send and receive signals from the SCB box, and the amplifier that supplies voltage to the drive coil, is shown in Figure 12.

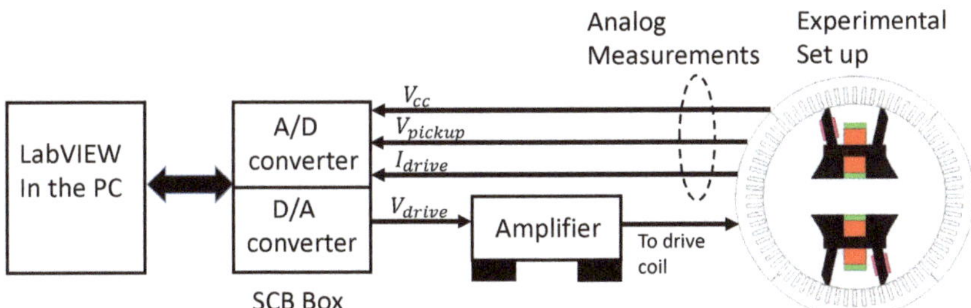

Figure 12. Schematics showing the overall experimental set-up and its main components.

4.2. Procedure for the Estimation of BH and Loss Curves of the Oriented Steel

The equations discussed in Section 3 are used to calculate the values of the core losses and MMF drops of X_i and T_i at selected values of flux density and frequencies, as shown in Table 1.

Table 1. Values of $\hat{B}_{limb}$ and the frequency used to perform the experiments.

Supply Frequency (in Hz)	Levels of $\hat{B}_{limb}$ (in T)
50	0.2, 0.4, 0.6, 0.8, 1.0, 1.2, 1.4, 1.5
100	0.2, 0.4, 0.6, 0.8, 1.0, 1.2, 1.4
150	0.2, 0.4, 0.6, 0.8, 1.0, 1.2, 1.3

The core loss for each field intensity (H) and orientation angle is calculated using the core loss and MMF drop values of X_i and T_i, combined with the data supplied by the manufacturer for the rolling and transverse directions. Section 2 shows that X_1 represents the properties of 90° away from the rolling direction, and X_2 represents the properties of 85° away from the rolling direction, and so on. For a given value of flux density and frequency, the % change in core loss of X_i with X_i is as follows:

$$\Delta x_i = \frac{Y_{X_i} - Y_{X_1}}{Y_{X_1}} \cdot 100 \tag{10}$$

Since X_1 is oriented at 90° away from the rolling direction, the core loss density (in W/m^3) in the transverse direction for the same value of flux density and frequency, denoted by $Y_{transverse}$, is used to calculate the core loss corresponding to the angle of orientation of X_i. For example, the value of $Y_{85°}$ is calculated using the following:

$$\Delta x_2 = \frac{Y_{X_2} - Y_{X_1}}{Y_{X_1}} \cdot 100 = \frac{Y_{85°} - Y_{transverse}}{Y_{transverse}} \cdot 100 \tag{11}$$

Therefore, the estimated value of core loss (in W/m^3) for the 85° angle away from the rolling direction, at a given value of flux density and frequency, is as follows:

$$Y_{85°} = \frac{\Delta x_2 + 100}{100} \cdot Y_{transverse} \tag{12}$$

The same process is repeated for all the values of X_i, at all values of flux densities and frequencies, in order to obtain the core losses for the values of orientation from 85° to 45°. Moreover, using the MMF drop values of X_i from the process discussed above, and using the value of H (in A/m) from the transverse direction, flux data are generated. Finally, the core loss and MMF drop values of T_i with T_1 are used to estimate the core loss and BH curves for all the orientation angles between 0° and 40°.

4.3. Experimental Results

The estimated BH and loss curves were obtained from the core loss and MMF drop values of X_i and T_i using the analysis discussed in Section 4.2. It was observed that at 55° away from the rolling direction had highest core losses, while the lowest losses were in the rolling direction. This is a long-known result and proves the correctness of the proposed method. However, for the FEA-interpolated curves, the rolling direction had the lowest losses, and the transverse direction had the highest losses. The core loss curves obtained from the analysis of the experimental results and the FEA-interpolated core loss curves with a supply frequency of 50 Hz are shown in Figure 13.

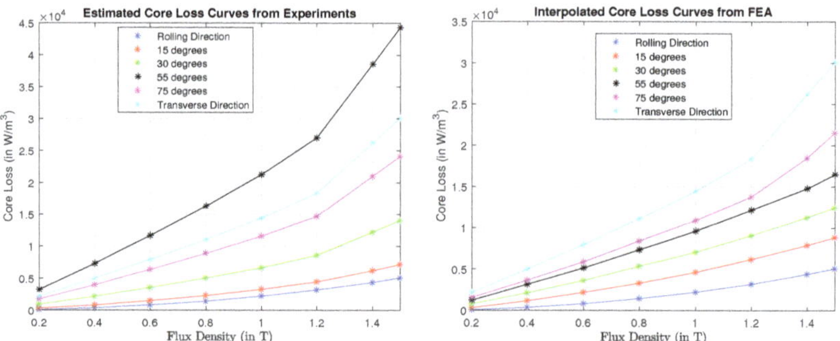

Figure 13. Comparison of the core losses obtained from the analysis of experimental results and the FEA-interpolated core loss curves using the built-in function in the JMAG software for different angles away from the rolling direction, at a supply frequency of 50 Hz.

The variation of the core loss curves, with the orientation angle obtained from the analysis of experimental results and the FEA-interpolated core loss curves for 1.5 T and a supply frequency of 50 Hz, is shown in Figure 14. This variation shows the difference between the interpolated properties of the oriented steel in the FEA (the conventional method) compared to the actual properties of the oriented steel obtained from the experiments. This shows the importance of the proposed method in obtaining the characteristics of oriented steel, for the correct modelling of the oriented steel modular PMSM.

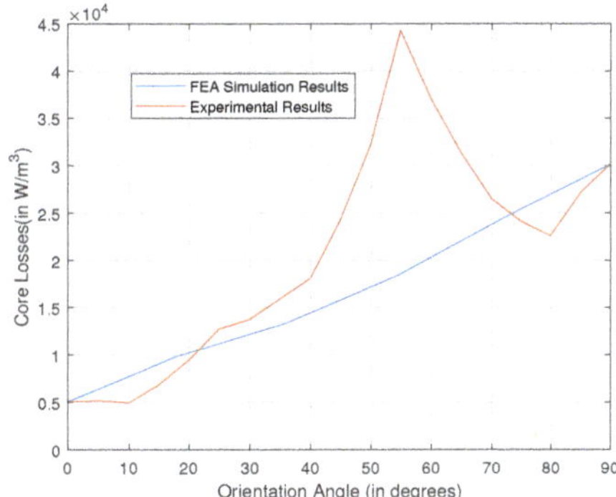

Figure 14. The variation of core losses, with an orientation angle obtained from the analysis of experimental results and simulations using the JMAG FEA software at 1.5 T and a supply frequency of 50 Hz.

5. Analysis of Oriented Steel Modular Stator PMSM

In this section, the BH and loss curves that were obtained from the analysis of experimental results are used in the piecewise isotropic model, discussed in Section 2, are used to model an oriented steel segmented stator. The oriented steel has the lowest core loss and the highest permeability along the rolling direction, while losses are the highest and permeability the lowest along the 55° direction, as shown in Figure 14.

This implies that building the whole stator with a single sheet of oriented steel will not lead to the proper utilization of its magnetic properties. Therefore, a segmented-stator PMSM with different numbers of segments, where segmentation is performed along the slot in the circumferential direction, is used for analysis. Moreover, the effect of increased cut edges, discussed in [2], on the performance of the machine is not studied in this work. The analysis is performed on the 12-pole, 72-slot PMSM. The selected orientation direction of each segment is along the general direction of the teeth, i.e., the center tooth is at 0°. This orientation direction is selected because the flux density in the teeth is higher than in the back iron. One problem with segmented stators is the presence of unavoidable gaps between the segments, also known as parasitic gaps. Parasitic gaps reduce the total flux in the machine, leading to a reduced average torque, and the effect increases with the number of segments. Conversely, as the number of segments increases, the magnetic properties of the teeth improves on average. The purpose of this analysis is to find the best possible number of segments in order to minimize the core losses and maximize the average torque.

The performance of the anisotropic modular stator PMSM is compared with modular and non-modular non-oriented steel PMSMs. The BH and loss curves in oriented and non-oriented steel are shown in Figure 15. The rolling direction has both higher permeability and lower losses compared to the non-oriented steel. Moreover, the non-oriented steel has slightly better permeability and lower losses compared to the transverse direction in the oriented steel. The design is analyzed using 4, 8, 9, 12, and 18 segments. The effect of the size of the parasitic gap is also considered. Laser-cutting tolerances result in a maximum possible parasitic gap of 0.2 mm. Therefore, in this study, the value of the parasitic air gap is increased from 0 mm, which is the ideal case, to 0.2 mm in steps of 0.05 mm. In this work, all the gaps are assumed to be uniform and equal between all the segments. The comparison of the three models—modular oriented steel stator (A_{ori}), modular non-oriented steel stator (A_{nonori}), and non-modular non-oriented steel stator ($A_{conventional}$)—is discussed in the following sub sections.

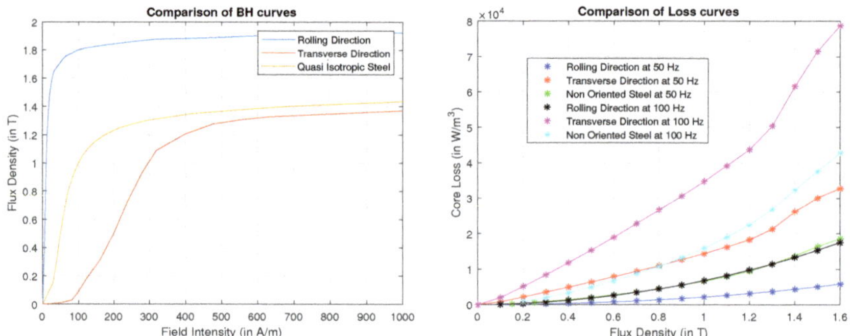

Figure 15. Comparison of the BH and core loss curves for the rolling, transverse direction and the quasi-isotropic steel.

5.1. Variation of λ_d and λ_q

In this section, a comparison of the average torque and core losses for $A_{conventional}$, A_{nonori}, and A_{ori} at point O, with the stator current = 100 A, the power angle = 120°, and speed = 1000 RPM, is presented.

To explain the effect of oriented and non-oriented steel on core loss and average torque, the variation of the λ_d and λ_q linkage must be explained first. The values of λ_d and λ_q in the IPM are explained using the first reluctance path, the second reluctance path, and the magnet path, as proposed in [20] and shown in Figure 16. In [20], it is stated that the primary and secondary reluctance flux paths contribute towards λ_q, while the magnet flux path contribute towards λ_d. The presence of parasitic gaps between the segments in the stator leads to an increase in the reluctance in the direction of both the *d*- and *q*-axis flux paths. The reluctance in the *d*-axis flux path in the non-modular stator is due to the magnet, machine air gap, and steel present in both the stator and the rotor. On the other hand, the reluctance in the q-axis flux path in the non-modular stator is only due to the machine air gap and steel present in both stator and rotor. The reluctance of rare earth magnets is close to that of air. Therefore, segmentation leads to a higher change in the reluctance of the *q*-axis as compared to the *d*-axis, and also causes the decrease in the *q*-axis flux to be more dominant than the change in the *d*-axis flux. Moreover, the change in flux linkage due to the improved quality of steel is more prominent in the *q*-axis as compared to the *d*-axis. Figures 17 and 18 show the variation of λ_d and λ_q at point O for A_{nonori} and A_{ori}, respectively, with the parasitic gaps and the number of segments supporting the claims made in this section.

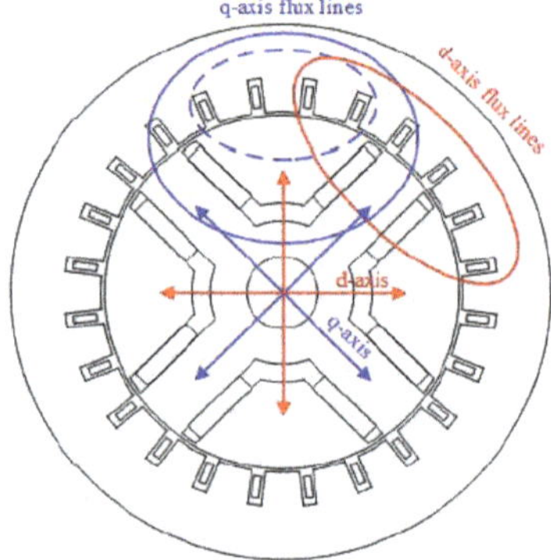

Figure 16. IPM flux paths: first reluctance path (solid blue), second reluctance path (dashed blue), and magnet path (solid red) [20].

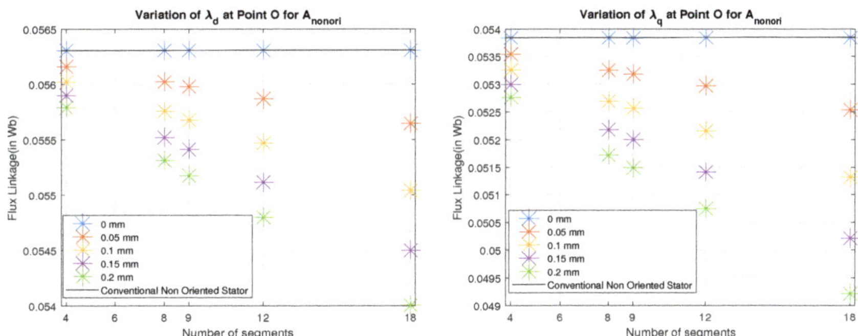

Figure 17. Variation of λ_d and λ_q for A_{nonori} at $I_s = 100$ A, $\delta = 120°$, and speed = 1000 RPM.

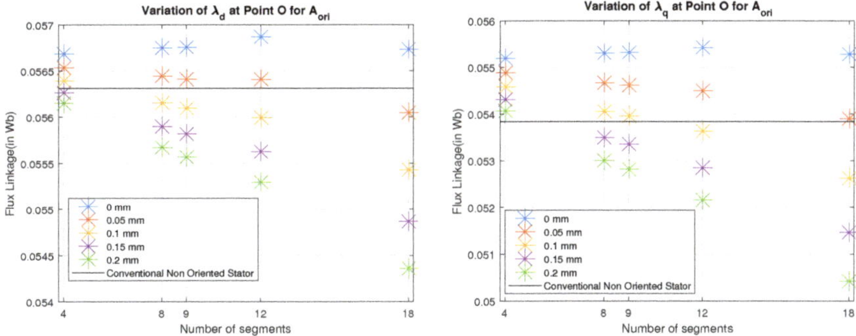

Figure 18. Variation of λ_d and λ_q for A_{ori} at $I_s = 100$ A, $\delta = 120°$, and speed = 1000 RPM.

5.2. Comparison of Core Loss

The three teeth closest to the rolling direction have almost identical magnetic properties, which are approximately the same as the properties in the rolling direction. There is an increase in the core losses from the fourth teeth onwards, i.e., more than 12.5° from the rolling direction, as shown in Figure 19.

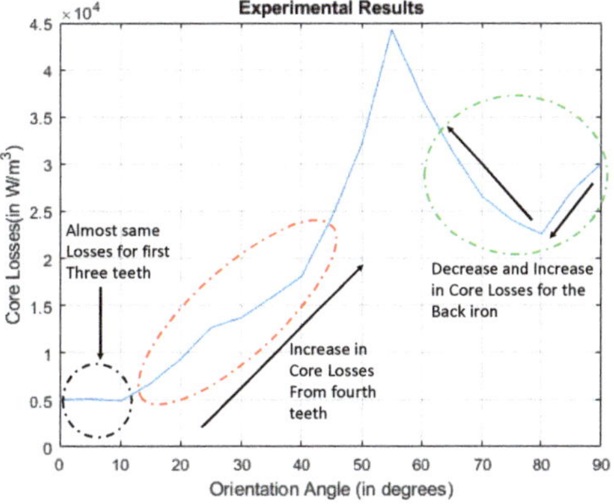

Figure 19. The variation of core losses with the orientation angle obtained from the analysis of experimental results, at 1.5 T and a supply frequency of 50 Hz. This shows the core loss variation for the teeth and the back iron from the line of symmetry.

The core loss in the back iron decreases as the orientation angle changes from 90° to 80°, and then increases, as shown in Figure 19. This implies that the back iron has the best properties for the first three divisions. Therefore, when the number of teeth and back iron divisions are 6 in each segment, which means $N_s = 12$, the segment has the best possible magnetic properties when the value of the parasitic gap is 0 mm. However, the higher decrease in the flux linkage values for A_{ori} from $N_s = 12$ to $N_s = 18$ when the parasitic gap is 0.2 mm, shown in Figure 18, results in the decrease of the core loss from $N_s = 12$ to $N_s = 18$. Figure 20 shows the variation of the core loss of the complete stator, with the number of segments for A_{ori} at point O for parasitic gap values of 0 mm and 0.2 mm.

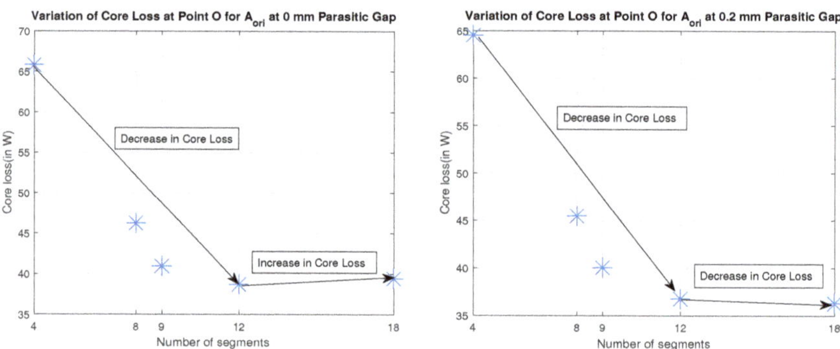

Figure 20. Core loss variation of the complete stator for A_{ori} at $I_s = 100$ A, $\delta = 120°$, and speed = 1000 RPM at a parasitic gap of 0 mm and 0.2 mm.

Moreover, due to the superior magnetic properties of the teeth of the A_{ori}, the losses are lower compared to the A_{nonori} and $A_{conventional}$ when N_s is 9, 12, and 18. The losses of A_{nonori} and $A_{conventional}$ are almost similar due to same material being used in both machines. For the non-zero values of the parasitic gaps, the value of the core loss is lower for A_{nonori} than $A_{conventional}$ due to the decrease in flux in the machine. Figure 21 shows the variation of the core loss, with the number of segments for all machines at point O, and a value of 0.2 mm for the parasitic gap.

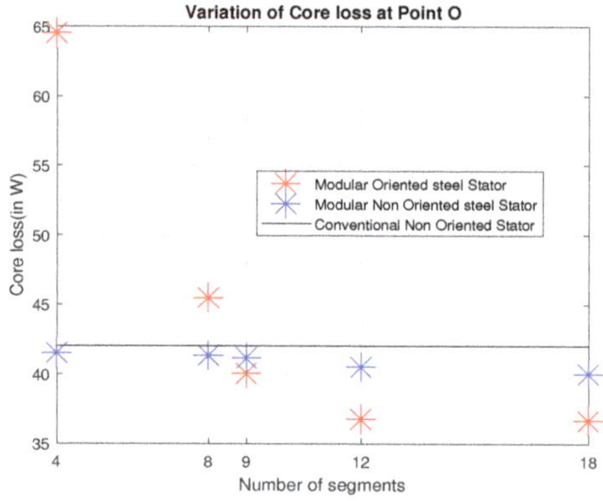

Figure 21. Core loss variation for all machines at $I_s = 100$ A, $\delta = 120°$, and speed = 1000 RPM, and a parasitic gap value of 0.2 mm.

5.3. Comparison of Average Torque

There is an improvement in the average torque in the A_{ori} compared with the A_{nonori} and $A_{conventional}$ because of the higher permeability of the oriented steel when the value of the parasitic gaps between the segments is 0 mm. The value of the average torque is maximum when the number of segments is 12 for A_{ori}, as both λ_d and λ_q are maximized at $N_s = 12$ for a 0 mm parasitic gap, as shown in Figure 18. However, when the parasitic gap is 0.2 mm, the values of both the d and q axis flux linkages are less as compared to the $A_{conventional}$ machine, and hence the average torque is also lower. The only exception is when A_{ori} has four segments. As shown in Figure 18, the value of λ_q is slightly higher for A_{ori} as compared to $A_{conventional}$ when the parasitic gap is 0.2 mm, which overshadows the lower values of λ_d, leading to a slightly higher torque for A_{ori}. Moreover, the average torque for A_{ori} is higher than A_{nonori} due to the better quality of the stator steel at all values of the parasitic gap, with same value of N_s. Finally, the average torque value for A_{nonori} is lower compared to the $A_{conventional}$ due to the decrease in flux linkage when the values of the parasitic gaps are non-zero. Figure 22 shows the variation of the average torque, with the number of segments for all machines and for parasitic gap values of 0 and 0.2 mm.

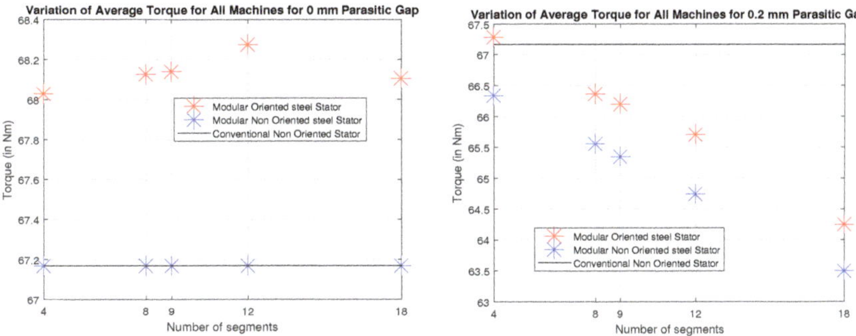

Figure 22. Variation of average torque with the number of segments for all machines and parasitic gaps of 0 and 0.2 mm.

6. Selection of the Number of Segments for the Best Possible Performance of the Machine

The selection criterion considers the worst possible scenario, i.e., when the parasitic gaps between the segments is 0.2 mm. Figure 23 shows the % change in the values of average torque and core loss from the reference value of $N_s = 18$, which shows the minimum average torque and the minimum core losses at point O.

- Selection of Number of Segments for the Oriented Steel Model
  First, it is clear from Figure 23 that the increase in average torque is much less than the increase in core loss as the number of segments is decreased from 18 to 4. Second, the increase in core loss from $N_s = 18$ to $N_s = 12$ is negligibly small; however, there is a significant increase in the average torque. Third, for 8, 9, and 4 segments, there is an increase in the average torque as compared to $N_s = 12$. However, the increase in core loss is far greater than the increase in the average torque. The value of the lower torque for $N_s = 12$ can be compensated by injecting a slightly higher current, which will lead to slightly higher core and copper losses. An analysis of the machine with an increased current is part of future work, and beyond the scope of this paper. Nevertheless, considering the two objectives of average torque and core losses, it can be inferred that for A_{ori}, the best possible performance is obtained from a machine consisting of 12 segments.
- Selection of Number of Segments for the Non-Oriented Model
  Figure 23 shows the almost-linear trend of the increase in core losses and average torque for the A_{nonori} machine, from $N_s = 18$ to $N_s = 4$. This is due to the fact that as the number of segments decreases, the flux in the machine increases, which then leads to a decrease in both core loss and average torque. Similar to the A_{ori} case, the lost torque can be increased by injecting slightly more current. Only an analysis

that considers both core loss and copper loss at higher values of current for same torque can provide details for the best possible selection of the number of segments for A_{nonori}.

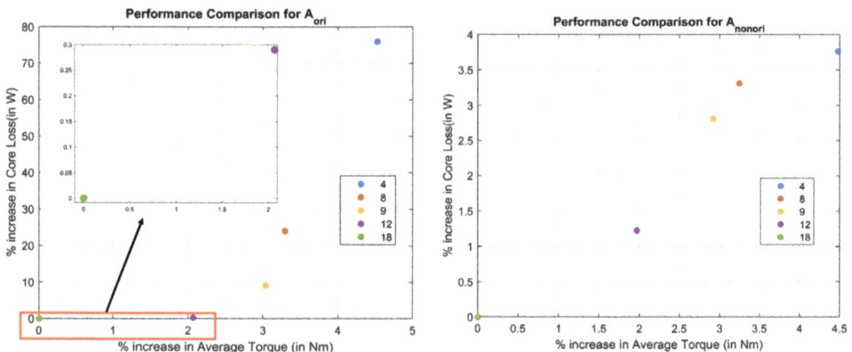

Figure 23. Comparison of the performance of different numbers of segments for oriented and non-oriented segmented stators at I_s = 100 A, δ = 120°, and speed = 1000 RPM, with a parasitic gap value of 0.2 mm.

7. Conclusions

In this paper, a method for estimating the magnetic characteristics of oriented steel was proposed. The method utilized oriented steel segments and flux-injecting probes to obtain the magnetic characteristics of the oriented steel. This data were then used to model a modular stator consisting of oriented steel. It was shown that the experimentally-extracted magnetic characteristics of the oriented steel were different from the FEA method that utilized the properties of both the rolling and transverse directions to interpolate the magnetic characteristics. The obtained magnetic characteristics of the oriented steel were used to model the modular machine, and the oriented steel modular stator was compared to the non-oriented modular and non-modular stators.

The analysis showed that the number of segments in the oriented-steel modular-stator machine can be selected to minimize losses, although this selection is likely design-specific. For example, this was 12 segments for the unloaded machine and 18 segments for the loaded machine. Additionally, for the sample machine, the oriented steel modular stator machine had less loss for all values of the segments as compared to the non-modular and modular non-oriented steel machines under loaded and unloaded conditions. An improvement in the values of the average torque was also shown for the oriented steel modular stator machine as compared to the non-modular and modular non-oriented stator machines. However, the improvement was overshadowed at higher values of the number of segments due to the presence of parasitic gaps. The only exception to this variation was the improvement of the average torque for the oriented-steel machine with four segments. Moreover, the average torque of the non-oriented modular stator machine was lower compared to the non-modular non-oriented steel machine due to the presence of the parasitic gaps. Finally, it was shown that the oriented steel machine performs best when the number of segments is 12, while taking into account both objectives: core losses and average torque.

Future work should include an analysis of the oriented steel modular stator and the homogeneous steel modular stator machines at slightly higher currents in order to compare the core losses and copper losses for the same average torque. Finally, the stator will be redesigned for the oriented modular stator machine to achieve the desired objectives of core loss minimization and average torque maximization.

Author Contributions: Conceptualization, A.A.; methodology, A.A.; software, A.A.; validation, A.A.; formal analysis, A.A.; investigation, A.A.; resources, A.A.; data curation, A.A.; writing—original draft preparation, A.A. and M.M.; writing—review and editing, A.A., M.M., E.S. and J.A.; visualization, A.A.; supervision, E.S.; project administration, J.A.; funding acquisition, J.A. All authors have read and agreed to the published version of the manuscript.

Funding: This research was funded by GM Global Technical Center, Warren, MI, USA grant number GAC 3146.

Institutional Review Board Statement: Not applicable.

Informed Consent Statement: Informed consent was obtained from all subjects involved in the study.

Conflicts of Interest: The authors declare no conflict of interest.

References

1. Li, G.J.; Zhu, Z.Q.; Foster, M.; Stone, D. Comparative Studies of Modular and Unequal Tooth PM Machines Either with or without Tooth Tips. *IEEE Trans. Magn.* **2014**, *50*, 1–10. [CrossRef]
2. Aggarwal, A.; Strangas, E.G.; Karlis, A. Review of Segmented Stator and Rotor Designs for AC Electric Machines. In Proceedings of the 2020 International Conference on Electrical Machines (ICEM), Gothenburg, Sweden, 23–26 August 2020; pp. 2342–2348.
3. Huang, S.; Strangas, E.G.; Aggarwal, A.; Li, K.; Niu, F. Robust Inter-turn Short-circuit Detection in PMSMs with Respect to Current Controller Bandwidth. In Proceedings of the 2019 IEEE Energy Conversion Congress and Exposition (ECCE), Baltimore, MD, USA, 29 September–3 October 2019; pp. 3897–3904.
4. Huang, S.; Aggarwal, A.; Strangas, E.G.; Li, K.; Niu, F.; Huang, X. Robust Stator Winding Fault Detection in PMSMs With Respect to Current Controller Bandwidth. *IEEE Trans. Power Electron.* **2021**, *36*, 5032–5042. [CrossRef]
5. Aggarwal, A.; Strangas, E.G.; Agapiou, J. Robust Voltage based Technique for Automatic Off-Line Detection of Static Eccentricity of PMSM. In Proceedings of the 2019 IEEE International Electric Machines Drives Conference (IEMDC), San Diego, CA, USA, 12–15 May 2019; pp. 351–358.
6. Aggarwal, A.; Allafi, I.M.; Strangas, E.G.; Agapiou, J.S. Off-Line Detection of Static Eccentricity of PMSM Robust to Machine Operating Temperature and Rotor Position Misalignment Using Incremental Inductance Approach. *IEEE Trans. Transp. Electrif.* **2021**, *7*, 161–169. [CrossRef]
7. Aggarwal, A.; Strangas, E.G. Review of Detection Methods of Static Eccentricity for Interior Permanent Magnet Synchronous Machine. *Energies* **2019**, *12*, 4105. [CrossRef]
8. Li, G.J.; Zhu, Z.Q.; Foster, M.P.; Stone, D.A.; Zhan, H.L. Modular Permanent-Magnet Machines With Alternate Teeth Having Tooth Tips. *IEEE Trans. Ind. Electron.* **2015**, *62*, 6120–6130. [CrossRef]
9. Tomida, T.; Sano, N.; Hinotani, S.; Fujiwara, K.; Kotera, H.; Nishiyama, N.; Ikkai, Y. Application of fine-grained doubly oriented electrical steel to IPM synchronous motor. *IEEE Trans. Magn.* **2005**, *41*, 4063–4065. [CrossRef]
10. Sugawara, Y.; Akatsu, K. Characteristics of a Switched Reluctance Motor using Grain-Oriented electric steel sheet. In Proceedings of the 2013 International Conference on Electrical Machines and Systems (ICEMS), Busan, Korea, 26–29 October 2013; pp. 18–23. [CrossRef]
11. Ma, J.; Li, J.; Fang, H.; Li, Z.; Liang, Z.; Fu, Z.; Xiao, L.; Qu, R. Optimal Design of an Axial-Flux Switched Reluctance Motor With Grain-Oriented Electrical Steel. *IEEE Trans. Ind. Appl.* **2017**, *53*, 5327–5337. [CrossRef]
12. Chwastek, K.R.; Baghel, A.P.S.; de Campos, M.F.; Kulkarni, S.V.; Szczygłowski, J. A Description for the Anisotropy of Magnetic Properties of Grain-Oriented Steels. *IEEE Trans. Magn.* **2015**, *51*, 1–5. [CrossRef]
13. Liu, J.; Basak, A.; Moses, A.; Shirkoohi, G. A method of anisotropic steel modelling using finite element method with confirmation by experimental results. *IEEE Trans. Magn.* **1994**, *30*, 3391–3394. [CrossRef]
14. Lin, D.; Zhou, P.; Badics, Z.; Fu, W.; Chen, Q.; Cendes, Z. A new nonlinear anisotropic model for soft magnetic materials. *IEEE Trans. Magn.* **2006**, *42*, 963–966. [CrossRef]
15. Di Napoli, A.; Paggi, R. A model of anisotropic grain-oriented steel. *IEEE Trans. Magn.* **1983**, *19*, 1557–1561. [CrossRef]
16. Gao, L.; Zeng, L.; Yang, J.; Pei, R. Application of grain-oriented electrical steel used in super-high speed electric machines. *AIP Adv.* **2020**, *10*, 015127. [CrossRef]
17. Maraví-Nieto, J.; Azar, Z.; Thomas, A.; Zhu, Z. Utilisation of grain-oriented electrical steel in permanent magnet fractional-slot modular machines. *J. Eng.* **2019**, *2019*, 3682–3686. [CrossRef]
18. Nazrulla, S.; Strangas, E.G.; Agapiou, J.S.; Perry, T.A. A Device for the Study of Electrical Steel Losses in Stator Lamination Stacks. *IEEE Trans. Ind. Electron.* **2014**, *61*, 2217–2224. [CrossRef]
19. Jeong, K.; Ren, Z.; Yoon, H.; Koh, C. Measurement of Stator Core Loss of an Induction Motor at Each Manufacturing Process. *J. Electr. Eng. Technol.* **2014**, *9*, 1309–1314. [CrossRef]
20. Hayslett, S.; Strangas, E. Analytical Design of Sculpted Rotor Interior Permanent Magnet Machines. *Energies* **2021**, *14*, 5109. [CrossRef]

 *energies*

Article

Analytical Design of Sculpted Rotor Interior Permanent Magnet Machines

Steven Hayslett [1],* **and Elias Strangas** [2]

[1] Department of Mechanical Engineering, Michigan State University, East Lansing, MI 48824, USA
[2] Department of Electrical and Computer Engineering, Michigan State University, East Lansing, MI 48824, USA; strangas@egr.msu.edu
* Correspondence: hayslet4@egr.msu.edu

Abstract: A computationally efficient design of interior permanent magnet (IPM) motor rotor features is investigated utilizing analytical methods. Over the broad operating range of IPM machines, interactions of MMF sources, permeances, and currents result in torque harmonics. The placement of traditional rotor features along with sculpt features are utilized to minimize torque ripple and maximize average torque. We extend the winding function theory to include the IPM rotor's primary and secondary reluctance paths and the non-homogeneous airgap of the rotor sculpt features. A new analytical winding function model of the single-V IPM machine is introduced, which considers the sculpted rotor and how this model can be used in the design approach of machines. Results are validated with finite elements. Rotor feature trends are established and utilized to increase design intuition and reduce dependency upon the lengthy design of experiment optimization processes.

Keywords: electric motor; interior permanent magnet; reluctance; MMF-permeance; winding function; torque ripple

Citation: Hayslett, S.; Strangas, E. Analytical Design of Sculpted Rotor Interior Permanent Magnet Machines. *Energies* **2021**, *14*, 5109. https://doi.org/10.3390/en14165109

Academic Editor: Athanasios Karlis

Received: 30 June 2021
Accepted: 17 August 2021
Published: 19 August 2021

Publisher's Note: MDPI stays neutral with regard to jurisdictional claims in published maps and institutional affiliations.

1. Introduction

The IPM motor is increasingly being utilized throughout industry as a primary source of propulsion due to its good efficiency, torque and power density. Examples include the development of battery electric vehicle traction motors [1,2], plugin hybrid electric vehicles [3], and hybrid electric vehicles [4]. Ideally, the traction machine provides an average torque produced from a sinusoidal distribution of the airgap flux density. In reality, embedding the magnet within the salient structure of the rotor lamination and distributing windings in discrete locations result in airgap flux density harmonics. These harmonics result in increased torque ripple, radial forces, losses, and other unwanted phenomena.

In this paper, an approach to minimize torque ripple with rotor features is presented, based upon analytically modeling the machine features. The analytical modeling approach enables efficient use of computational resources, without the sacrifice of harmonic content, prior to the use of more expensive finite element methods. The calculation of the IPM machines' spatially-dependent torque harmonics is performed through the extension of the winding function method. New to the winding function framework is a method to model the equipotential nature of the rotor's salient features and rotor surface modifications. The non-homogeneous airgap of rotor surface modifications is included in the model through an additional MMF term. Unique to this analytical method, both the constituents and aggregates of the torque harmonics are found. A detailed investigation into the rotor geometry design space to minimize torque harmonics while managing average torque design trade-off is presented.

Inherent to the design of IPM machines, torque ripple is a persistent problem. Design choices to increase torque density or decrease manufacturing cost are often at odds with minimizing the torque ripple [5]. Rotor features, including surface modifications or sculpt features, are utilized to minimize the torque harmonics. The ability to reduce ripple by design must consider the speed, current, and control angle ranges of the IPM machine.

Analytical expressions for the airgap and torque harmonics are developed for the IPM in [6,7]. The synchronous reluctance of torque harmonics presented in [8] is extended to the IPM machine in [7]. The expressions are useful in setting the stator slot and rotor barrier counts but do not model the machine.

Analytical models better relate the physical geometry of the machine to its airgap and flux density harmonics. Directly solving the Laplacian–Poisson is difficult [9,10]. Subdomain models break the model into pieces in which the Laplacian–Poisson can be more readily solved [11,12]. Magnetically Equivalent Circuits (MEC) divide the geometry into smaller manageable pieces [13]. Methods depending on winding functions allow for the geometry and harmonics to be described, but the second reluctance path can be difficult to model. The airgap harmonics of the salient pole permanent magnet synchronous machine are presented in [14] but does not address torque ripple or the secondary reluctance path of the IPM machine. The rotor permeance path is approximated in [15] to determine the torque ripple of the machine under study, but does not fully describe an IPM machine. The double V shaped IPM is presented in [16], in which flux densities are calculated through an MEC model and described with a Fourier series. The single V IPM presented in [17] considers the pole cap effect but does not consider torque ripple harmonics. The single V, delta and double V IPM rotor configurations are shown in Figure 1. Moreover, the airgap harmonics in permanent magnet synchronous machines were calculated in [18,19], but the effect of the second reluctance path on the airgap harmonics was not included in the calculations.

Recently, researchers have investigated rotor modifications to alter the airgap, modify airgap flux, and improve torque harmonics. The first feature type is pole shaping, which creates a small airgap near the d-axis and an increased airgap in the region of the q-axis. The torque ripple was reduced for the single magnet flat magnet IPM and optimized with a differential evolution algorithm and finite elements [20]. A surface-mount PM pole-shaped machine was studied with an analytical solution to the field in [21]. The 2D solution was confirmed both by finite element and testing. The pole-shaped single flat magnet IPM was optimized with a response surface method within FE [22]. This included the use of rotor core modifications as well; both FE and experimental results were presented. The flat magnet IPM pole shape was optimized, along with the creation of design rules for the ratio of q-axis and d-axis airgap length in [23]. The single V magnet-shape IPM was improved with pole shaping using finite elements in [24]. Cogging torque and back emf were measured. A third harmonic was added to the pole shape in [25], which studied the machine in finite elements. A second feature type is in the rotor core, which creates a small hole in the rotor core near the airgap in order to redirect flux. Holes in the rotor core's second reluctance path of the single magnet IPM were shown to decrease torque ripple using finite elements in [26]. The double V magnet IPM machine with improved torque ripple, due to holes in rotor iron core and rotor surface sculpt features, was shown to improve torque ripple but lower average torque in [27]. The delta magnet IPM shape included modified internal rotor features to improve for average torque and decrease iron loss in [28].

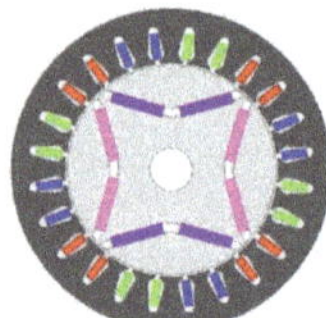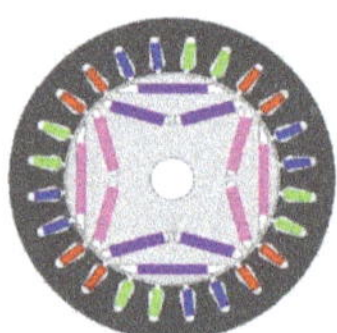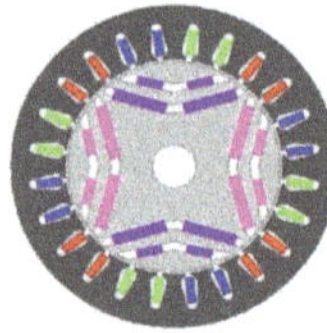

Figure 1. IPM rotor types: single V (**left**), delta (**center**), double V (**right**).

The third and final feature type is sculpting the rotor surface at the airgap to redirect flux. The single flat magnet IPM machine cogging torque was reduced in [29] and experimentally verified. A grid on/off optimization of the rotor surface was conducted on the single flat magnet IPM using finite elements in [30,31], resulting in an asymmetric rotor surface with reduction in torque ripple and maintaining average torque. The double V-magnet IPM torque ripple was minimized with both rotor core and surface sculpted features in [27]. The delta-magnet IPM machine torque ripple was minimized with rotor surface sculpt features in [32]. Then, a general analytical expression for torque harmonics was developed and utilized to optimize the solution with finite elements.

This paper presents a detailed analytical model of the sculpted rotor IPM machine. The model allows for a break down of flux and torque into magnet, primary reluctance, secondary reluctance, and sculpt features. Multiple sculpt features configurations are demonstrated to achieve similar torque harmonic reductions. Results are validated with finite elements and utilized to improve the torque harmonic characteristics of an existing industrialized machine. This is because finite elements accurately predict experimental results across a broad range of machines [33–36] and have been utilized to evaluate and compare machine types [37–39] and validate analytical solutions [40–50]. Section 2 introduces the topics relevant to the design of IPM motor construction and control. Section 3 provides details on how to model an IPM motor magneto motive force (MMF), permeance, and linear current density in order to model the machine geometry, flux, and torque harmonics. The model developed in Section 3 is applied to that of a well-known industrial IPM machine in Section 4. Design features are explored in Section 5. Contributions of this paper include a novel analytical winding function-based IPMSM model, the analytical description of rotor sculpt features, and modeling of magnet and reluctance torque component alignment due to asymmetric sculpt features. In addition, this paper demonstrates the torque effects of reluctance path pole arc, sculpt feature type (symmetrical/asymmetrical), sculpt feature location, and sculpt feature depth and sculpt feature width.

2. Flux Distribution and Control of IPM Machines

Performance, harmonics, and control are all dependent upon the distribution of flux density within the machine. Permanent magnets provide a constant source of flux density, which enable efficient torque production but can limit high speed operation. Reluctance features provide a source of torque dependent upon armature current at high current angles, useful for extending operation at high speeds. The ratio of magnet and reluctance torque is balanced to enable the machine to stay within its operation constraints while efficiently using the voltage and current available at the terminals. This section provides a brief overview of the machine's construction, design features, and flux paths. In addition, the necessary framework for control is introduced.

2.1. Flux Distribution

Figure 2 shows a two-dimensional illustration of a four pole IPM machine. The IPM motor is fundamentally constructed of a stator and a rotor. The stator is the mechanically grounded part of the machine. It is constructed of slots, teeth, a yoke, and the three phase windings. The stator teeth and yoke are constructed of a magnetically permeable iron alloy. The teeth and yoke allow for easy flow of magnetic flux to and from the airgap of the machine. The slots allow space for the copper windings. The windings are distributed within the slots to produce a current dependant magneto motive force (MMF), which in turn creates the radial magnetic flux density. The placement of the windings also creates a current density along the bore of the stator, resulting in a tangential component of flux density. When arranged and controlled properly, the currents in the windings produce a rotating set of fields to produce torque.

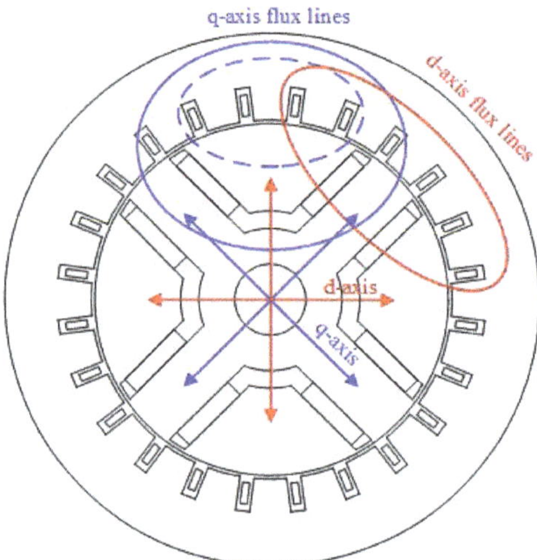

Figure 2. IPM flux paths: first reluctance path (solid blue), second reluctance path (dashed blue), and magnet path (solid red).

The rotor, the mechanically rotating part, is constructed of iron ribs, air-filled barriers, iron bridges, and permanent magnets. The iron core is constructed of ribs and bridges. The ribs control the distribution of flux density while the bridges mechanically couple all parts together. Barriers provide air pockets, which assist the ribs in directing flux, and also contain embedded magnets. The permanent magnets are embedded within the rotor and produce an MMF, which is independent of current.

From the perspective of the rotor, a direct axis (*d*-) and a quadrature axis (*q*-) of the machine are electromagnetically aligned to the rotor characteristics. The *d*- axis is the primary axis of which the permanent magnet flux density flows. The magnet flux density flows through the magnet into the central rib of the magnet pole, into the airgap, through the stator teeth and yoke and returns into the adjacent opposite magnet pole. This permanent magnet flux density path is shown in Figure 2 as a red ellipse. The *q*- axis is the axis in which the armature-induced flux flows through the rotor. This flux is produced from the armature MMF and the reaction of the rotor/stator permeance. Two armature-induced paths result; one through the primary reluctance path, and a second through the secondary reluctance path. The primary reluctance path flux is shown as the solid blue ellipse, and the second reluctance path flux is shown as the dashed blue ellipse in Figure 2.

2.2. Control of IPM Machine

Control must be considered in the design of the IPM motor. The steady state torque and voltage equations are shown in Equations (1)–(3). These are the fundamental starting points to develop the necessary analysis for the control of electric machines. The equations are based upon *d*- and *q*- axis voltages, v_d and v_q, currents, i_d and i_q, inductance λ_d and λ_q, magnet flux linkage, λ_m, and phase resistance R_s. Magnet offset, δ, as shown in Figure 3, is included to account for magnet alignment relative to the reluctance path, which may be caused by rotor sculpting features [51,52]. Traditional IPM alignment would feature $\delta = 0°$, with the *d*-axis aligned to the maximum of magnet flux linkage. For purposes of this paper, the *q*-axis remains aligned to the minimum reluctance of the rotors first reluctance path. By inspection, the resistance or loss terms do not have an effect on the torque and only affect the voltage and electrical power. It may be useful to assume the phase resistance is negligible.

$$\tau = \frac{1}{2}\frac{3}{2}\left((l_d - l_q)i_d i_q + \lambda_m\left(\cos(\delta)i_q - \sin(\delta)i_d \right) \right) \tag{1}$$

$$v_d = R_s i_d + -\omega l_q i_q - \omega \lambda_m \sin(\delta) \tag{2}$$

$$v_q = R_s i_q + \omega l_d i_d + \omega \lambda_m \cos(\delta) \tag{3}$$

Performance assessments require including an analysis of the phase current constraint, I_{max}, and phase voltage V_{max}. For a wye connected machine, I_{max} is equal to the phase current, I_{ss}, and is limited by the power devices of the inverter and the electric machines thermal capability. The voltage limit is the maximum phase voltage that the inverter can apply, limited by the specific pulse width modulation (PWM) technique used. The voltage limit for space vector PWM is $V_{max} = \frac{V_{dc} \cdot MI}{\sqrt{2} \cdot \sqrt{3}}$ and six step PWM is $V_{max} = \frac{V_{dc} \cdot MI \cdot}{\sqrt{2} \cdot \pi}$. The maximum modulation index is set to $MI = 0.95$ to account for cable and device voltage drops.

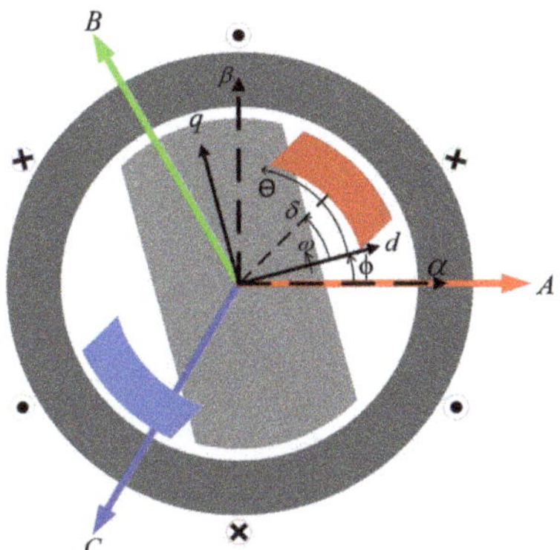

Figure 3. Vector diagram of IPM rotor with unaligned magnet, with variables: ϕ rotor position, θ rotor spatial coordinate, δ magnet alignment, ω rotor speed.

3. Analytical Model: MMF, Permeance, Flux, and Torque

This section develops the necessary analytical winding function model for the IPM machine idealized to focus on the effects of the rotor geometry. Assumptions include closed stator slots and no saturation leading to infinite permeability, leaving the permeability of the airgap assumed to be that of free space $\mu_\circ = 4\pi 10^{-7}\ \frac{H}{m}$. Stator conductors are modeled by discrete current sheets along the stator bore inner diameter and phases are assumed to be wye connected.

Focused on the MMF interaction with the second reluctance path, the analytical model describes MMF and permeance functions. These winding function-based MMFs, $F_x(\theta, \phi)$, and permeance functions, $\Lambda(\theta, \phi)$, express the harmonic content of stator and rotor features as Fourier series. Relationships between the rotor spatial coordinate θ and rotor position ϕ and current angle β are included. Flux densities are computed using Equation (4), and contributions of the stator and rotor harmonic interactions to the torque ripple are determined.

$$B_r(\theta, \phi) = 2\Lambda_r(\theta, \phi)F_r(\theta, \phi) \tag{4}$$

3.1. Permeance Functions

The general form of the permeance functions Λ_x, shown in Figure 4, can be written as Equation (5), where the amplitude Y and phase γ define the location of features relative to the d-axis of the machine.

$$\Lambda_x(\theta) = \sum_{n=0,2,4,6...}^{\infty} Y_x(n)\cos(n\theta + \gamma_x(n)) \tag{5}$$

Permeance functions, along with rotor and stator construction, are shown in Figure 4. The salient features of the rotor begin with the definition of the primary reluctance path, which assumes a small airgap l_g, aligned with the minimum reluctance of the q-axis, and a large airgap considering barrier and magnet dimensions, l_m. Above the magnet, a secondary reluctance path exists, which reacts as equipotential salient iron to the armatures MMF. The permanent magnet permeance path describes the total amount of air the magnet must push its flux through, including its thickness in the same region of the second reluctance path. Coefficients of Equation (5) are derived from the local definition of the permeance Equation (6).

$$\Lambda = \frac{\mu_\circ}{g} \tag{6}$$

The defining airgaps of the permeance functions are listed in Table 1.

Table 1. Permeance functions and related minimum and maximum airgaps.

Permeance Term	Minimum Airgap	Maximum Airgap
First Reluctance Path	l_g	$l_g + l_m$
Second Reluctance Path	l_g	∞
Permanent Magnet Path	$l_g + l_m$	∞

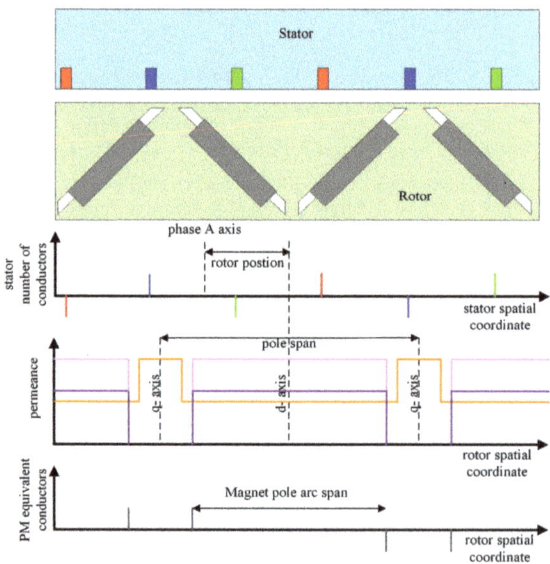

Figure 4. Conductor locations and permeability functions, (red) phase A conductor, (blue) phase B conductor locations, (green) phase C conductor location, (gold) first reluctance path permeability function, (lavender) second reluctance path permeability functions, (purple) magnet path permeability function, (gray) equivalent permanent magnet conductor locations.

3.2. Magneto Motive Forces (MMF)

Equation (7) gives the general form of the winding function composed of the turns function $n(\theta, \phi)$ and its mean $< n(\phi, \theta) >$, where the turns function is a result of the closed path integral of the conductor [53]. MMF functions, $F(\theta, \phi)$, are comprised of the turns function multiplied by a current.

$$N(\theta, \phi) = n(\theta, \phi) - < n(\phi, \theta) > \tag{7}$$

$$< n(\theta, \phi) > = \frac{1}{2\pi} \int_0^{2\pi} n(\theta, \phi) d\theta \tag{8}$$

Expanded into a Fourier series, the winding takes the form as follows (9):

$$N_x(\theta, \phi) = \sum_{n=1,3,5,7\ldots}^{\infty} Y_{wf_x}(n) \cos(n\theta + \gamma_{wf_x}(n) + n\phi_i + n\phi) \tag{9}$$

where Y_{wf_x} is the coefficient, $\gamma_{wf_x}(n)$ is the phase, mechanical order n, rotor position ϕ, and initial rotor position ϕ_i.

3.2.1. Stator

Stator MMF is formed from the interaction of the phase currents (I_a, I_b, I_c) and the phase winding functions (N_a, N_b, N_c). The summation of the three phases creates a rotating MMF, F_{abc}, which directly acts upon the primary reluctance path.

$$F_{abc}(\theta, \phi) = N_a(\theta, \phi) I_a(\phi, \beta) + N_b(\theta, \phi) I_b(\phi, \beta) + N_c(\theta, \phi) I_c(\phi, \beta) \tag{10}$$

3.2.2. Magnet

The permanent magnet is represented as an equivalent current I_{PM}, which is related to the remnant flux B_r, permeability of free space u_o, relative permeability μ_r, and magnet thickness. Magnet MMF for the IPM machine interacts with the magnet path permeance, where τ_m is the magnet's salient iron pole pitch, and w_m is the combined width of the magnets for a single pole.

$$I_{PM} = \frac{B_r}{\mu_0 \mu_r} l_m \tag{11}$$

$$F_{PM}(\theta, \phi) = N_{PM}(\theta) I_{PM} \frac{\tau_m}{w_m} \tag{12}$$

3.2.3. Second Reluctance Path Modification

The equipotential nature of the rotor's second reluctance path reacts only to the regional harmonics of the stator MMF. In this case, all looping flux that enters the second reluctance path pole arc through the airgap must exit through the same airgap. Modification to the armature MMF by removing its mean satisfies this condition and is made possible through Equation (13). The symbols $F_{<abc>}$ and $< F_{abc}(\theta, \phi) N_{PM}(\theta) >$ represent the modified MMF, which interacts with the second reluctance path and the mean of the MMF across this same boundary. As a matter of convenience, the permanent magnet winding function N_{PM} is also used to consider the stator MMF in the region of the second reluctance path, invert it, remove the mean, and revert to the original polarity.

$$F_{<abc>}(\theta, \phi) = \left(F_{abc}(\theta, \phi) N_{PM}(\theta) - < F_{abc}(\theta, \phi) N_{PM}(\theta) > \right) N_{PM}(\theta) \tag{13}$$

$$< F_{abc}(\theta, \phi) N_{PM}(\theta) > = \frac{\frac{1}{2\pi} \int_0^{2\pi} F_{abc}(\theta, \phi) N_{PM}(\theta) d\theta}{\frac{1}{2\pi} \int_0^{2\pi} | N_{PM}(\theta) | d\theta} \tag{14}$$

3.3. Sculpt Feature Description: Equivalent Magnetizing Dipole Current

MMF-permeance methods allow for flux density for the IPM smooth rotor homogeneous airgap to be calculated. Rotor sculpting affects both magnet flux distribution and reluctance flux distribution. This smooth rotor MMF-permeance theory does not adequately describe stator slots and rotor sculpt features as it has been developed with a constant airgap dimension [53]. In this section, an extension of winding function theory is developed, which can be used in the description of both slots and sculpts based upon equivalent magnetic currents (EMC) [54] and the equivalent magnetic dipole [55]. These non-homogeneous airgap features are represented by additional MMF terms utilizing the description of a magnetic dipole and its equivalent magnetic currents. The redistribution of flux density and MMF is possible with the use of the equivalent dipole concept.

The magnetic dipole in free space is formed by a loop of radius b and current of I. The solution at far fields, when $R >> b$, solved in spherical coordinates, using the magnetic vector potential, A, is shown in Equation (15), where the magnetic dipole moment m is written as $m = a_z I \pi b^2$ [55].

$$A = \frac{\mu_0 m \times a_R}{4\pi R^2} \tag{15}$$

This dipole in free space can be used to explain the magnetism at the atomistic level, where small circulating currents are formed by the process of magnetization. This magnetization aligns the individual atomic dipoles and modifies the orbital spin of the electrons for each atom.

The macroscopic volume density of magnetization, M, with units of A/m, is computed through a sum of the individual microscopic dipoles. Shown in [55], the magnetization vector M is equivalent to both a volume current density, J_m with units of $\frac{A}{m^2}$, and a surface current density J_{ms} with units of $\frac{A}{m}$.

$$J_m = \nabla \times M \tag{16}$$

$$J_{ms} = M \times a_n \tag{17}$$

Given M, the flux density B can be found by computing both J_m and J_{ms}. These values are used to determine the magnetic vector potential A. Uniform M within a magnetic material will result in no volume current density and only a surface current density J_{ms} on its borders. If space variations of M exist within a material, a net volume current density will exist. Hence, a magnetic dipole inside a material with constant magnetization M can be represented by a current loop in the air, formed at the exterior boundary material.

Figure 5 illustrates the process of analyzing the rotor sculpt feature effects. Rather than account for the changing flux density over the sculpt feature, the assumption of homogeneous flux density holds when breaking the geometry into smaller discrete dipoles (i) of fixed width. For purposes of this analysis, it is assumed the sculpt features have a constant depth l_{ms}. A sufficient number of points (i) must be defined in order to hold the assumption of homogeneous flux density. For each point (i) contained within the sculpt feature, the first current $I_{ms1}(i)$ is applied based upon prior analysis of the magnetic dipole.

$$I_{ms}(i) = \frac{l_{ms}(i) * B_{ms}(i)}{\mu_\circ} \tag{18}$$

A second dipole counter current in the adjacent point $I_{ms2}(i+1) = -I_{ms1}(i)$. The net effective dipole current for each point, $I_{ms3(i)}$ is formed through summation dipole currents $I_{ms3(i)} = I_{ms1(i)} + I_{ms2(i)}$. A third dipole current I_{ms3}, or summed current, becomes the current-turns function for the sculpted feature, in which an equivalent MMF for the rotor surface features can be determined through the use of a winding function. It should be noted that the sculpt features analyzed with this process do not create flux but only distribute flux away from the sculpt feature.

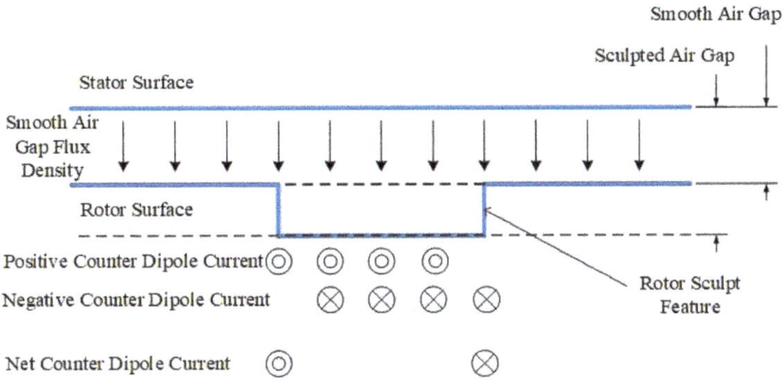

Figure 5. Sculpted Rotor Reluctance Counter Dipole Current.

3.4. Flux Density

Total airgap flux density, B_{tot}, is a result of the flux creating sources and flux distributing features. Primary reluctance path, secondary reluctance path, and the permanent magnets create and distribute flux. The sculpt and slotting features serve to redistribute flux, with the assumption of small features. Flux densities can be computed for each individual component or in summation.

$$B_{tot} = 2\Lambda_{R1}F_{abc} + 2\Lambda_{R2}F_{<abc>} + 2\Lambda_{PM}F_{PM} + 2\Lambda_{sculpt}F_{sculpt} + 2\Lambda_{slot}F_{slot} \tag{19}$$

3.5. Torque

The Maxwell stress tensor, Equation (20), allows for torque computation given the radial flux density and the conductor's linear current density. Torque due to individual components of flux density can be separated or combined for the net effects.

$$T = \int_{\theta=0}^{2\pi} r^2 l \cdot B_{tot}(\theta)K(\theta)d\theta \tag{20}$$

3.6. Rotation and Convolution

In this section, a computationally efficient method to evaluate the phenomenon of rotation and the ensuing interaction of the time and space domain harmonics is reviewed. To simplify analysis of the magnetic fields, the rotor is left frozen while the stator rotates in the counter direction. The Fourier spatial coefficients of the permeance and winding functions are determined at the initial position of the rotor through use of the Fast Fourier Transform (FFT). For each rotor position, $\phi(i)$, a complex rotation, Equation (21), is used to form a rotation vector, Equation (22). The complex rotation vector is used to transform the FFT coefficients of permeance and winding functions at each position.

$$\epsilon = (\cos(\phi) + i\sin(\phi)) \tag{21}$$

$$\Phi = \begin{bmatrix} \epsilon^0 & \epsilon^1 & \epsilon^2 & \dots & \epsilon^{\frac{n}{2}-1} & \epsilon^0 & \overline{\epsilon^1} & \overline{\epsilon^2} & \dots & \overline{\epsilon^{\frac{n}{2}-1}} \end{bmatrix}^{\top} \tag{22}$$

In place of the convolution of the Fourier coefficients to determine flux density, the permeance and winding function is reconstructed with the inverse fast Fourier transform.

3.7. Comparative Analysis to Recent Analytical Methods

This section will compare the analytical methods developed in this paper to three recent methods [45,49,50].

The analytical MEC model is used to design a reduced magnet cost single V consequent pole (CP) machine with the same average torque as single V IPM in [45]. The models are developed based on zones and regions, allowing for an assumption of the open circuit flux density distribution. Two flux sources (magnets) and six reluctance network paths are used to create the open circuit single V IPM model. Open circuit flux density is assumed to take a trapezoidal waveform where the flux density is determined spatially from the fluxes in the MEC reluctance network. Similarly, the single V CP IPM network consists of one magnet flux source and five reluctance paths. In both cases, the rotors reluctance path reaction to the armature loading is not computed. The open circuit flux densities are used to quickly determine an equivalent single V CP IPM fundamental to that of the single V IPM fundamental. In order to guide design, the method is used to find an equivalent consequent pole open circuit flux density fundamental to the traditional IPM flux density. Finite elements are relied on to complete the study of the torque performance.

Multi-barrier synchronous reluctance and Permanent Magnet Assisted Synchronous Reluctance Machines (PMSynRM) are modeled using conformal mapping and magnetic equivalent circuits in [50]. Hyperbolic shaped flux barriers are assumed. Conformal transformations are employed to the rotors flux barrier geometry to compute the magnetic reluctance. The reluctance values calculated from conformal mapping are subsequently used in the reluctance network values of the MEC model. The MEC model considers MMF sources of both the armature and magnet. Loaded and open circuit flux densities, average torque, and torque ripple are compared to finite elements with reasonable accuracy.

The slotless U-type IPM machine open circuit flux density is analytically modeled with a subdomain method solving Laplace's and Poisson's equations in [49]. Analytical equations are derived and presented for each subdomain. Results are validated against finite elements. The model is divided into four regions, which consist of the airgap and magnets. The governing system of partial differential equations is developed, along with simplifications, interface and boundary conditions. A separation of variables is used to develop the general solutions of the PDEs, and they are written as a Fourier series. The system of equations is solved and compared to FE. Strong agreement of the radial and tangential flux density is shown between the FE and the subdomain methods.

The analytical models discussed were developed for multiple purposes. The MEC method is used in [45] to quickly estimate the open circuit flux density fundamental of the single-V IPM and single-V CP IPM machines. The armature reaction of the reluctance features is not considered by the model, and finite elements are used to finish the designs. Conformal mapping is used in [50] to determine the reluctances of a multi-barrier PMSynRM and further evaluated using a MEC network. Both open circuit and loaded conditions are evaluated for airgap flux densities and torque performance and compared to finite elements. The analysis is not extended to the design. The subdomain methodology is employed in [49] for the analysis of the U-shape IPM machine open circuit conditions. Both tangential and radial flux densities are shown to match finite element results. The model requires further extension to consider the torque performance due to a loaded armature. The single-V sculpted rotor IPM winding function model developed in this paper considers both open circuit and loaded conditions. The model is constructed such that computational efficiency is possible without the sacrifice of spatially dependent harmonic content to drive the design. The choice of which analytical model to develop is dependent upon the application, computational resources, and intended purpose. In this case, the new winding factor IPM model was developed.

4. Application of Analytical Model to Example Machine

The method developed in Section 3 is validated with a well-known industrialized IPM machine. Details of the 2004 Toyota Prius traction motor are included in Table 2, and the geometry is modeled analytically within Matlab and finite elements within Ansys Maxwell.

Table 2. Example motor parameters.

Parameter	Value	Unit
Pole Pairs	4	
Stator Slots	48	
Number of Phases	3	
Stack Length	83	mm
Rotor Diameter	161.15	mm
Airgap Length	0.75	mm
Magnet Pole Arc % of Pole Pitch	63.8	%
Barrier Type	Single V	
Magnet Thickness	6.48	mm
Magnet Width	16×2	mm
Permanent Magnet Remnant	1.19	T
Permeability of Iron	∞	H/m
Permeability of Bridge Features	$4\pi \cdot 10^{-7}$	H/m

Focusing on the effects of rotor, the stator geometry has been idealized with no slots. Both sinusoidally distributed stator windings and the production configuration of discretely placed windings are modeled. Only the stator winding harmonics interactions with rotor geometry harmonics are considered. With the assumption of infinite permeability, the bridge features are omitted. Airgap fringing in the region of the magnet barrier is not considered.

Design parameters are studied within this section using the analytical winding function model previously validated. Rotor sculpt features are included along with their additional MMF term developed in Section 3.

4.1. Sculpting Geometry

The sculpted rotor IPM machine geometry design space to be explored is shown on a single pole of the example machine in Figure 6. Rotor primary and secondary reluctance paths are shown in green with no bridge features. The stator, shown in gray, continues to have omitted its slot features, and distributed windings, orange, are placed within the airgap.

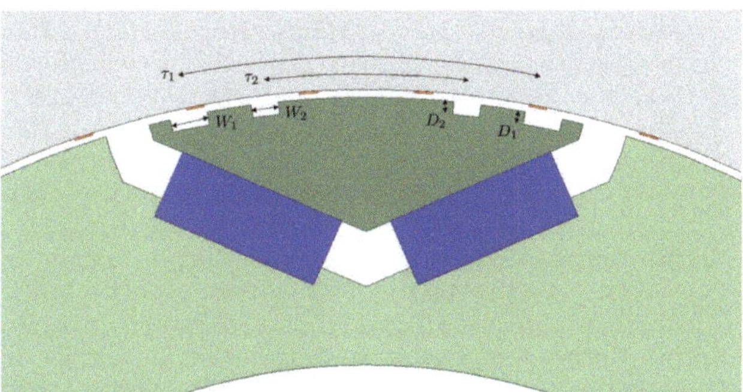

Figure 6. Rotor sculpt features.

The primary design parameters are centered on the rotor effects, which include the ratio of primary and secondary reluctance path and sculpt features. The magnet pole arc width is varied. Up to two sets of symmetrically placed sculpt features are placed on the second reluctance path. The symmetrical feature span locations τ_1 and τ_2 define the symmetrical location of the feature in terms of its percentage of the magnet pole arc span. Single asymmetrical features are described with a similar parameter but with only one sculpt feature on the pole. In this single asymmetrical case, the feature location, τ, is set to be positive for right hand side placement and negative for the left hand side placement. The depths D_1 and D_2 are measured from the outer surface of the rotor to the root of the sculpt feature. The widths W_1 and W_2 are measured in terms of a single feature percentage of the pole span.

A Fourier analysis of design feature effects on torque harmonic analysis is presented. The phase of the torque harmonics is set with respect to the negative zero crossing of phase A back EMF, i.e., the north pole of the machine. Torque components are separated in terms of total, primary reluctance, secondary reluctance and magnet torque with both the torque amplitude and its corresponding phase.

4.2. Model Implementation

Implemented in Matlab, the analytical model in Section 3 has been utilized to explore the design space. Rotor permeances of primary reluctance, secondary reluctance, and magnet and the permanent magnet winding function are modeled spatially as the rectangular waves. Stator phase winding functions and linear turn densities are spatially modeled, and the spatial harmonic coefficients are determined with the FFT. For each rotor position, the complex rotation vector is created and applied to the stator winding and linear turn functions. With the current applied, the modified secondary reluctance path is determined from Equation (13). Radial flux density is now determined for the primary reluctance, second reluctance, and magnet paths. A broad range of phase current, control angle, and rotor sculpting features are studied. Maxwell's stress tensor in conjunction with Equation (19) provides the ability to separate components of torque.

The analytical model is implemented utilizing both the FFT and the inverse FFT algorithms to efficiently move between frequency and spatial domains. Rotation is best performed within the frequency domain, Equation (22), and convolution is performed within the spatial domain. This provides the most efficient use of resources by a factor of five times. Results are computed over an entire electrical cycle with a sufficient number of points to provide a smooth torque waveform. The analytical model is executed within Matlab in 6.7 s, and the finite element results are executed in 20 min. Ansys Maxwell was also used to perform the finite element analysis.

4.3. Model Validation: Radial Flux Density

This section compares the radial flux density results of the analytical model and finite elements while varying: (1) winding type, (2) current, (3) control angle, and (4) rotor sculpt features. Figures 7–10 plot the flux density along the rotors spatial coordinate, θ, over a single pole pair. Both sinusoidal and distributed windings are compared. The q-axis, which is aligned to the rotor minimum reluctance, occurs at $\theta_{elec} = 90°$ and $\theta_{elec} = 270°$. The d-axis is aligned, which is aligned to the smooth rotors permanent magnet maximum flux linkage, occurs at $\theta_{elec} = 0°$, $\theta_{elec} = 180°$, and $\theta_{elec} = 360°$. In all results, the finite element and the analytical model result in comparable flux densities.

Flux densities shown for sinusoidal windings, Figure 7, illustrate the changing airgap flux density harmonics with current and control angle. When the phase current is set to zero, only the permanent magnet field is present. As current and current angle increases, the flux density becomes more jagged, with the case of a fully negative d-axis current displaying the most harmonic content. It is clear that as the negative d-axis current becomes dominant, so do the reluctance path harmonics. Harmonic effects of the discretely distributed windings are shown in Figure 8. As current increases, so do the airgap reluctance harmonics. In all cases, the analytical model and finite element results agree with reasonable accuracy.

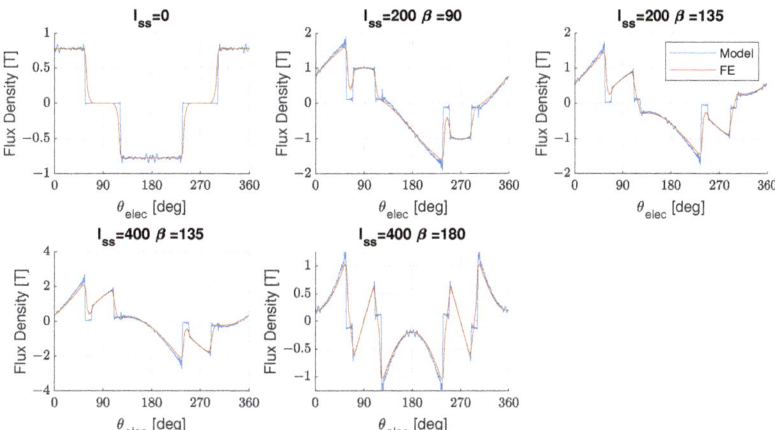

Figure 7. Radial flux densities with sinusoidally distributed windings ($N = 200$) at various currents and control angles.

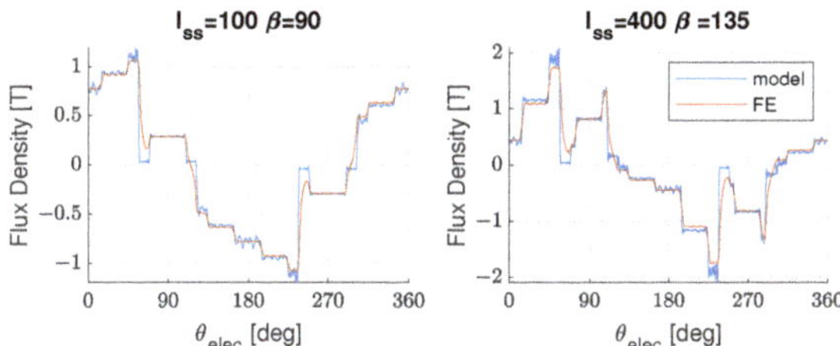

Figure 8. Radial flux densities with distributed windings (2 SPP) at various currents and control angles.

The effects of rotor sculpt features on the second reluctance path are shown in Figures 9 and 10. A symmetrical pair of sculpt features are shown with distributed windings in Figure 9. The flux densities of the smooth rotor and sculpted rotor are plotted. Reduced flux density in the region of the sculpt features is observed, and sculpt features are located approximately at $\theta_{elec} = 30°, 150°, 210°, 270°$. This flux density from the sculpt features is conserved and redistributed across the regions of the second reluctance path.

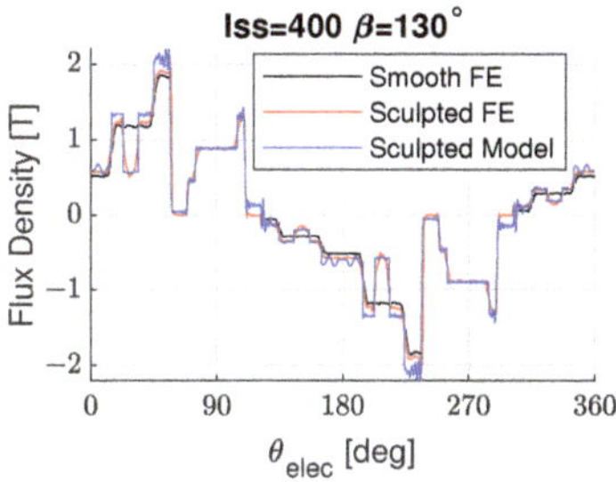

Figure 9. Radial flux densities with single symmetrical sculpt feature located at $\tau_1 = 50\%$, $W_1 = 10\%$, and $D_1 = 1.2$ mm.

The reluctance flux density single sculpt feature is plotted in Figure 10. Similar to the symmetrical sculpt features, the flux density is reduced in the region of the sculpt feature. In all cases, the analytical model and finite element results agree with reasonable accuracy.

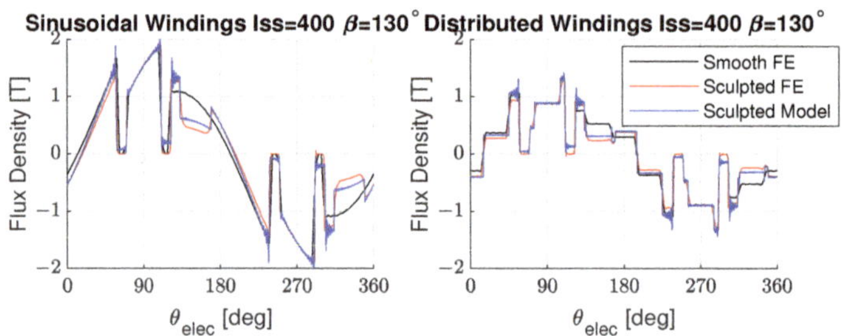

Figure 10. Radial flux densities (reluctance only) with single asymmetrical sculpt feature located at $\tau = 50\%$, $W = 20\%$, and $D = 1.2$ mm.

4.4. Model Validation: Torque Ripple

Torque ripple of the smooth rotor IPM, Table 2, is compared between finite element and the analytical model in Figure 11. Good agreement between the finite element and analytical models is observed. The torque ripple effects of two symmetrical rotor sculpt features are demonstrated in Figure 12, directly calculated by the analytical model, whereas the finite element model requires two runs, once with and once without sculpt features, to determine the sculpt feature effects.

Good correlation between the model and finite elements is demonstrated and shown.

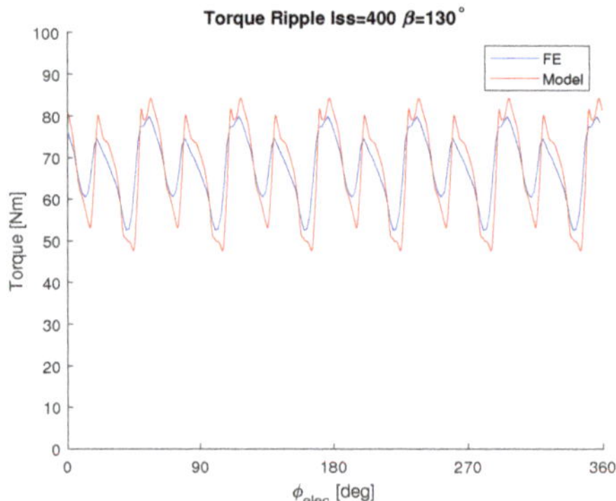

Figure 11. Smooth rotor IPM torque ripple.

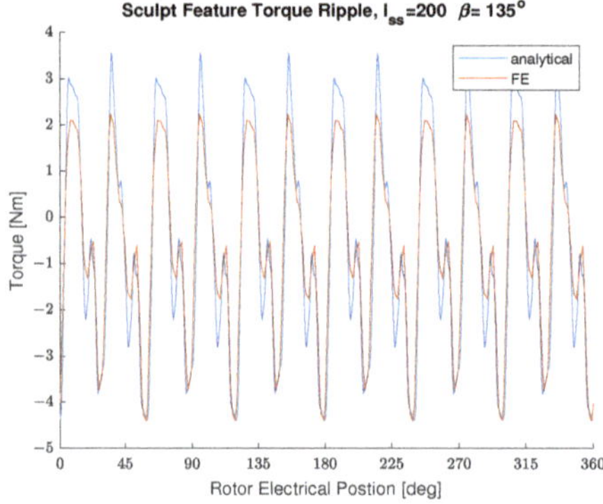

Figure 12. Sculpt feature torque ripple.

4.5. Torque Ripple Components

In this section, the torque ripple results of the analytical model are studied while varying current and control angle. Figure 13 plots the torque ripple for a complete electrical cycle of the example machine. The torque components for the first reluctance path, second reluctance path, total reluctance torque, magnet torque, and total machine torque are plotted. Magnet torque and its harmonics are dominant at lower currents, but the reluctance paths cannot be ignored. As current is increased, the reluctance torque increases relative to the magnet torque. The stronger field weakening currents cause the contribution of the reluctance features torque to increase. The dominant torque harmonic orders are the 6th and 12th electrical orders. In the design, both the torque harmonic amplitudes and phases of each of the components need to be considered as the sculpt feature design will provide the counter torque at the counter phase.

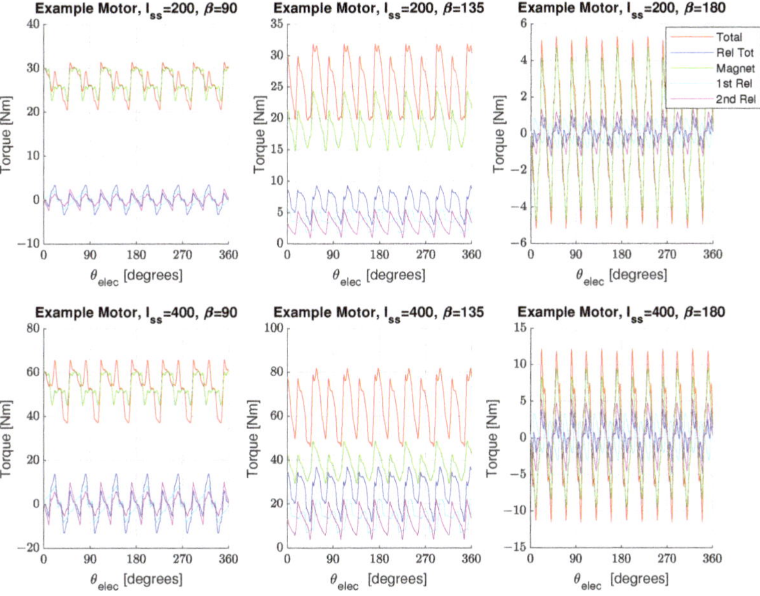

Figure 13. Analytical model torque ripple components at various control angles.

5. Investigation of Design Features

In this section, the effects of design features are demonstrated to influence both torque harmonic amplitudes and phases. Carefully applied, these effects are used to design counter torque harmonics. Second reluctance path pole arc, sculpt feature type (symmetrical/asymmetrical), sculpt feature location, sculpt feature depth, and sculpt feature width can all be used to design an appropriate counter torque to reduce the machine's torque harmonics. While mildly affecting average torque, the second reluctance path pole arc, τ_p, strongly affects the phase of the 12th electrical order torque harmonic. A single pair of symmetrical sculpt features placed upon the second reluctance path pole arc reduce the average torque. Feature position provides 12th electrical order torque harmonic phasing, and the feature width and depth directly affect the torque harmonics amplitude. The single asymmetrical feature is shown to increase average torque when placed on a specific side of the second reluctance path pole arc. The asymmetrical feature placement can also be used to modify the phase of both 6th and 12th electrical order torque harmonics. Finally, feature phasor summation is shown to be effective in combining the effects of multiple design features, further providing the ability to design both the amplitude and phase of these minimizing torque harmonics. These relationships provide the necessary intuition to reduce computationally intensive design steps.

5.1. Magnet Pole Arc

Magnet pole arc span, τ_m, effects upon the torque harmonics, without rotor sculpt features, are explored. Figure 14 shows the average, 6th, and 12th harmonics of torque as a function of τ_m. For this case, magnet torque is dominant. Although not always the case, it is just as important to follow the trends of individual torque components. Total, first reluctance, and magnet average machine torque are reduced as the pole arc is increased, and only the second reluctance torque increases the average torque. The 12th order torque harmonic is dominant, with primary contributions from the magnet and the second reluctance path, whereas the 6th order torque harmonic is mostly contributed to by the primary reluctance and magnet. Rotor geometry has a strong influence on the phase of the 12th order torque harmonics, whereas the 6th harmonic is less affected by rotor geometry.

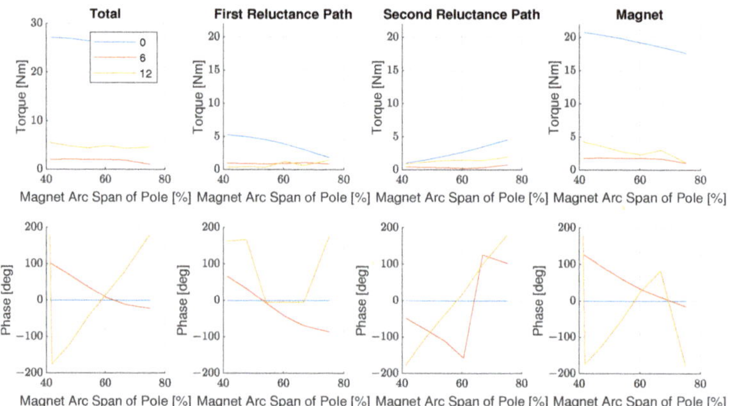

Figure 14. Effects of magnet pole arc τ_p loaded at $I_{ss} = 200$ and $\beta = 135°$.

5.2. Single Pair Symmetrical Rotor Sculpt Feature

A single symmetrical rotor sculpt feature torque is studied in Figures 15–17. Only the effects of the sculpt feature torque are plotted. In Figures 15 and 16, a single sculpt feature position is varied, with fixed width, W_1, and fixed depth, D_1, along the magnet pole arc. Rotor sculpt features have a negative effect upon average torque, as the phase is 180° out of phase with the smooth rotor average torque. Figure 15 compares the analytical model to finite elements and shows precise agreement with the phase and matching trends for torque amplitude. Using a wider feature width, D_1, Figure 16 translates amplitude and phase plots to a phasor representation. The 12th harmonic is clearly the dominant torque in both amplitudes and choice of phase.

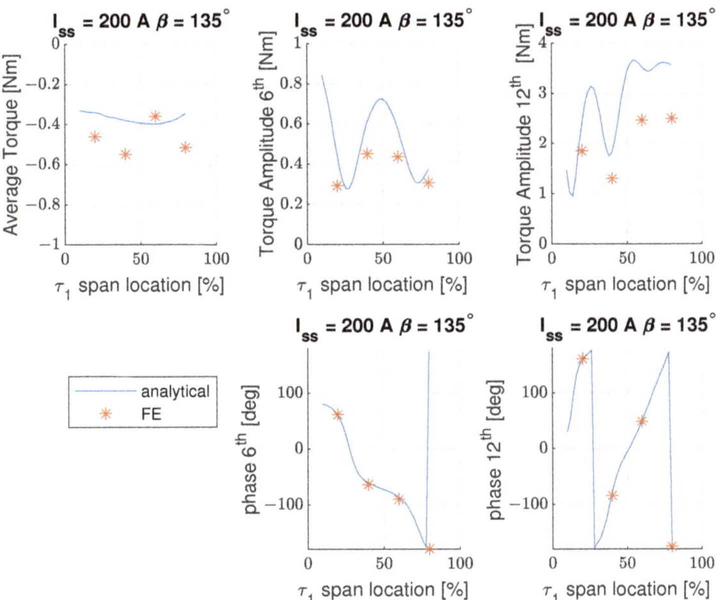

Figure 15. Symmetrical sculpt feature effects compared to finite elements: $D_1 = 1.2$ mm, $W_1 = 5\%$.

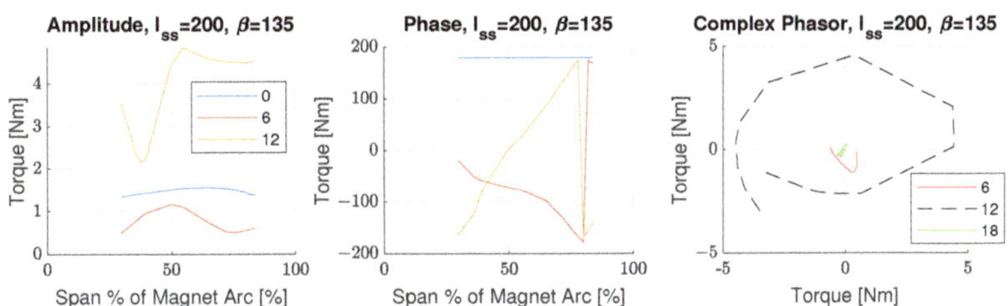

Figure 16. Symmetrical sculpt feature effects with phasor diagram: $D_1 = 1.2$ mm, $W_1 = 9\%$.

Figure 17 includes the sculpt features width and depth effects. Increasing sculpt feature width, W_1, and/or the sculpt features depth, D_1, increases the amplitude of the sculpt features torque harmonic. The primary influence on the torque harmonic amplitude is the width of the sculpt feature. Sculpt feature width and depth have no effect on the torque harmonics phase.

5.3. Single Asymmetrical Rotor Sculpt Feature

A single sculpt feature resulting in asymmetrical placement upon the rotor surface is studied in this section. Similar parameters D_1, W_1, and τ_1 are used to describe the features width, depth, and location. In the asymmetrical case, the location, τ_1, is described with the same location parameter, where in this case, a positive τ_1 results on the right side of Figure 6 and a negative τ_1 results in sculpt feature placement on the left hand side. Figure 18 compares the model to finite element and shows precise agreement with phase and matching trends for torque amplitude. In the asymmetric sculpt feature case, a torque improvement is possible due to the aligned axis effect from $\tau_1 > 0$. The placement of the sculpt feature allows for placement of the torque harmonic phase angle across a broad range of phases. Negative values of τ_1 result in the largest amplitudes of the 12th electrical torque harmonic.

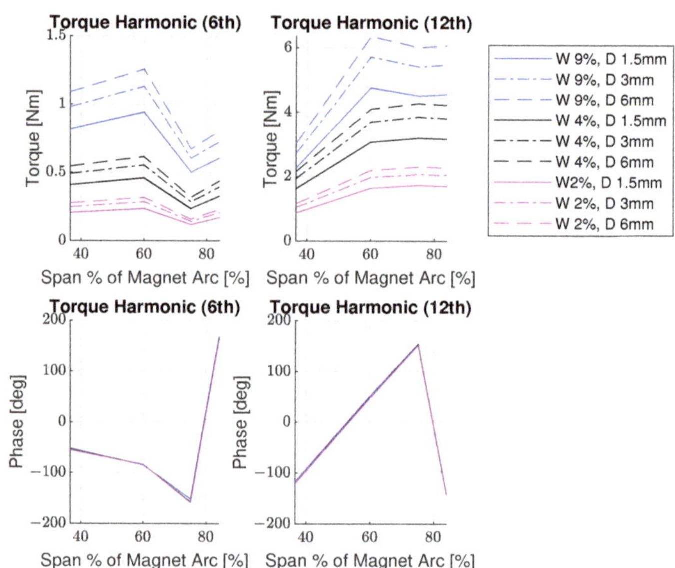

Figure 17. Sculpt feature effects $N = 1$.

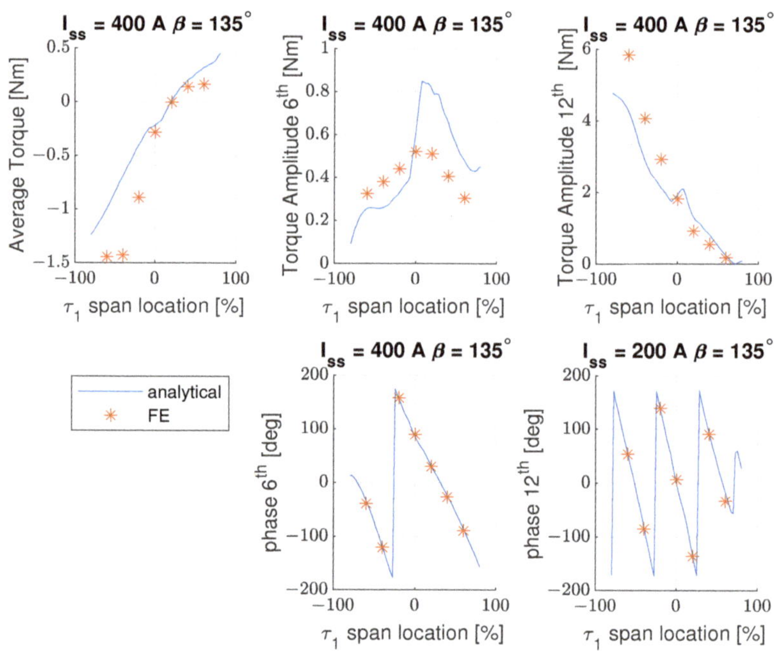

Figure 18. Single asymmetrical sculpt feature effects with phasor diagram: $D_1 = 1.2$ mm, $W_1 = 5\%$.

5.4. Two Symmetrical Rotor Sculpt Features

More than one symmetrical rotor sculpt feature can be used. In this section, it is shown that the components of a first symmetrical feature can be combined with that of the second symmetrical rotor sculpt feature. The MMF-permeance model is validated by comparing to finite element results in Figure 12.

To illustrate this concept, the parameters of the two sculpt feature sets of Figure 6 are shown in Table 3. Through vector summation, the two vectors were used to create a 12th order counter torque with a phase of $-116°$.

Table 3. Two sculpt feature parameters.

Feature	Value	Unit
τ_1	82	% of magnet pole arc
τ_2	46.5	% of magnet pole arc span
W_1	5.5	% of pole span
W_2	5.5	% of pole span
D_1	1.2	mm
D_2	1.2	mm

Figure 19 illustrates the first (red) and second 12th order electrical torque (green) phasors. The two phasors combine to create the effective total phasor (blue). This phasor summation is plotted along with the torque complex mapping of the previous single feature design sweep. These symmetrical rotor sculpt features are designed to mitigate the 12th order electrical torque harmonics to near zero. A single feature or multiple features can be designed to minimize the torque ripple. The sculpt features are not without consequence, as the average torque is negatively affected.

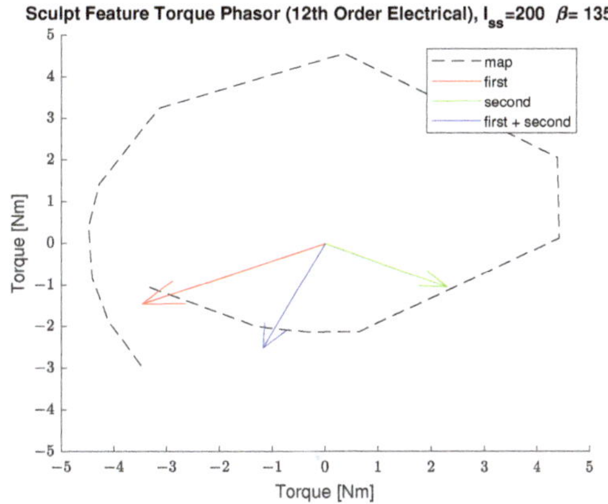

Figure 19. Two sculpt features of 12th order electrical torque phasor plot.

6. Conclusions

This paper has presented an analytical modeling and design approach to reduce torque ripple with rotor sculpt features. By carefully placing rotor sculpt features and rotor barrier features, average torque can be maintained while minimizing torque harmonics. Contributions of this paper include:

- A new analytical winding factor modeling approach for the single V IPM machine relating the rotor's first reluctance feature, second reluctance feature, permanent magnet features, sculpt features, and stator windings to the resulting torque harmonics;
- An analytical modeling approach accounting for both symmetrical and torque aligning asymmetrical rotor sculpt features;
- Results from the analytical model providing valuable insights for identifying rotor feature design improvements;

- Design approach for placement of rotor sculpt features to minimize torque ripple while maintaining average torque;
- Demonstration of close agreement of radial flux density and torque harmonics results between the analytical model and that of finite element results.

These results enable better design insight and an efficient design process through use of an analytical model.

Author Contributions: Investigation, S.H.; supervision, E.S. Both authors have read and agreed to the published version of the manuscript.

Funding: This research received no external funding.

Institutional Review Board Statement: Not applicable.

Informed Consent Statement: Not applicable.

Data Availability Statement: Data sharing not applicable.

Conflicts of Interest: The authors declare no conflict of interest.

References

1. Momen, F.; Rahman, K.; Son, Y.; Savagian, P. Electrical propulsion system design of Chevrolet Bolt battery electric vehicle. In Proceedings of the 2016 IEEE Energy Conversion Congress and Exposition (ECCE), Milwaukee, WI, USA, 18–22 September 2016; pp. 1–8.
2. Namiki, K.; Murota, K.; Shoji, M. High Performance Motor and Inverter System for a Newly Developed Electric Vehicle. *SAE Tech. Pap. Ser.* **2018**. [CrossRef]
3. Jurkovic, S.; Rahman, K.; Bae, B.; Patel, N.; Savagian, P. Next generation chevy volt electric machines; design, optimization and control for performance and rare-earth mitigation. In Proceedings of the 2015 IEEE Energy Conversion Congress and Exposition (ECCE), Montreal, QC, Canada, 20–24 September 2015; pp. 5219–5226.
4. Kanayama, T.; Yanagida, E.; Kano, S.; Geller, B.; Nakao, Y.; Fukao, M. Development of New Hybrid System for Mid-Size SUV. *SAE Tech. Pap. Ser.* **2020**. [CrossRef]
5. Jensen, W.R.; Pham, T.Q.; Foster, S.N. Comparison of Multi-objective Optimization Methods Applied to Electrical Machine Design. In *Evolutionary Multi-Criterion Optimization*; Deb, K., Goodman, E., Coello Coello, C.A., Klamroth, K., Miettinen, K., Mostaghim, S., Reed, P., Eds.; Springer International Publishing: Cham, Switzerland, 2019; pp. 719–730.
6. Han, S.; Jahns, T.M.; Soong, W.L. Torque Ripple Reduction in Interior Permanent Magnet Synchronous Machines Using the Principle of Mutual Harmonics Exclusion. In Proceedings of the 2007 IEEE Industry Applications Annual Meeting, New Orleans, LA, USA, 23–27 September 2007; pp. 558–565. [CrossRef]
7. Pellegrino, G.; Guglielmi, P.; Vagati, A.; Villata, F. Core Losses and Torque Ripple in IPM Machines: Dedicated Modeling and Design Tradeoff. *IEEE Trans. Ind. Appl.* **2010**, *46*, 2381–2391. [CrossRef]
8. Vagati, A.; Pastorelli, M.; Francheschini, G.; Petrache, S.C. Design of low-torque-ripple synchronous reluctance motors. *IEEE Trans. Ind. Appl.* **1998**, *34*, 758–765. [CrossRef]
9. Zhu, Z.; Howe, D.; Bolte, E.; Ackermann, B. Instantaneous magnetic field distribution in brushless permanent magnet DC motors. I. Open-circuit field. *IEEE Trans. Magn.* **1993**, *29*, 124–135. [CrossRef]
10. Zhu, Z.; Howe, D.; Chan, C. Improved analytical model for predicting the magnetic field distribution in brushless permanent-magnet machines. *IEEE Trans. Magn.* **2002**, *38*, 229–238. [CrossRef]
11. Tang, C.; Shen, M.; Fang, Y.; Pfister, P.D. Comparison of Subdomain, Complex Permeance, and Relative Permeance Models for a Wide Family of Permanent-Magnet Machines. *IEEE Trans. Magn.* **2021**, *57*, 1–5. [CrossRef]
12. Mi, C.; Filippa, M.; Liu, W.; Ma, R. Analytical method for predicting the air-gap flux of interior-type permanent-magnet machines. *IEEE Trans. Magn.* **2004**, *40*, 50–58. [CrossRef]
13. Tariq, A.R.; Nino-Baron, C.E.; Strangas, E.G. Iron and Magnet Losses and Torque Calculation of Interior Permanent Magnet Synchronous Machines Using Magnetic Equivalent Circuit. *IEEE Trans. Magn.* **2010**, *46*, 4073–4080. [CrossRef]
14. Dajaku, G.; Gerling, D. Air-Gap Flux Density Characteristics of Salient Pole Synchronous Permanent-Magnet Machines. *IEEE Trans. Magn.* **2012**, *48*, 2196–2204. [CrossRef]
15. Koo, B.; Nam, K. Analytical Torque Ripple Prediction Using Air-Gap Permeance and MMF Functions in PM Synchronous Motors. In Proceedings of the 2018 21st International Conference on Electrical Machines and Systems (ICEMS), Jeju, Korea, 7 October 2018; pp. 302–307. [CrossRef]
16. Pina, A.J.; Xu, L. Modeling of synchronous reluctance motors aided by permanent magnets with asymmetric rotor poles. In Proceedings of the 2015 IEEE International Electric Machines Drives Conference (IEMDC), Coeur d'Alene, ID, USA, 10–13 May 2015; pp. 412–418. [CrossRef]
17. Li, Q.; Fan, T.; Wen, X. Armature-Reaction Magnetic Field Analysis for Interior Permanent Magnet Motor Based on Winding Function Theory. *IEEE Trans. Magn.* **2013**, *49*, 1193–1201. [CrossRef]

18. Aggarwal, A.; Strangas, E.G.; Agapiou, J. Analysis of Unbalanced Magnetic Pull in PMSM Due to Static Eccentricity. In Proceedings of the 2019 IEEE Energy Conversion Congress and Exposition (ECCE), Baltimore, MD, USA, 29 September–3 October 2019; pp. 4507–4514.
19. Aggarwal, A.; Strangas, E.G. Review of Detection Methods of Static Eccentricity for Interior Permanent Magnet Synchronous Machine. *Energies* **2019**, *12*, 4105. [CrossRef]
20. Du, Z.S.; Lipo, T.A. Torque Ripple Minimization in Interior Permanent Magnet Machines Using Axial Pole Shaping. In Proceedings of the 2018 IEEE Energy Conversion Congress and Exposition (ECCE), Portland, OR, USA, 23–27 September 2018; pp. 6922–6929. [CrossRef]
21. Jang, S.; Park, H.; Choi, J.; Ko, K.; Lee, S. Magnet Pole Shape Design of Permanent Magnet Machine for Minimization of Torque Ripple Based on Electromagnetic Field Theory. *IEEE Trans. Magn.* **2011**, *47*, 3586–3589. [CrossRef]
22. Lee, S.; Kang, G.; Hur, J.; Kim, B. Stator and Rotor Shape Designs of Interior Permanent Magnet Type Brushless DC Motor for Reducing Torque Fluctuation. *IEEE Trans. Magn.* **2012**, *48*, 4662–4665. [CrossRef]
23. Evans, S.A. Salient pole shoe shapes of interior permanent magnet synchronous machines. In Proceedings of the The XIX International Conference on Electrical Machines—ICEM 2010, Rome, Italy, 6–8 September 2010; pp. 1–6. [CrossRef]
24. Seo, U.; Chun, Y.; Choi, J.; Han, P.; Koo, D.; Lee, J. A Technique of Torque Ripple Reduction in Interior Permanent Magnet Synchronous Motor. *IEEE Trans. Magn.* **2011**, *47*, 3240–3243. [CrossRef]
25. Wang, K.; Zhu, Z.Q.; Ombach, G.; Chlebosz, W. Optimal rotor shape with third harmonic for maximizing torque and minimizing torque ripple in IPM motors. In Proceedings of the 2012 XXth International Conference on Electrical Machines, Marseille, France, 2–5 September 2012; pp. 397–403. [CrossRef]
26. Kioumarsi, A.; Moallem, M.; Fahimi, B. Mitigation of Torque Ripple in Interior Permanent Magnet Motors by Optimal Shape Design. *IEEE Trans. Magn.* **2006**, *42*, 3706–3711. [CrossRef]
27. Shimizu, Y.; Morimoto, S.; Sanada, M.; Inoue, Y. Reduction of torque ripple in double-layered IPMSM for automotive applications by rotor structure modification. In Proceedings of the 2017 IEEE 12th International Conference on Power Electronics and Drive Systems (PEDS), Honolulu, HI, USA, 12–15 December 2017; pp. 429–434. [CrossRef]
28. Yamazaki, K.; Kumagai, M.; Ikemi, T.; Ohki, S. A Novel Rotor Design of Interior Permanent-Magnet Synchronous Motors to Cope with Both Maximum Torque and Iron-Loss Reduction. *IEEE Trans. Ind. Appl.* **2013**, *49*, 2478–2486. [CrossRef]
29. Kang, G.; Son, Y.; Kim, G.; Hur, J. A Novel Cogging Torque Reduction Method for Interior-Type Permanent-Magnet Motor. *IEEE Trans. Ind. Appl.* **2009**, *45*, 161–167. [CrossRef]
30. Liang, J.; Parsapour, A.; Moallem, M.; Fahimi, B. Asymmetric Rotor Surface Design in Interior Permanent Magnet Synchronous Motors for Torque Ripple Mitigation. In Proceedings of the 2019 IEEE International Electric Machines Drives Conference (IEMDC), San Deigo, CA, USA, 15 May 2019; pp. 727–732. [CrossRef]
31. Liang, J.; Parsapour, A.; Yang, Z.; Caicedo-Narvaez, C.; Moallem, M.; Fahimi, B. Optimization of Air-Gap Profile in Interior Permanent-Magnet Synchronous Motors for Torque Ripple Mitigation. *IEEE Trans. Transp. Electrif.* **2019**, *5*, 118–125. [CrossRef]
32. Yamazaki, K.; Utsunomiya, K. Mechanism of Torque Ripple Generation by Time and Space Harmonic Magnetic Fields in Interior Permanent Magnet Synchronous Motors. In Proceedings of the 2020 International Conference on Electrical Machines (ICEM), Virtual, 23–26 August 2020; Volume 1, pp. 232–238. [CrossRef]
33. Islam, R.; Husain, I.; Fardoun, A.; McLaughlin, K. Permanent-Magnet Synchronous Motor Magnet Designs with Skewing for Torque Ripple and Cogging Torque Reduction. *IEEE Trans. Ind. Appl.* **2009**, *45*, 152–160. [CrossRef]
34. Cao, R.; Mi, C.; Cheng, M. Quantitative Comparison of Flux-Switching Permanent-Magnet Motors with Interior Permanent Magnet Motor for EV, HEV, and PHEV Applications. *IEEE Trans. Magn.* **2012**, *48*, 2374–2384. [CrossRef]
35. Wu, D.; Zhu, Z.Q. Design Tradeoff Between Cogging Torque and Torque Ripple in Fractional Slot Surface-Mounted Permanent Magnet Machines. *IEEE Trans. Magn.* **2015**, *51*, 1–4. [CrossRef]
36. Liu, X.; Chen, H.; Zhao, J.; Belahcen, A. Research on the Performances and Parameters of Interior PMSM Used for Electric Vehicles. *IEEE Trans. Ind. Electron.* **2016**, *63*, 3533–3545. [CrossRef]
37. Pellegrino, G.; Vagati, A.; Guglielmi, P.; Boazzo, B. Performance Comparison Between Surface-Mounted and Interior PM Motor Drives for Electric Vehicle Application. *IEEE Trans. Ind. Electron.* **2012**, *59*, 803–811. [CrossRef]
38. Pellegrino, G.; Vagati, A.; Boazzo, B.; Guglielmi, P. Comparison of Induction and PM Synchronous Motor Drives for EV Application Including Design Examples. *IEEE Trans. Ind. Appl.* **2012**, *48*, 2322–2332. [CrossRef]
39. Yang, Z.; Shang, F.; Brown, I.P.; Krishnamurthy, M. Comparative Study of Interior Permanent Magnet, Induction, and Switched Reluctance Motor Drives for EV and HEV Applications. *IEEE Trans. Transp. Electrif.* **2015**, *1*, 245–254. [CrossRef]
40. Zarko, D.; Ban, D.; Lipo, T. Analytical calculation of magnetic field distribution in the slotted air gap of a surface permanent-magnet motor using complex relative air-gap permeance. *IEEE Trans. Magn.* **2006**, *42*, 1828–1837. [CrossRef]
41. Wu, L.J.; Zhu, Z.Q.; Staton, D.; Popescu, M.; Hawkins, D. An Improved Subdomain Model for Predicting Magnetic Field of Surface-Mounted Permanent Magnet Machines Accounting for Tooth-Tips. *IEEE Trans. Magn.* **2011**, *47*, 1693–1704. [CrossRef]
42. Kim, B.; Lipo, T.A. Analysis of a PM Vernier Motor with Spoke Structure. *IEEE Trans. Ind. Appl.* **2016**, *52*, 217–225. [CrossRef]
43. Dajaku, G. Open-Circuit Air-Gap Field Calculation of a New PM Machine Having a Combined SPM and Spoke-Type Magnets. *IEEE Trans. Magn.* **2020**, *56*, 1–9. [CrossRef]
44. Dajaku, G. Analytical Analysis of Electromagnetic Torque and Magnet Utilization Factor for Two Different PM Machines with SPM and HUPM Rotor Topologies. *IEEE Trans. Magn.* **2021**, *57*, 1–9. [CrossRef]

45. Kwon, J.W.; Li, M.; Kwon, B.I. Design of V-Type Consequent-Pole IPM Machine for PM Cost Reduction with Analytical Method. *IEEE Access* **2021**, *9*, 77386–77397. [CrossRef]
46. Hoang, K.D. Simplified Analytical Model for Rapid Evaluation of Interior PM Traction Machines Considering Magnetic Nonlinearity. *IEEE Open J. Ind. Electron. Soc.* **2020**, *1*, 340–354. [CrossRef]
47. Hu, W.; Zhang, X.; Lei, Y.; Du, Q.; Shi, L.; Liu, G. Analytical Model of Air-Gap Field in Hybrid Excitation and Interior Permanent Magnet Machine for Electric Logistics Vehicles. *IEEE Access* **2020**, *8*, 148237–148249. [CrossRef]
48. Ghahfarokhi, M.M.; Amiri, E.; Boroujeni, S.T.; Aliabad, A.D. On-Load Analytical Modeling of Slotted Interior Magnet Synchronous Machines Using Magnetic Islands Method. *IEEE Access* **2020**, *8*, 95360–95367. [CrossRef]
49. Hajdinjak, M.; Miljavec, D. Analytical Calculation of the Magnetic Field Distribution in Slotless Brushless Machines with U-Shaped Interior Permanent Magnets. *IEEE Trans. Ind. Electron.* **2020**, *67*, 6721–6731. [CrossRef]
50. Mirazimi, M.S.; Kiyoumarsi, A. Magnetic Field Analysis of SynRel and PMASynRel Machines with Hyperbolic Flux Barriers Using Conformal Mapping. *IEEE Trans. Transp. Electrif.* **2020**, *6*, 52–61. [CrossRef]
51. Hayslett, S.; Strangas, E. Design and Analysis of Aligned Axis Interior Permanent Magnet Machines Considering Saturation. In Proceedings of the 2019 IEEE International Electric Machines Drives Conference (IEMDC), San Diego, CA, USA, 12–15 May 2019; pp. 686–692. [CrossRef]
52. Thike, R.; Pillay, P. Mathematical Model of an Interior PMSM with Aligned Magnet and Reluctance Torques. *IEEE Trans. Transp. Electrif.* **2020**, *6*, 647–658. [CrossRef]
53. Toliyat, H.A. *Electric Machines: Modeling, Condition Monitoring, and Fault Diagnosis*; CRC Press: Boca Raton, FL, USA, 2017.
54. Taghipour Boroujeni, S.; Zamani, V. A Novel Analytical Model for No-Load, Slotted, Surface-Mounted PM Machines: Air Gap Flux Density and Cogging Torque. *IEEE Trans. Magn.* **2015**, *51*, 1–8. [CrossRef]
55. Cheng, D.K. *Field and Wave Electromagnetics*; Addison-Wesley: Boston, MA, USA, 1989.

 *energies*

Article

Parasitic Effects of PWM-VSI Control Leading to Torque Harmonics in AC Drives [†]

Juriy Plotkin [1,*], Nurgul Almuratova [2], Assel Yerzhan [3] and Victor Petrushin [4]

[1] Department of Cooperative Studies, Berlin School of Economics and Law, 10315 Berlin, Germany
[2] Electrical Machines and Electric Drive Department, Almaty University of Power Engineering and Telecommunications, Almaty 050013, Kazakhstan; nur0577@mail.ru
[3] Telecommunications and Innovative Technologies Department, Almaty University of Power Engineering and Telecommunications, Almaty 050013, Kazakhstan; a.erzhan@aues.kz
[4] Electrical Machines Department, Odessa National Polytechnic University, 65044 Odessa, Ukraine; victor_petrushin@ukr.net
* Correspondence: juriy.plotkin@hwr-berlin.de
† This paper is an extended version of our paper published in 35th Annual Conference of IEEE Industrial Electronics, Porto, Portugal, 3–5 November 2009; pp. 1368–1372.

Abstract: Precise torque control without pulsations is one of the major quality issues in pulse-width modulated voltage-source inverter (PWM-VSI) drives. Theoretically, it could be postulated that at frequencies of some kHz, the machine's inertia absorbs switching frequency torque harmonics, and the resulting torque becomes smooth; though, in reality, parasitic effects in voltage source inverters may cause additional torque harmonics of low order. In particular, first, second and sixth torque harmonics are observed. Such torque harmonics are especially dangerous for normal drive operation, since they may be amplified by drive train resonances at corresponding rotational velocities. New parasitic effects in PWM-VSI control, leading to torque harmonic of low order, are described in the paper, and recommendations for their compensation are given.

Keywords: torque ripple; 6th harmonic; induction motor; AC machine; PWM inverter; space phasor modulation

Citation: Plotkin, J.; Almuratova, N.; Yerzhan, A.; Petrushin, V. Parasitic Effects of PWM-VSI Control Leading to Torque Harmonics in AC Drives . *Energies* **2021**, *14*, 1713. https://doi.org/10.3390/en14061713

Academic Editor: Federico Barrero

Received: 8 February 2021
Accepted: 10 March 2021
Published: 19 March 2021

Publisher's Note: MDPI stays neutral with regard to jurisdictional claims in published maps and institutional affiliations.

1. Introduction

Modern industry applications require inverter-fed pulse-width modulated (PWM) drives with a high standard of performance. Excellent dynamic characteristics have been achieved in recent years with the development of power electronic devices and sophisticated control strategies. However, there is still improvement potential concerning torque ripple reduction methods. Torque harmonics may cause mechanical oscillations, which are particularly dangerous on resonance frequencies of the drive train. In addition, they may produce audible noise.

Drive train resonance frequencies are typically within the range of a couple hundred Hertz. That is why avoiding low-order torque harmonics is so important; such torque harmonics could excite the mechanical resonance oscillations hitting the resonance frequencies of the drive train. With variable drive rotational frequency also, the torque harmonic frequencies vary; during the start-up they could successively coincide with the drives' resonance frequency, as is shown in the measurement in Figure 1.

First, second and sixth torque harmonics in PWM-VSI-fed drives were observed in the literature [1–6], the sixth torque harmonic is reported to be usually the dominating one [3–6]. The origin of these torque harmonics especially in pulse-width modulated voltage-source inverter (PWM-VSI)-fed drives has not been studied sufficiently. This paper summarizes the authors' own preceding research upon parasitic effects as a reason for low-order torque harmonics [1,3,6–9]—it is mainly based on the conference paper [3].

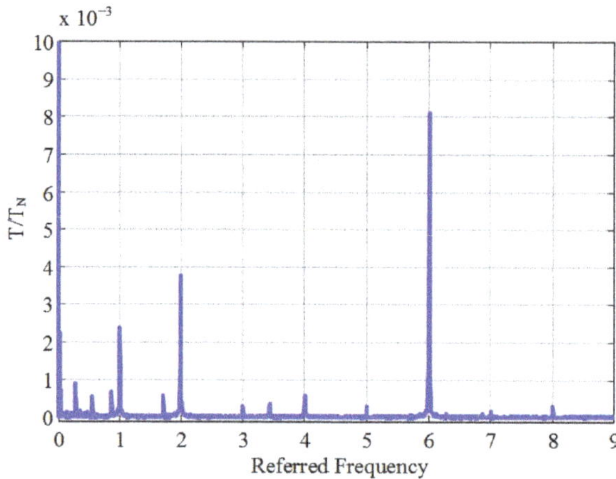

Figure 1. Torque spectrum of a 220 kW asynchronous drive [3].

Torque harmonics in alternating current (AC) machines may occur also at perfectly sinusoidal power supplies: they may be caused by intrinsic electrical machine properties, by its control technique or by the load characteristics. This paper focuses, however, solely on torque harmonics due to the possible parasitic effects of the PWM's supply voltage in a kHz range. Investigation of current and torque harmonics for PWM supply with small switching frequencies can be found in the literature [10–14].

Figure 1 shows measured torque harmonics of a 220 kW induction test-bench motor drive [3,6]. It can be observed that the sixth torque harmonic is the predominating one. The induction machine ratings are given in Table 1. It is a special purpose drive with an untypically rated frequency of 67 Hz. Further investigation and explanations will be given in the example of 50 Hz supply frequency—for instance, in the case of the investigated 2.2 kW induction machine in Section 2.

Table 1. Tested motor ratings.

Power	220 kW
Voltage	400 V
Current	390 A
Frequency	67 Hz
Rotational speed	3985 rpm

Possible reasons for torque harmonics due to parasitic effects are given in the following sections.

2. First Torque Harmonic

Rotating mass eccentricity is one possible reason for the first torque harmonic appearance in drives. For PWM-VSI-fed drives, the direct current (DC) component could be a further reason for first torque harmonic emergence [6–9,15–17].

In the presence of a current control loop, any current measurement offset could lead to real offset in machines currents. Usually only a very small measurement offset can be expected, which would lead to an unnoteworthy current offset. Another possible reason for the DC component in drives with and without current control loops could be control circuit- and power switch parameter dispersion [9].

In Figure 2, phase currents of the PWM-inverter-fed ohmic-inductive load at switching frequency of 15 kHz are shown. They contain a DC component, though no DC voltage offset was commanded. PWM control signals within this inverter are transmitted by means of plastic optical fiber coupling to IGBT gate drivers. Such optical coupling brings an advantage of insensitivity to electromagnetic noises and of galvanic isolation between the circuits.

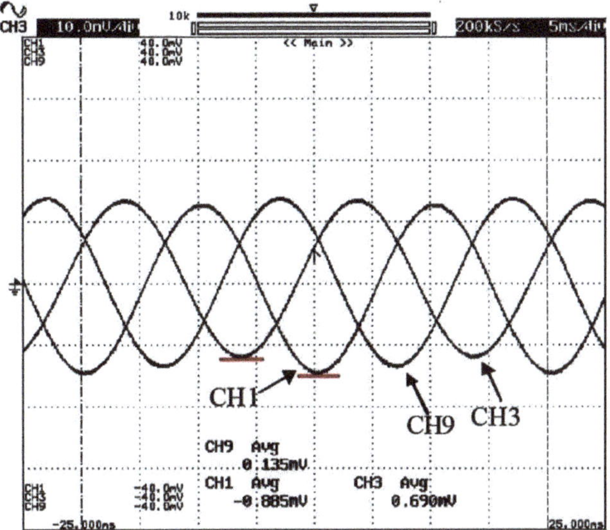

Figure 2. Currents in the three-phase ohmic-inductive load at 15 kHz switching frequency [9].

Control signals for transistors of one inverter phase are usually complementary to each other and are transmitted by only one signal. Control signal inversion with necessary dead time for the switching between the upper and lower transistors is realized within the gate unit. The driver circuit for the optical transmitter is implemented according to the producer recommendations given in the application notes, see Figure 3.

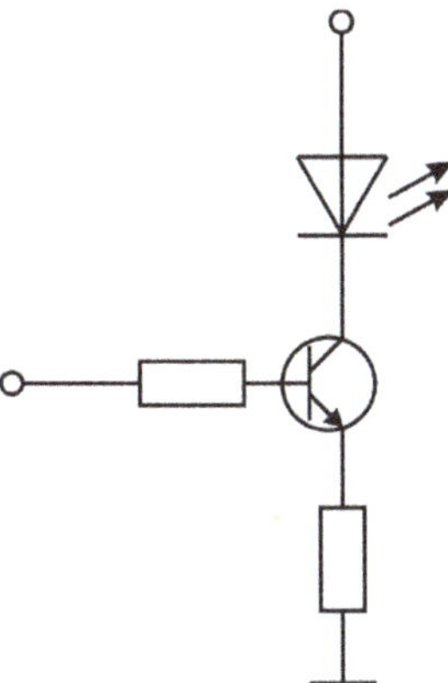

Figure 3. Emitting diode driver circuit [16].

To determine the source of DC offset in the inverter output voltage, control signals are investigated upon the propagation delay time. For this aim, the same pulse sequence is commanded to all inverter phases simultaneously. As expected, control signals measured directly at the output port of the micro-controller are perfectly simultaneous without any lag-time relative to each other.

On the contrary, control signals measured at the output of the optical link, behind the optical receiver, have delay time relative to each other. As is shown in Figure 4, the rising edges show time delays of more than 100 ns, and the falling edges are close to each other. That means that, with each switching cycle, an additional non-commanded DC voltage offset between the phases arises, and its magnitude is proportional to the switching frequency and time delay between the signals.

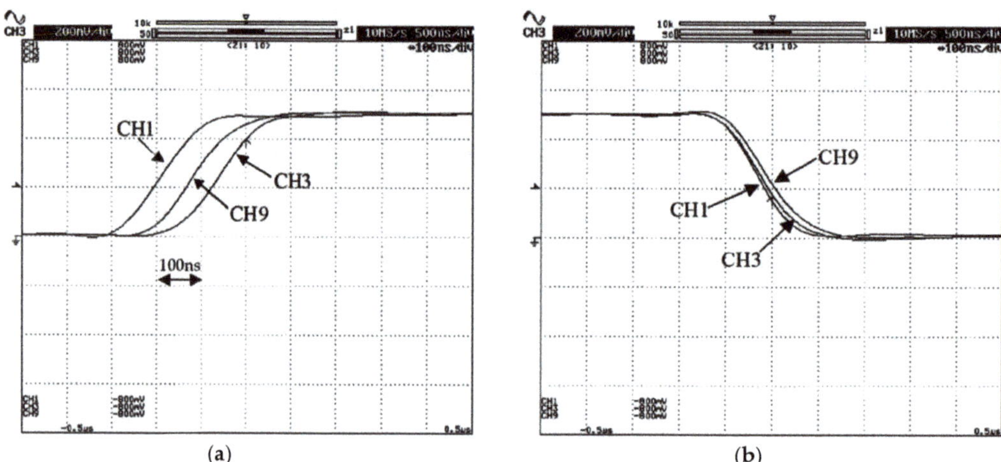

Figure 4. Control signals for three inverter legs measured after the optical transmission [16]: (**a**) rising edge—light off; (**b**) falling edge—light on [9].

The DC offset in the currents of an ohmic-inductive load, see Figure 2, corresponds to a measured delay in the control signals. Since phase A (CH1 in the Figures 2 and 4) has the biggest negative DC voltage offset compared to the other phases, it has consequently also the biggest negative current offset.

The light-emitting diode of the optical transmitter is driven by the bipolar transistor. It is known, that transistors within the same production series show some dispersion in their characteristics. Among other parameters, the switching on/off time delays may differ sufficiently. Whereby the switching on times are usually much smaller than the switching off times; therefore, their variation may be considered as of no consequence. The rising edge of the voltage in the Figure 4a corresponds to bipolar transistor switching off in the emitting diode driver circuit. This edge shows the strongest time deviation from phase to phase because of the switching-off delay time variation of the given bipolar transistors. The time deviation for the transistors switching on shows a much lower dispersion level, which could be neglected for DC offset generation (Figure 4b).

Measurement directly on the emitter side circuit showed the same time deviations between the voltages, as on the optical receiver side. Such small time deviations in the control signals play a role only for high switching frequencies of some kHz. Even comparatively small resulting DC voltage offsets may lead to considerable DC currents, since the active resistance of the driven load is usually small, as in the case of large AC machines.

We also measured the fact that the light dispersion, which occurs due to improper optical fiber contacting with the receiver, leads to a similar effect of different switching delays and, therefore, to the DC voltage offset. Attention must be paid to proper contacting and clean surfaces of optic fiber endings.

Another possible reason for a potentially much higher DC current component is the malfunction of dead time compensation because of an offset in a current measurement [8]. Even the minimal measurement offset level may lead to a considerable DC current shift. Dead time voltage due to interlock times between switching transistors of the same phase may be represented as rectangular voltage, which polarity is solely dependent upon current direction. The most common method for its compensation is to add to the reference's rectangular voltage of the same magnitude but of the opposite polarity. If the current measurement shows an offset, then an error will occur in generating the compensating voltage reference, as is shown in Figure 5. Such error leads to unequal duration of compensating voltage positive and negative half-waves. Such asymmetry leads to a nonzero mean value in compensating voltage and correspondingly to a DC current offset. Due to the small DC resistance of the connected AC machines significant DC current level may appear even at the lowest bit offset in the current measurement.

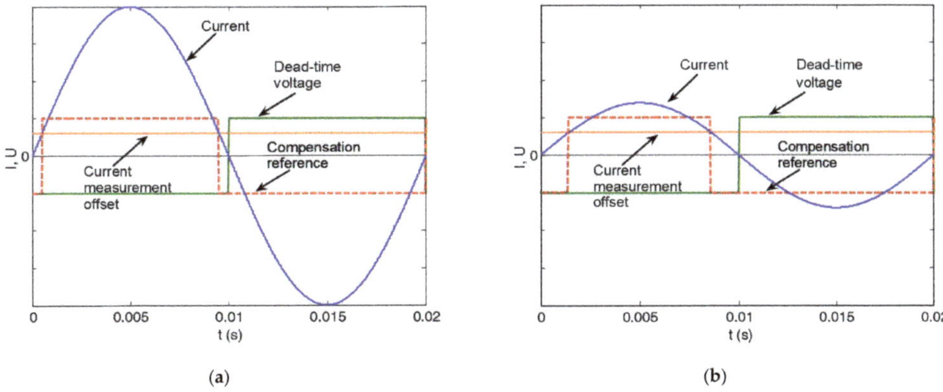

Figure 5. Dead time voltage compensation malfunction caused by current measurement offset: (**a**) larger current magnitude; (**b**) smaller current magnitude [3].

The longer phase current polarity is wrongly detected because of the measurement offset—larger asymmetry and resulting DC offset in the compensating reference voltage will result. As is shown in Figure 5, the duration of false current polarity detection depends upon the current slope near its zero crossings and, consequently, upon its magnitude at a given frequency. At small current magnitudes, the error increases and the resulting DC component in the compensating voltage rises. Figure 6 shows the measured DC current offset with 23% of the rated current in a phase of the induction machine operated in field weakening mode caused by just 0.8% DC offset in the current measurement [8]. At no load operation magnetizing current component is the prevailing one in the phases of the induction machines; hence, the total current magnitude becomes strongly reduced in the field-weakening mode, which provokes large DC offset due to faulty generated dead-time-compensating voltage.

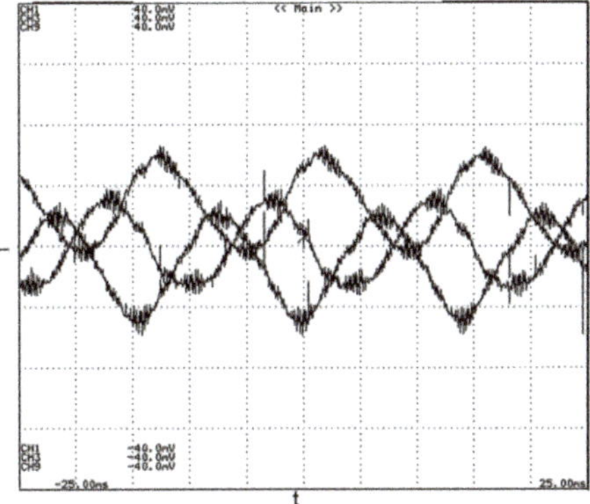

Figure 6. DC current offset in 2.2 kW induction machine phase currents in field-weakening region at no load due to faulty dead time compensation [8].

3. Second Torque Harmonic

The second torque harmonic can be provoked by amplitude error in the current measurement. Through asymmetry in the current measurement, current control would produce asymmetry in real currents, containing a negative sequence. An interaction of this negative sequence with the positive sequence would lead to the second torque harmonic.

A further reason for the second torque harmonic could also be the consequence of the faulty dead time voltage compensation described in the previous section. Dead time compensation voltage here is asymmetric due to the offset in the current measurement; the resulting DC component is discussed above. Another important consequence of rectangular voltage asymmetry is the emergence of second harmonic component, as is shown in Figure 7 [3]. Slip for all higher harmonics is very high, which is why the current components became disproportionally high despite the relatively small harmonic voltage magnitude. Interaction between DC and second harmonic currents leads to second torque harmonic.

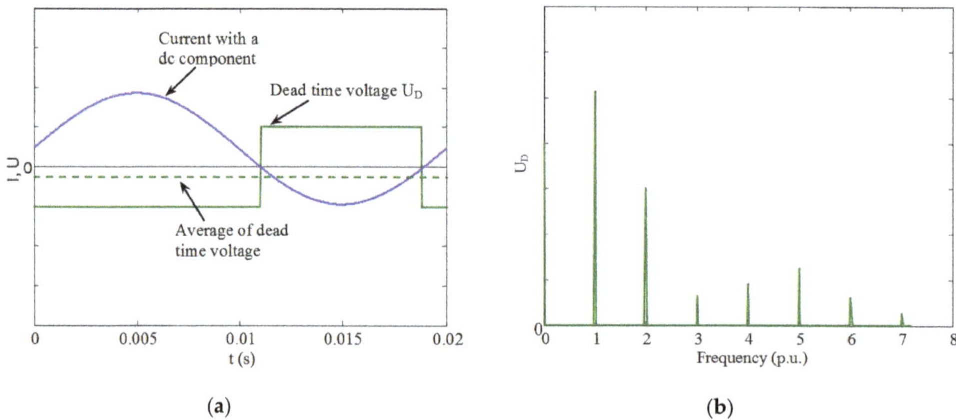

Figure 7. Dead time voltage in a presence of a DC component in a phase current: (**a**) time course; (**b**) spectrum [1].

A further reason for the second and higher odd harmonics is saturation due to DC current component [17].

The second harmonic, belonging to negative sequence, results in an interaction with the fundamental flux to a third torque harmonic.

4. Sixth Torque Harmonic

Uncompensated rectangular dead time voltage contains fifth and seventh harmonics. The fifth harmonic belongs to the negative sequence and the seventh harmonic belongs to the positive sequence, which is why both of them lead to an interaction with fundamental voltage to a sixth torque harmonic [2]. Dead time voltage magnitude does not vary as the output frequency changes. The magnitude of the dead time voltage is solely dependent on DC link voltage, switching frequency and interlock time between transistors switching [1]. Its influx will increase with lower output frequency and consequently lower fundamental voltage. That is why the sixth torque harmonic caused by dead time voltage will increase at lower VSI output frequency.

At 220 kW test bench drive, a vice versa behavior of the sixth torque harmonic was observed in [4]—its magnitude increased with higher rotational speeds (Figure 8). The origin of the sixth torque harmonic in that case could be explained by parasitic effects in the PWM-VSI control by operation in the nonlinear area of the machine's magnetization curve [1,3,6].

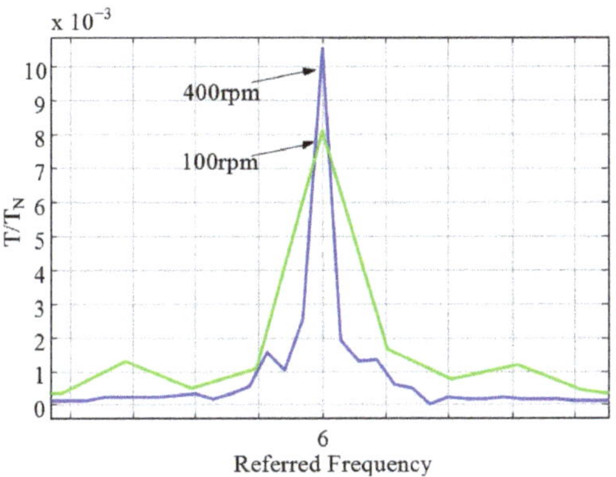

Figure 8. Measured torque harmonic at a load of 150 Nm [6].

For space phasor modulation of two-level VSI, eight voltage phasors corresponding to eight possible switching states are available. Six of them are active phasors and two are zero phasors, when all phasors switched to the upper or lower DC link bus. Any voltage space phasors within the hexagon are modulated by switching between these eight phasors (Figure 9). In order to obtain the desired circular trajectory, zero phasors must be utilized. Switching between active and zero phasors produces torque vibration. The duration of zero phasors increases in the vicinity of each active phasor and decreases in the middle of sectors formed by active phasors, so it varies six times per period (Figure 8). With the zero-phasor duration variation also, the magnitude of torque vibration of switching frequency will vary six times a period.

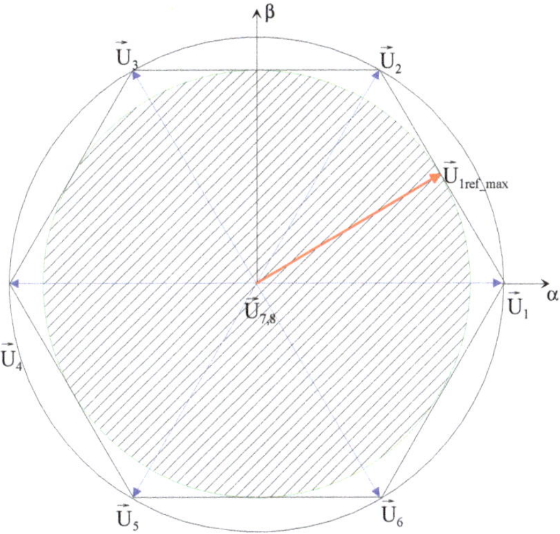

Figure 9. Space phasors of two-level voltage-source inverter (VSI) [6].

In order to demonstrate torque pulsation of switching frequency of an induction motor drive, simulation was performed in Matlab Simulink. The machine model has been realized as a set of differential equations in α-β-coordinates in the form of S-Function. Voltage PWM pulse pattern was generated by sine-triangle comparison, as also was the case of the 220 kW test bench drive under investigation. The pulse pattern included the pulse dropping effect; ideal switching neglecting transistor dynamics was implemented in the model. Simulation results confirmed the theoretical assumption given above; resulting torque pulsation of the sixth order is shown in Figure 10a. Measuring such torque pulsation at higher switching frequencies is not possible, because torque pulsation with frequency of some kHz would be absorbed by rotor inertia even in small power machines. Therefore, it cannot be measured on the machines shaft, but it can be demonstrated through simulation.

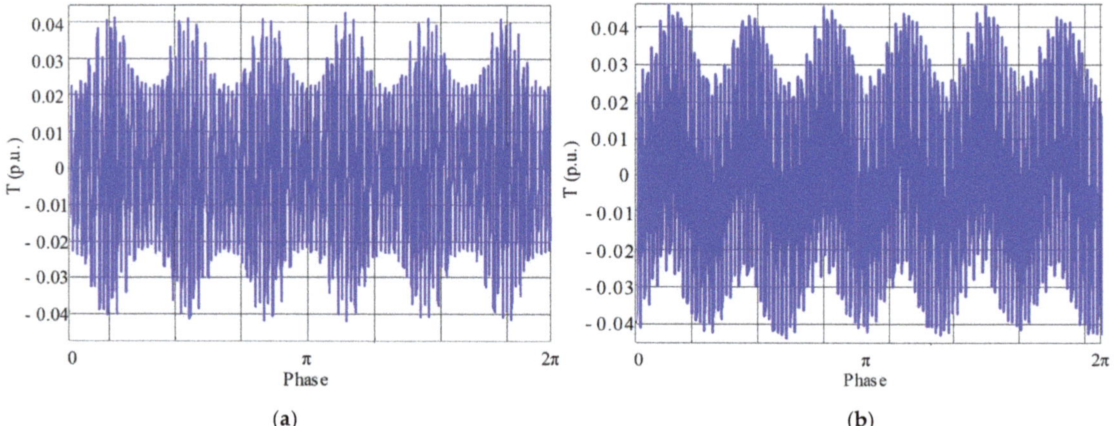

Figure 10. Simulated torque of pulse-width modulated voltage-source inverter (PWM-VSI)-fed induction machine: (**a**) unsaturated; (**b**) at the "knee" of the magnetization characteristic [6].

By operation in the nonlinear area of the machine's magnetization characteristic, pulsation will get distorted by its nonlinearly. The upper half-waves of the pulsation will be weakened due to saturation; they will not compensate the lower half-waves anymore. The higher the pulsation magnitude becomes, the lower the average torque due to the saturation effect. Therefore, operating in the beginning of the magnetic saturation area will transform torque pulsation with switching frequency to torque harmonic of the sixth order with minima at $n{\cdot}60°$ ($n = 0; +/-1\ldots$). The sharper the saturation curve, the higher the magnitude of the sixth torque harmonic that can be expected.

To prove this assumption, the measured magnetic saturation curve of a 220 kW induction machine (Table 1) was incorporated into the Matlab model by a look-up table. Simulated torque for operating point on a knee of the magnetization characteristic is given in Figure 10b. The simulation result clearly shows the appearance of the sixth torque harmonic. Torque pulsation of the switching frequency is now a carrier frequency for the machine's sixth torque harmonic. Carrier torque frequency will be absorbed by drive inertia as occurred in the previous case, but the sixth torque component will be passed through to the machine's shaft.

By increasing the voltage reference, zero phasor variation, relative to the active phasors, and, consequently, the torque pulsation amplitude, increased. That means that, unlike for the sixth torque harmonic due to dead time voltage, the sixth torque harmonic due to saturation would rise with higher output voltage of VSI. With field weakening, the sixth torque harmonic would significantly decrease [6] or completely disappear.

5. Corrective Measures

Compensating the torque harmonics due to parasitic effects in PWM-VSI may be performed by control software changes in the majority of cases.

To avoid the first and second torque harmonic, a DC current component must be prevented. It can be achieved by current measurement refinement. A DC offset in the measurement can be easily detected prior to the start by blocked transistor control signals. It should be subtracted from the current measurement during the further operation.

During the drive operation, temperature drift in the current measurement may arise, so, in each idle mode, the zero-point calibration should be repeated. Furthermore, current measurement accuracy should be improved. In the 2.2 kW drive given in the second section, only 10-bit analog–digital converters for current measurement were applied, which is obviously not precise enough. Analog–digital converters with higher resolution would help to avoid the parasitic effects described above, which result in torque harmonics.

In the case of driver circuits for optical fiber coupling, the simplest way to reduce the measured DC voltage offset owing to the transistor tolerances would be to replace bipolar transistors by MOSFETs, which have much lower switching off delay times and, therefore, narrower tolerances.

Avoiding DC current component insertion is important, not only for machine drives but also for grid-side inverters.

Precise compensation of dead time voltage would prevent the sixth torque harmonic appearance. The major difficulty for its implementation lies in precise current polarity detection, which may vary due to multiple zero crossings at high current ripple.

Sixth torque harmonic due to PWM in the nonlinear magnetization area can be compensated by torque reference with the opposite phase. Minima and maxima of the sixth torque component are provided by the voltage angle, with a high degree of accuracy reference voltage phasor angle that can be used for compensation aim. The magnitude of the occurring sixth torque harmonic can be measured once during the drive commissioning for various voltage values and stored in a look-up table; it can be used for the compensation reference during drive operation. Solely, phase delay of torque control loop must be considered for the optimal implementation of this method. Due to production tolerances, machine parameters may vary significantly [5], and it is advisable to perform such measurements for each machine individually.

6. Conclusions

Parasitic effects in PWM-VSI controls are found to be the reason for low-order torque harmonics in AC drives operated at switching frequencies of some kHz. Low-order torque harmonics could coincide with drive train resonance frequency leading to strong vibrations, depending on the grade of mechanical damping. This may disturb the technological process and decrease production quality.

Knowing the origin of such low-order torque harmonics enables one to meet countermeasures. Not only are torque harmonic origins explained, but also ways to compensate them are presented in this paper. Mostly compensation measures are restricted to control software modification only. However, precise current measurement hardware is advisable in any case.

Author Contributions: Conceptualization, J.P. and V.P.; methodology, J.P.; validation, J.P., N.A. and A.Y.; investigation, J.P.; writing—original draft preparation, J.P.; writing—review and editing, V.P. and N.A. All authors have read and agreed to the published version of the manuscript.

Funding: This research received no external funding.

Conflicts of Interest: The authors declare no conflict of interest.

References

1. Plotkin, J. *Parasitic Effects in PWM-Converters with DC Voltage Link and Their Impact onto the Operation of Connected Loads*; Shaker Verlag: Aachen, Germany, 2008.
2. Barro, R. Torque Ripple Compensation of Induction Motors Under Field Oriented Control. In Proceedings of the APEC 97—Applied Power Electronics Conference, Atlanta, GA, USA, 27 February 1997; Volume 1, pp. 527–533.
3. Plotkin, J.; Schaefer, U.; Hanitsch, R. Torque Ripple in PWM-VSI-fed Drives due to Parasitic Effects in the Inverter Control. In Proceedings of the 35th Annual Conference of IEEE Industrial Electronics, Porto, Portugal, 3–5 November 2009; pp. 1368–1372.
4. Hofmeyer, D. Harmonic Torque Pulsations of Induction Machines—Analysis and Compensation Techniques Using PWM Inverter. In Proceedings of the EPE 10th European Conference on Power Electronics and Applications, Toulouse, France, 2–4 September 2003.

5. Holtz, J. Identification and Compensation of Torque Ripple in High- Precision Permanent Magnet Motor Drives. *IEEE Trans. Ind. Electron.* **1996**, *43*, 309–320. [CrossRef]
6. Plotkin, Y.; Stiebler, M.; Hofmeyer, D. Sixth Torque Harmonic in PWM Inverter-fed Induction Drives and Its Compensation. *IEEE Trans. Ind. Appl.* **2005**, *41*, 1067–1074. [CrossRef]
7. Plotkin, J.; Schaefer, U.; Hanitsch, R. High First Torque Harmonic Due to Insufficient Function of Dead Time Compensation in PWM Inverters. In Proceedings of the EPE 12th European Conference on Power Electronics and Applications, Aalborg, Denmark, 2–5 September 2007.
8. Plotkin, J.; Schaefer, U.; Hanitsch, R. Malfunction of a Dead-Time Compensation in PWM-Converters Leading to a High DC Current Offset. In Proceedings of the IEEE IECON, Orlando, FL, USA, 10–13 November 2008.
9. Plotkin, J.; Centner, M.; Schulz, D.; Hanitsch, R. DC Offset at PWM-Inverter Output due to Tolerances of the Optical Link Junction for IGBT-Control Signals. In Proceedings of the International Conference on Electrical Machines (ICEM), Chania, Greece, 2–5 September 2006.
10. Enjeti, P.N.; Ziogas, P.D.; Lindsay, J.F. Programmed PWM Techniques to Eliminate Harmonics: A Critical Evaluation. *IEEE Trans. Ind. Appl.* **1990**, *26*, 418–430. [CrossRef]
11. Holtz, J. Pulsewidth Modulation—A Survey. *IEEE Trans. Ind. Electron.* **1992**, *39*, 410–420. [CrossRef]
12. Jenni, F.; Wuest, D. *Control Methods for Selfcommutated Converters*; B.G. Teubner: Stuttgart, Germany; vdf Hochschulverlag: Zürich, Switzerland, 1995.
13. Michel, M. *Fundamentals of Power Electronics*; Springer: Berlin, Germany, 1995.
14. Patel, H.S.; Hoft, R.G. Generalized Techniques of Harmonic Elimination and Voltage Control in Thyristor Inverters. *IEEE Trans. Ind. Appl.* **1973**, *3*, 310–317. [CrossRef]
15. Antic, D.; Klaassens, J.B.; Deleroi, W. Side Effects in Low-speed AC Drives. In Proceedings of the PESC'94 Power Electronics Specialists Conference, Taipei, Taiwan, 20–25 June 1994; Volume 2, pp. 527–533.
16. Chen, S.; Namuduri, C.; Mir, S. Controller-induced Parasitic Torque Ripples in a PM Synchronous Motor. *IEEE Trans. Ind. Appl.* **2002**, *38*, 1273–1281. [CrossRef]
17. Zhang, P.; Du, Y.; Habetler, T.G.; Lu, B. Magnetic Effects of DC Signal Injection on Induction Motors for Thermal Evaluation of Stator Windings. *IEEE Trans. Ind. Electron.* **2010**, *58*, 1479–1489. [CrossRef]

Article

A General and Accurate Measurement Procedure for the Detection of Power Losses Variations in Permanent Magnet Synchronous Motor Drives

Massimo Caruso [1], Antonino Oscar Di Tommaso [1], Giuseppe Lisciandrello [1], Rosa Anna Mastromauro [2], Rosario Miceli [1,*], Claudio Nevoloso [1], Ciro Spataro [1] and Marco Trapanese [1]

[1] Department of Engineering, University of Palermo, Viale Delle Scienze, Parco D'Orleans, 90128 Palermo, Italy; massimo.caruso16@unipa.it (M.C.); antoninooscar.ditommaso@unipa.it (A.O.D.T.); giuseppe.lisciandrello@unipa.it (G.L.); claudio.nevoloso@unipa.it (C.N.); ciro.spataro@unipa.it (C.S.); marco.trapanese@unipa.it (M.T.)

[2] Department of Information Engineering, University of Florence, 50139 Florence, Italy; rosaanna.mastromauro@unifi.it

* Correspondence: rosario.miceli@unipa.it

Received: 12 October 2020; Accepted: 2 November 2020; Published: 4 November 2020

Abstract: The research of innovative solutions to improve the efficiency of electric drives is of considerable interest to challenges related to energy savings and sustainable development. In order to successfully validate the adoption of new and innovative software or hardware solutions in the field of electric drives, accurate measurement procedures for either efficiency or power losses are needed. Moreover, high accuracy and expensive measurement equipment are required to satisfy international standard prescriptions. In this scenario, this paper describes an accurate measurement procedure, which is independent of the accuracy of the adopted instrumentation, for the power losses variations involved in electrical drives, namely $\Delta\Delta P$, useful to detect the efficiency enhancement (or power losses reduction) due to the real-time modification of the related control algorithm. The goal is to define a valuable measurement procedure capable of comparing the impact of different control algorithms on electric drive performance. This procedure is carried out by experimentally verifying the action of different control algorithms by the use of a Field Oriented Control (FOC) with different values of the direct-axis current component (i.e., $I_d = 0$ A and $I_d = -1$ A) applied for fixed working conditions in terms of speed and load torque. Two different measurement systems of power losses, each one characterized by different accuracy and cost, are taken into account for the validation of the proposed method. An investigation is, then, carried out, based on the comparison between the measurements acquired by both instrumentations, for different working conditions in terms of load and speed, highlighting that the uncertainty generated by systematic errors does not affect the $\Delta\Delta P$ measurements. The results reported in this work demonstrate how the $\Delta\Delta P$ parameter can be used as a valuable index for the characterization of the power drive system, which can also be evaluated even with low-accuracy instrumentation.

Keywords: power loss minimization; speed control drive systems; efficiency measurement; IPMSM

1. Introduction

In recent years, the challenges related to the reduction of environmental pollution, energy savings and sustainable development have aroused the interest not only of the scientific and industrial communities but also of society and policies of the countries. In this scenario, the field of electrical drives plays a fundamental role, since it represents the biggest consumer related to global energy consumption in industrial applications [1,2]. Therefore, the efforts of the scientific community have been recently dedicated to on the research and development of innovative solutions concerning the design and control of high-performance power converters and electric motors, in order to obtain a relevant improvement of the electrical drives both in terms of efficiency and adaptability for challenging operating conditions. In particular, a significant increase in the adoption of electrical drives equipped with Permanent Magnet Synchronous Motors (PMSMs) has been detected, due to their high power factor, high torque density and high efficiency, especially for e-mobility [3,4], aircraft [5,6] and marine propulsion applications [6–8].

A possible solution for the energy savings purpose is the design and development of control algorithms for PMSM electric drives, with the aim of minimizing the power losses, maximizing, thus, the related overall efficiency. For instance, the topic of the design and development of maximum torque per ampere control algorithms for PMSM is continuously discussed [9–11] in the recent literature, as well as for the so-called loss model algorithms (LMAs), which involve the real-time determination of the optimal value of the magnetization level (or any other loss variable) for the power losses minimization of the motor [12–14] in any working operation in terms of speed and applied load. In order to validate the effectiveness of these types of control algorithms, it is needed to perform an accurate measurement of the PMSM electric drive efficiency [15–17].

In the last decade, the IEC 60034-2 [18–22] treated standardized methods in order to determine the efficiency of electrical machines fed by the electrical grid and converter-fed motors with adequate accuracy, repeatability and reproducibility. These international standards provide accurate prescriptions for induction machines, dc machines, and wound-field synchronous machines; however, specific prescriptions regarding the PMSMs efficiency measurement are not provided. Moreover, these international standards do not cover the electric machines specifically adopted for traction applications. The latest international standards IEC 61800-9, issued in 2017 [21,22], which provide the methodologies needed for the determination of the efficiency of each part of the electric drive, such as the electric motor, Complete Drive Modules (CDMs) and Power Drive Systems (PDS), are a valid reference for the electric drive efficiency measurement. These standards introduce the definition of a conventional measurement methodology to energy efficiency standardization for any extended product by defining the guidance of the Extended Product Approach (EPA), which considers not only the efficiency of the motor, but also the efficiency of the whole electric drive, including of its load.

In this context, the scope of this work concerns the definition and validation of an innovative and general measurement procedure of power losses variations involved in electrical drives, which can be applied independently from the accuracy of adopted instrumentation. Indeed, it can be generally stated that the measurement of the efficiency controlled by different control algorithms requires the PDS efficiency measurement for each control algorithm with accurate and expensive measurement instruments. Nevertheless, the proposed and discussed methodology addresses the accurate estimation of the power losses variations, namely $\Delta\Delta P$, which can be determined by real-time changes of the algorithms capable of controlling the electrical drive to maximize the related performance, even by the use of low-accuracy and cheap measurement equipment. Therefore, it will be demonstrated that, for real-time control, the $\Delta\Delta P_{PDS}$ index represents an optimal parameter for the characterization of the control system.

For this purpose, two different measurement systems, each one characterized by different accuracy and cost, have been taken into account of the proposed methodology. More specifically, an extended experimental investigation has been carried out on a low-power PMSM electric drive by changing the working conditions in terms of load and speed and by varying the magnetization level of the motor under test by means of the applied control algorithm. This work demonstrates that the $\Delta\Delta P$ index is independent of the uncertainty given by systematic errors and, therefore, the $\Delta\Delta P$ measurements can be carried out even with low-accuracy instrumentation.

This work is structured as follows: Section 2 describes the methodologies suggested by the recent standards regarding the determination of the power losses in a power drive system, Section 3 reports the mathematical description of the proposed methodology, Section 4 provides a brief description of the test bench set-up for the experimental tests, Section 5 discusses the acquisition of experimental results and their analysis, Section 6 summarizes the validation of the proposed method and provides the uncertainty analysis that validates the proposed measurement approach.

2. Methodologies for the Power Losses Measurements in Electrical Drives

Figure 1 shows a schematical representation of the PDS and its three main sections (single-phase PDS input section, no. 1, three-phase input motor section, no. 2 and mechanical PDS output section, no. 3). From the reported scheme, it is possible to define the PDS efficiency that can be measured with the direct method, by using the following relationship:

$$\eta_{PDS} = \frac{P_M}{P_{PDS}} \tag{1}$$

where P_{PDS} and P_M are the active power flows through Sections 1 and 3, respectively.

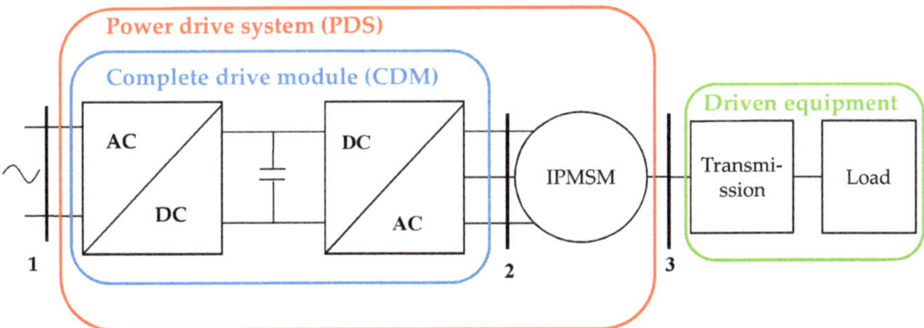

Figure 1. Schematic representation of a PDS.

Otherwise, by referring to the Standard IEC 60034-2-1, the efficiency can also be indirectly determined from the computation of the power losses involved in the system, namely ΔP, by adopting the following equation:

$$\eta = \frac{P_{PDS} - (\Delta P_{CDM} + \Delta P_M)}{P_{PDS}} \tag{2}$$

where ΔP_{CDM} and ΔP_M are the total loss of the CDM and the total loss of the motor, respectively. They can be also defined as:

$$\Delta P_{CDM} = P_{PDS} - P_{Mot} \tag{3}$$

$$\Delta P_M = P_{Mot} - P_M \tag{4}$$

where P_{Mot} is the power flow through Section 2.

Therefore, it appears evident that the indirect method involves the computation of the power losses involved in the section of both converter and motor. The indirect measurement methodology requires a huge amount of tests for the determination of each power loss component of both the converter and the motor. Therefore, this methodology is a high time-consuming measurement approach. In this context, the standards provide to apply the indirect method only for high-power motors with a power greater than 1 kW [18].

An interesting study [23] addresses the uncertainty evaluation for both direct and indirect determinations of efficiency, demonstrating that the use of the direct method with the adoption of technologically advanced instrumentation for the measurement of torque and speed, also for high-power motor, provides simpler and straight efficiency determination with uncertainties even lower than those obtained by means of the indirect method. Nevertheless, no specific standard is dedicated towards PMSMs: the standard IEC 60034-2-3 is limited only to induction machines, dc machines and wound-field synchronous machines, while the IEC 61800-9 is directed towards the characterization of the whole drive, including CDM and motor [21,22]. The reference for this topology of motors is the IEEE Std 1812-2014 [24], which provides the guidelines for the determination of ΔP, composed by the iron losses ΔP_{fe}, stator copper losses ΔP_{cu}, friction and windage losses ΔP_m and additional losses ΔP_{add} including the additional losses due to inverter voltage harmonics. The short-circuit and load tests are practically identical to those described in the standard IEC 60034-2-1 referred to traditional synchronous motors, while the no-load test presents some differences. Firstly, the rotor flux of a PMSM cannot be controlled and, therefore, the friction losses cannot be computed with the procedure suggested in the standard IEC 60034-2-1. In any case, the determination of ΔP_{fe} and ΔP_m (which cannot be easily separated like a traditional motor) leads to undoubted practical difficulties, since the rotor must be replaced with an identical rotor without PMs. Besides, the guide suggests other procedures like the direct measurement of the mechanical power transmitted by a tared machine method, or with retardation or coast-down test. Moreover, the magnetic field produced by the PMs is temperature-sensitive; therefore, during different working conditions, the PMSM temperature changes could cause variations of the rotor flux, affecting, consequentially, the produced power.

In recent literature, a small number of papers addressed the indirect determination of the PMSMs efficiency, obtained by computing the sum of the power losses. More specifically, Deusinger et al. described a new indirect measurement method of the PMSM efficiency that requires an open circuit test, a removed-rotor test and a pure reactive current test [25]. Lateb et al. [26] proposed a suitable indirect efficiency measurement method for high-speed surface-mounted PMSMs by adopting over-fluxing test at near-zero power factor, in order to estimate the motor core losses close to load conditions. However, these studies do not address the determination of measurement uncertainties, whose aspect, instead, represents a relevant aspect for the evaluation of the accuracy of the proposed efficiency measurement approaches [27,28].

In conclusion, it can be stated that the direct method is a simpler procedure than the indirect method, requiring only the measurement of input power and output power of the system taken into account (motor, CDM, or PDS). As for the PDS, the most critical aspect is related to the measurement of the output mechanical power, which requires accurate transducers or very expensive instrumentation [29]. On the contrary, the indirect method requires the measurement of each power loss component, which leads to a more complex and time-consuming procedure. However, it provides the advantage of performing efficiency measurement with only electrical equipment, which is very accurate and relatively cheap.

3. Proposed Methodology for the Power Losses Measurements in Electrical Drives

This section describes the procedure adopted for the determination of a new index, namely $\Delta\Delta P$, which represents a useful tool for the measurement of the power losses variations involved in the PDS, demonstrating the fact that the measurement of the $\Delta\Delta P$ index can be provided even with low-cost instrumentation, without compromising the accuracy of the measurement.

This study was applied to a whole PDS, taking into consideration the measurements achieved with the two following measurement systems (deeply discussed in Section 4):

- A highly accurate measurement system, composed of a NI 9225 acquisition module and a high precision A40B Fluke shunt resistor for the PDS input power measurement and another NI 9215 acquisition module for the measurement of mechanical quantities;
- A low-accuracy measurement system, composed of a NI 9225 acquisition module and a low precision i400 Fluke current probe for the PDS input power measurement and the previous measurement equipment for the measurement of mechanical quantities.

By defining $\Delta P_{PDS,1}$ and $\Delta P_{PDS,2}$ as the PDS power losses for two different control algorithms, namely 1 and 2, the PDS power losses are evaluated as the difference between the PDS input active power P_{PDS} and the PDS output mechanical power P_M, therefore with the following equations:

$$\Delta P_{PDS,1} = P_{PDS,1} - P_{M,1} \tag{5}$$

$$\Delta P_{PDS,2} = P_{PDS,2} - P_{M,2} \tag{6}$$

Given that there is no correlation between the electric and mechanical measurements, the uncertainty related to the power losses ΔP in the PDS is given by:

$$u(\Delta P_{PDS}) = \sqrt{u^2(P_{PDS}) + u^2(P_M)} \tag{7}$$

where $u(P_{PDS})$ is the uncertainty of the input power measurement, whereas $u(P_M)$ is the uncertainty in the mechanical measurement. Thus, the following two quantities are defined:

$$\Delta\Delta P_{PDS,shunt} = \left(\Delta P_{PDS,1} - \Delta P_{PDS,2}\right)_{shunt} \tag{8}$$

$$\Delta\Delta P_{PDS,probe} = \left(\Delta P_{PDS,1} - \Delta P_{PDS,2}\right)_{probe} \tag{9}$$

where $\Delta\Delta P_{PDS,shunt}$ is evaluated with the shunt resistor and $\Delta\Delta P_{PDS,probe}$ is evaluated with the Fluke current probe. These quantities define the reduction or enhancement of the power losses obtained by switching from Algorithm 1 to Algorithm 2.

Since a large part of PMSM control algorithms are designed to optimally elaborate the direct-axis current value in order to optimize the PMSM performance [30–32], a Field Oriented Control (FOC) strategy, shown schematically in Figure 2, has been adopted. The FOC block scheme presents a closed-loop control of the motor speed, where the corresponding speed error is processed by the use of a PI regulator that provides the reference value of q-axis current component I_q^*. Moreover, the magnetization level of the motor can be controlled by selecting the reference value direct axis current I_d^*. These values are compared with the real values of I_q and I_d, measured with LEM sensors of CDM; the correspondings errors are processed in order to obtain the reference values of the supply voltage Park components v_d^* and v_q^*. The dq voltage quantities are transformed in a three-phase reference frame and applied to the motor through a Pulse Width Modulation (PWM) technique. Therefore, the control strategy can be varied by selecting different I_d^* values.

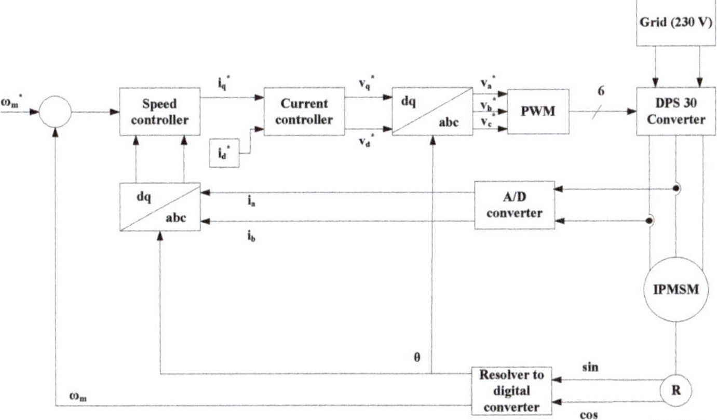

Figure 2. Block diagram of the control system taken into account.

It can be demonstrated that the $\Delta\Delta P_{PDS}$ measurement is not affected by uncertainty due to systematic errors. It is, indeed, well known that the uncertainty corresponding to a general quantity $y = f(x_1, x_2, \ldots . x_n)$, with $n \in N$, is given by:

$$u(y) = \sqrt{\sum_{i=1}^{n}\left(\frac{\partial f}{\partial x_i}\right) u^2(x_i) + 2\sum_{i=1}^{n-1}\sum_{j=i+1}^{n} \frac{\partial f}{\partial x_i}\frac{\partial f}{\partial x_j} r_{ij}u(x_i)u(x_j)} \tag{10}$$

where r_{ij} represents the coefficient of correlation between the i-th and the j-th quantities. For the $\Delta\Delta P_{PDS}$ quantity, the following formula is obtained:

$$y = \Delta\Delta P_{PDS} = \Delta P_{PDS,1} - \Delta P_{PDS,2} = P_{PDS,1} - P_{M,1} - P_{PDS,2} + P_{M,2} \tag{11}$$

Therefore:

$$\begin{aligned}
u(\Delta\Delta P_{PDS}) = \big[& u^2(P_{PDS,1}) + u^2(P_{M,2}) + u^2(P_{PDS,2}) + u^2(P_{M,2}) - 2r_{PDS,M}u(P_{PDS,1})u(P_{M,1}) \\
& -2r_{PDS,PDS}u(P_{PDS,1})u(P_{PDS,2}) + 2r_{PDS,M}u(P_{PDS,2})u(P_{M,2}) + 2r_{PDS,M}u(P_{PDS,2})u(P_{M,1}) \\
& -2r_{M,M}u(P_{M,1})u(P_{M,2}) - 2r_{PDS,M}u(P_{PDS,2})u(P_{M,2}) \big]^{\frac{1}{2}}
\end{aligned} \tag{12}$$

Since both the input electric power and mechanical power are measured with the same equipment and the measured values are similar when employing both the Algorithms 1 and 2, it is possible to assert that:

$$u(P_{PDS,1}) = u(P_{PDS,2}) = u(P_{PDS}) \tag{13}$$

$$u(P_{M,1}) = u(P_{M,2}) = u(P_M) \tag{14}$$

Let's suppose that the components of the uncertainty due to random errors are neglected. Then, by referring to the correlation coefficient, let's suppose to employ two different measurement equipment for independent acquisition of electrical quantities and mechanical quantities. Thus, it is possible to assert that $rij = 1$ if the uncertainty quantities involved in the double product are both electrical or both mechanical; otherwise, $rij = 0$. As a consequence, the uncertainty expression can be written as follows:

$$u(\Delta\Delta P_{PDS}) = \sqrt{2u^2(P_{PDS}) + 2u^2(P_M) - 2u^2(P_{PDS}) - 2u^2(P_M)} = 0 \tag{15}$$

The obtained result emphasizes that $\Delta\Delta P_{PDS,shunt}$ and $\Delta\Delta P_{PDS,probe}$ are not affected by the uncertainties due to systematic errors. Therefore, their measurements should provide the same results, although with a slight deviation due to random errors.

4. Test Bench

For the validation of the proposed methodology, it is needed to quantify the contribution of the random errors evaluating the short-term repeatability of the related measurements. For this purpose, a test bench has been set-up and it is composed by (Figure 3):

- A Complete Drive Module (CDM), consisting of a DPS 30-A power converter (Automotion Inc., Chantilly, VA, USA), which is composed of a diode bridge rectifier stage supplied by the electrical grid and an inversion stage. The IGBT bridge of the inverter is made by POWEREX, Model PM30CSJ060;
- A three-phase Interior Permanent Magnet Synchronous Motor (IPMSM) (Magnetic S.r.l., type BLQ-40, Italy) with SmCo permanent magnets (made by HITACHI Inc, type H-18B, maximum specific energy equal to 143 kJ/m^3). The nameplate data are summarized in Table 6;
- A Magtrol hysteresis brake (Model HD-715-8NA), connected to the shaft of the motor and used as a mechanical load for the IPMSM. The Magtrol can be controlled in real-time through a digital

dynamometer (model DSP6001) or software interface. The torque and speed measurement signals are acquired by the use of an acquisition board (in this case, the NI DAQ 9215). The main features of the brake are described in Table 1;

- A programmable data acquisition board NI cdaq-9172, suitable for the Labview® environment. This setup provides several advantages, such as a high degree of flexibility and acquisition of a large number of signals;
- A NI DAQ 9225 acquisition module, whose specifications are shown in Table 2;
- Two NI DAQ 9215 acquisition module, whose specifications are shown in Table 3;
- A Fluke i400 current probe (Table 4);
- A non-inductive Fluke A40B shunt resistor (Table 5).

Table 1. Main features of the Magtrol hysteresis brake.

Quantity	Value
Model	HD-715-8NA
Maximum torque	6.2 Nm
Maximum speed	25,000 rpm
Rated input inertia	1.449×10^{-3} kgm^2
Accuracy	Speed: 0.01% of reading from 10 rpm to 100,000 rpm TSC1: 0.02% of range (±1 mV) TSC2: 0.02% of range (±2 mV)
Maximum torque input	TSC1: ±5 V DC TSC2: ±10 V DC
Torque/Speed Output	Torque: ±10 V DC Speed: ±10 V DC

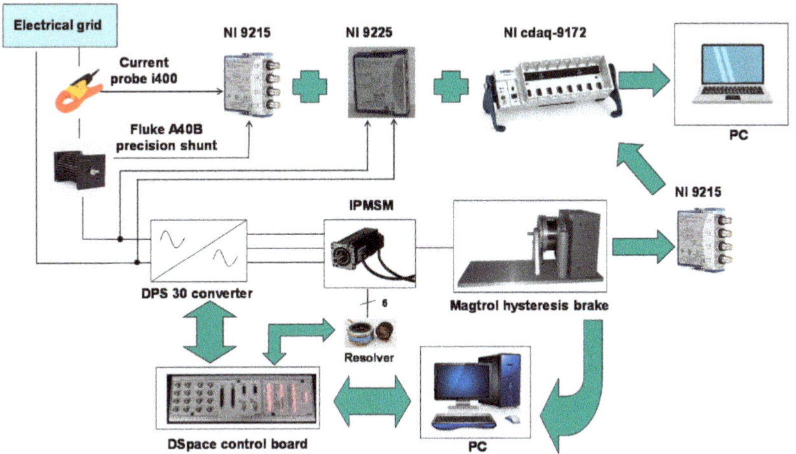

Figure 3. Schematic representation of the test bench.

Table 2. Technical data for the NI 9225 acquisition module.

Quantity	Value
Number of channels	4
ADC resolution	24 bit
Sampling	Simultaneous
Sample rate fs	$1.613 \div 50$ kS/s
Rated Voltage	300 V_{rms}
Bandwidth	0.453 fs
Accuracy THD	Gain error $\pm 0.05\%$

Table 3. Technical data for the NI 9215 acquisition module.

Quantity	Value	
Number of channels	4	
ADC resolution	16 bit	
Sampling	Simultaneous	
Sample rate fs	100 kS/s	
Input range	± 10 V	
Bandwidth	420 kHz	
Accuracy	Gain error	Offset error
	$\pm 0.02\%$	$\pm 0.0014\%$

Table 4. Technical data for the fluke i400current probe.

Quantity	Value
Reference temperature	$23 \pm 5\,^{\circ}$C;
Current range	1–400 A_{rms} or 1–40 A_{rms}
Output	1 mA/A
Accuracy	2% + 0.06 A, 45 Hz–400 Hz
Bandwidth	5–20,000 Hz

Table 5. Technical data for the A40 B shunt fluke.

Quantity	Value				
Rated current	20 A				
Rated resistance	0.04 Ω				
Bandwidth	0–100 kHz				
Accuracy [$\pm\mu$A/A] (confidence level 95%)	DC	1 kHz	10 kHz	30 kHz	100 kHz
	26	43	52	70	113
Maximum current	<5 s			Indefinitely	
	42 A			22 A	
Phase angle error [°]	1 kHz		10 kHz		100 kHz
	<0.013°		<0.125°		<1.250°
Work temperature range	13–33 $^{\circ}$C				

Table 6. Rated values and parameters of IPMSM under test.

Quantity	Value
Voltage [V]	132
Current [A]	3.6
Speed [rpm]	4000
Torque [Nm]	1.8
Pole pairs	3

The input power of the PDS is measured by adopting the scheme reported in Figure 3. The single-phase voltage is measured through a direct connection to the NI DAQ 9225 acquisition module. The current is sensed simultaneously by the use of a Fluke i400 current probe and through a non-inductive Fluke A40B shunt resistor, whose output signals are sent to the NI DAQ 9215 acquisition module. More in detail, the shunt presents high accuracy and cost especially regarding the phase angle error that present very low value. This feature allows the accurate measurement of active power. Instead, the current probe presents lower accuracy and cost, with respect to the shunt, and its angle error is not known. Therefore, as mentioned in the previous section, the two adopted transducers allow defining two measurement systems that perform the input power measurement with different accuracy and uncertainty. In this way, it is possible to compare the measurement results between a first measurement system with greater accuracy and cost (shunt resistor) able to meet the standard prescriptions [21,22]. and a second measurement system less accurate and not satisfying the standard prescriptions [21,22], but cheaper (current probe).

From the reported Tables, it can be stated that the bandwidth of the adopted components is adequate for the measurement of the electrical quantities involved in the CDM input section. Moreover, the NI 9225 is suitable for the acquisition of grid sinusoidal voltage (240 V_{rms}), since it can reach 300 V_{rms} of the input voltage. As for the current, it reaches peak values in the order of Ampère units. The shunt resistor has an already adequate value of rated range and it is chosen to arrange 4 turns on the pass-through current probe for operation close to the full-scale. It is easy to calculate the output voltages from the probe and the shunt resistance and verify that these values do not exceed the NI 9215 acquisition module full-scale value equal to 10 V.

Finally, the uncertainties introduced by the acquisition modules are negligible if compared to those given by the transducers.

The mechanical load of the motor consists of a Magtrol HD-715-8NA hysteresis brake. The torque offered by the brake can be adjusted in real-time by means of a Magtrol digital dynamometer model DSP6001, whose interface already provides the torque, speed and power values measured at the shaft motor. Furthermore, the digital dynamometer provides the torque and speed measurement signals can be sent to other acquisition systems. In this work, an additional acquisition module NI DAQ 9215 is used to acquire the torque and speed measurement signals. The accuracy specifications of Magtrol HD-715-8NA hysteresis brake satisfy the prescription of the standards [18–22]. Therefore, the adoption of the Magtrol HD-715-8NA hysteresis brake with the NI 9215 module allows performing the accurate measurement of mechanical power satisfactory for accurate direct measurement of the efficiency according to the standard [21,22].

5. Measurement of the PDS Power Losses

5.1. Data Acquisition Procedure

Based on the statements reported in the previous Sections and since the rated power of the IPMSM under test is lower than 1 kW, the direct method has been adopted in order to evaluate the efficiency of the PDS described in Section 3. Figure 4 shows a photograph of the test bench during a set of measurements of the PDS power losses.

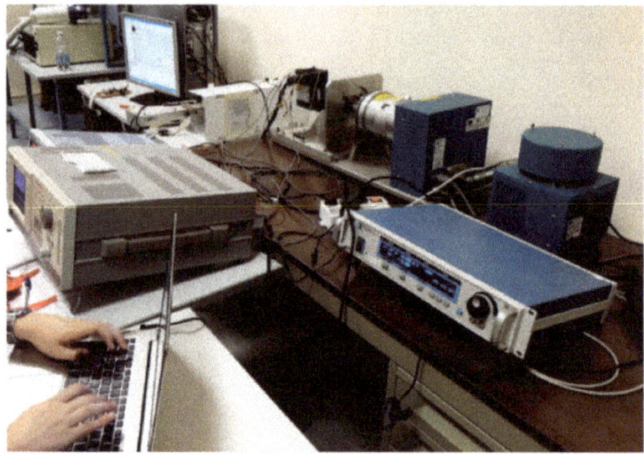

Figure 4. A photograph of the test bench during the power losses measurements.

The speed and torque transducers employed in the electrical drive under test satisfy the measurement requirements of the standard IEC 60034-2 [20].

The acquisition of both the electrical and mechanical quantities is carried out through the transducers (see Section 4) in a Labview environment. This acquisition is carried out with the Express Virtual Instrument (VI) DAQ Assistants, characterized by:

- Sampling frequency, namely f_s;
- Single acquisition time or measurement time, namely T_M.

The number of samples in one acquisition is given by:

$$N_s = T_M f_s \tag{16}$$

During a single acquisition time, the instantaneous values of electrical and mechanical quantities are acquired so that the instantaneous values of electrical power $p_{PDS}(t)$ and mechanical power $p_M(t)$ are computed by means of the following relationships:

$$p_{PDS}(t) = v_{PDS}(t) \cdot i_{PDS}(t) \tag{17}$$

$$p_M(t) = T_{em}(t) \cdot \omega_m(t) \tag{18}$$

where $v_{PDS}(t)$ is the instantaneous value of the PDS input voltage, $i_{PDS}(t)$ is the instantaneous value of the PDS input current, $T_{em}(t)$ is the instantaneous value of the electromagnetic torque and $\omega_m(t)$ is the instantaneous value of the mechanical speed. By means of the Labview VI library (e.g., VI BASIC DC RMS), the overall rms values of voltage and current quantities, as well as the mean value of instantaneous electric power and the mean value of mechanical quantities are computed with respect to time T_M. More specifically, the computation of the mean value of the instantaneous electric power allows considering the contribution, in terms of active power, of each product between isofrequential harmonic components of both voltage and current.

Consequently, in a period equal to T_M, it is possible to obtain one sample for each electrical and mechanical quantity, allowing the possibility of repeating this process and performing a very wide range of time window, without affecting the PC memory with a large number of samples.

Besides, the adoption of several measurement samples allows the evaluation of the average value and the measurement deviation for each quantity. The single measurement, which is characterized by T_M and f_s, is, then, implemented in a single iteration of a *for loop*. The number of iterations

N corresponds to the number of desired measurement samples. The total observation time or the total acquisition time T_w is equal to:

$$T_w = NT_M \qquad (19)$$

At the output of each loop iteration, the measurement sample is stored in an external array. At the end of the cycle, an array is obtained for each quantity of interest, consisting of scalar elements. The measurement arrays obtained are summarized in Table 7.

Table 7. Measurement arrays for the PDS Section.

Section	Quantity
PDS input section	• Rms value of the single-phase voltage, V_{PDS} [V]; • Current acquired with the Fluke probe, $I_{PDS,p}$ [A]; • Current acquired with the Fluke shunt resistor, $I_{PDS,sh}$ [A]; • Input PDS active power acquired with the Fluke current probe $P_{PDS,probe}$ [W]; • Input PDS active power acquired with Fluke shunt resistor, $P_{PDS,shunt}$ [W].
Mechanical section	• Output torque, T_{em} [Nm]; • Mechanical speed of the motor, ω_m [rad/s]; • Mechanical power, P_M [W].

The efficiency of the PDS is computed within the *for loop* by adopting the values of the active power evaluated in the input and mechanical sections and adopting the direct method. Therefore, the following efficiency arrays are defined and measured:

- PDS efficiency evaluated with the Fluke probe $\eta_{PDS\text{-}probe}$;
- PDS efficiency evaluated with the Fluke shunt resistor $\eta_{PDS\text{-}shunt}$;

In order to evaluate the average value, the VI standard Deviation and Variance are employed in the software. The sampling frequency values f_s and the acquisition time T_M adopted for the two DAQ Assistants are equal to 50 kHz and 60 s, respectively, satisfactory for the requirements of the PDS input section and the prescription of the standards [21,22].

In order to obtain a synchronous sampling between the measurement sections, the same measurement time was chosen.

Finally, the overall observation time T_w has been set equal to 40 min, which allows the repeatability of the measurements.

5.2. Results and Discussions

As suggested by the IEC 61800-9-2 standard, the IPMSM working points taken into consideration for the efficiency measurement, corresponding to different conditions of load and speed, are reported in Table 8.

Table 8. IPMSM Working Points.

Working Points	$\omega_m = \omega_n$	$\omega_m = 50\%\ \omega_n$	$\omega_m = 0$
$T_{em} = T_n$	(1) ω_n; T_n	(2) 50% ω_n; T_n	(3) 0; T_n
$T_{em} = 50\%\ T_n$	(4) ω_n; 50% T_n	(5) 50% ω_n; 50% T_n	(6) 0; 50% T_n
$T_{em} = 25\%\ T_n$	(7) ω_n; 25% T_n	(8) 50% ω_n; 25% T_n	(9) 0; 25% T_n

By defining T_n and ω_n as the rated IPMSM torque and speed values and according to Table 8, an additional point (no.7), which is not suggested by the standard, is included within the sets of measurement.

Concerning the $\omega_m = 0$, the standard defines this speed as a sufficiently low-speed corresponding to a supply frequency of the motor lower than 12 Hz. In this investigation, the low speed is equal to 200 rpm.

Moreover, in order to simulate the behavior of two different control algorithms, for each working point, the measurements are taken at two different values of the direct axis current I_d, equal to -1 A and 0 A, which have been set by means of the FOC strategy.

As already mentioned, 40 measurement samples were acquired for each of the quantities of interest previously defined. Each measurement sample presents a duration of 60 s and the overall observation window is equal to 40 min. In such a way, the related arrays of voltage and currents, in terms of rms and mean values of torque and speed, are created. For instance, the mean values for the first working point are reported in Table 9.

Table 9. Average values at the first working point for different I_d values.

	V_{PDS} [V]	$I_{PDS,p}$ [A]	$I_{PDS,sh}$ [A]	T_{em} [Nm]	ω_m [rad/s]
$I_d = 0$ A	230.441	6.402	6.368	1.801	420.61
$I_d = -1$ A	230.420	6.374	6.340	1.801	420.63

Furthermore, with regard to the repeatability of the measurement and the evaluation of the uncertainty, it is needed to evaluate the standard deviation for each quantity of interest. For instance, the average values of the standard deviation for the first working point are reported in Table 10.

Table 10. Standard deviations at the first working point for different I_d values.

	$\sigma(V_{PDS})$ [V]	$\sigma(I_{PDS,p})$ [A]	$\sigma(I_{PDS,sh})$ [A]	$\sigma(T_{em})$ [Nm]	$\sigma(\omega_m)$ [Rad/s]
$I_d = 0$ A	0.0162	0.004	0.0019	5.81×10^{-4}	8.2×10^{-4}
$I_d = -1$ A	0.0139	4.6×10^{-4}	4.8×10^{-4}	1.5×10^{-6}	7.3×10^{-4}

As reference examples, Figures 5 and 6 show the average values of the PDS input active power and the output mechanical power as a function of mechanical speed n_m [rpm] for $I_d = -1$ A.

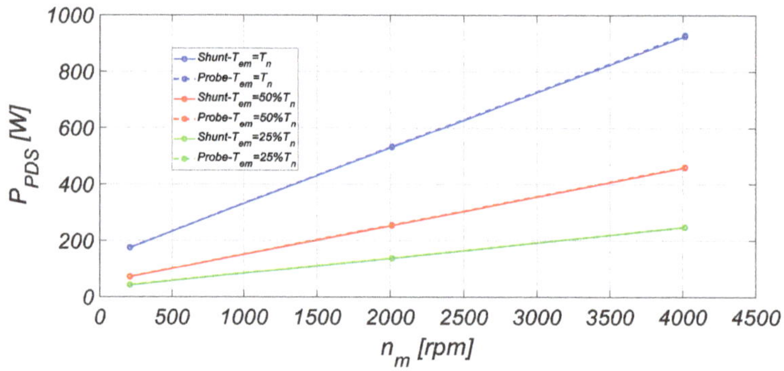

Figure 5. PDS input active power measured with the shunt resistor and fluke current probe at $I_d = -1$ A.

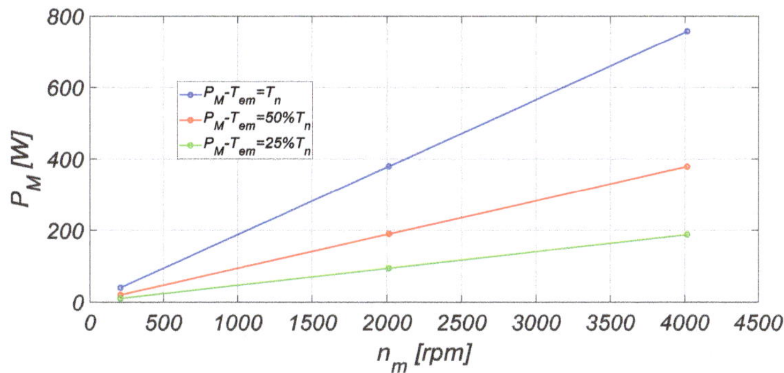

Figure 6. Mechanical power measured at $I_d = -1$ A.

From the acquired values of active power, it is possible to determine both the power losses and the efficiency of the entire PDS under test. For instance, Table 11 reports the efficiency detected for each working point at $I_d = -1$A.

Table 11. Efficiency detected for each working point at $I_d = -1$ A.

Working Points	$I_d = -1$ A	
	$\eta_{PDS-shunt}$	$\eta_{PDS-probe}$
1	0.23	0.2
2	0.72	0.68
3	0.83	0.79
4	0.3	0.26
5	0.76	0.71
6	0.84	0.8
7	0.26	0.21
8	0.71	0.66
9	0.78	0.73

It can be noticed that the highest efficiency values are detected for the load torque condition $T_{em} = 50\% \; T_n$. Moreover, for the speed condition $\omega_m = 0$, the PDS efficiency obtained at $T_{em} = 25\% \; T_n$ is higher than the one obtained at $T_{em} = 50\% \; T_n$. However, for speed conditions $\omega_m \geq 50\% \; \omega_n$, the PDS efficiency obtained at $T_{em} = 25\% \; T_n$ is lower than the one obtained at $T_{em} = 50\% \; T_n$. In any case, for all the proposed working points, the PDS efficiency values measured with the shunt resistor (continuous lines) are higher than those measured with the Fluke current probe (dot lines). These results highlight a not negligible difference between the efficiencies measured with the two measurement system. Similar behavior has been detected for $I_d = 0$ A.

The PDS efficiency uncertainty is affected by errors on power measurement both in the input single-phase section than in the output mechanical section. Regarding the PDS input single-phase section, the two current transducers adopted in this work present different accuracy. In particular, the uncertainty regarding the active power measurement in the single-phase sector with the shunt resistance has been already computed in [33], which is equal to 0.05%, corresponding to an extended uncertainty equal to 0.13% with a 99% of confidence level. With regards to the power measurement with the probe, the uncertainty is relatively high, due to an error equal to 2% ± 0.06 A and its angle error is not known. An appropriate angle error value is obtained by comparing the average specifications of the current probe transducer with those of the same precision class, reaching, therefore, the value of

0.03 rad. The expanded uncertainty on the measurement of PDS input single-phase power by Fluke current probe is equal to 9.3% (99% confidence interval).

The mechanical power measurement is mainly affected by the error introduced by the dynamometer, while the one introduced by the NI 9215 is negligible. From the data provided by the dynamometer datasheet, the expanded uncertainty on the measurement of mechanical power is equal to 0.17% with a confidence interval of 99%.

The mechanical power can also be measured by directly reading on the dynamometer interface. In this case, the related error is obtained from the datasheet of the used brake, equal to 0.01% for the speed and to 0.5% for the torque. The expanded uncertainty on the measurement of mechanical power with this second method is equal to 0.74% with a confidence interval of 99%.

Finally, the uncertainty concerning the efficiency measurement with the first system (shunt resistance and acquisition of torque and speed) is equal to 0.21% with a confidence interval of 99%, whereas the second system (Fluke current probe and acquisition of torque and speed) is characterized by an uncertainty equal to 9.4% with a confidence interval of 99%. The uncertainties considerations are valid for each case of study ($I_d = -1$ A, $I_d = 0$ A).

The PDS efficiency measurement performed with the first measurement system complies with the IEC61800-9 uncertainty prescriptions and, therefore, can be used for the PDS energy classification. On the contrary, the second measurement system is not suitable for the PDS energy classification. This fact underlines the need for high accuracy and high-cost instrumentation for the efficiency measurement of electric drives.

6. Experimental Validation of the Proposed Measurement Methodology

The purpose of this Section is the experimental validation of the methodology of measurement described in Section 3 by means of the experimental results carried out from the measurements discussed in Section 4. More specifically, a variation of the I_d current is applied to the FOC system, determining a variation of the power losses $\Delta\Delta P$ involved in the PDS. This variation is measured by means of the two measurement systems reported in Section 3 (shunt resistor and current probe) and the related values are compared.

Thus, in order to evaluate the power losses variations due to a variation of the magnetizing current ($I_d = 0$ A and $I_d = -1$ A) in the FOC system, the $\Delta\Delta P_{PDS,shunt}$ and $\Delta\Delta P_{PDS,probe}$ indexes are computed for each IPMSM working condition. These values are obtained by processing the data from the PDS input power array acquired with the Fluke shunt resistor, $P_{PDS,shunt}$, the PDS input power array acquired with the current probe, $P_{PDS,probe}$, and the PDS output mechanical power array, namely P_M. More in detail, the PDS power losses evaluated with the first measurement system $\Delta P_{PDS,shunt}$ and with the second measurement system $\Delta P_{PDS,probe}$ are calculated by adopting the following relationships, respectively:

$$\Delta P_{PDS,shunt} = P_{PDS,shunt} - P_M \tag{20}$$

$$\Delta P_{PDS,probe} = P_{PDS,probe} - P_M \tag{21}$$

Therefore, the measurement of power losses with $I_d = 0$ A and with $I_d = -1$ A has been carried out, and, consequently, the power losses variations have been determined with the first measurement system ($\Delta\Delta P_{PDS,shunt}$) and with the second measurement system ($\Delta\Delta P_{PDS,probe}$).

Figure 7 depicts the values of $\Delta\Delta P_{PDS}$ measured with two measurement systems (dot and continuous lines) as a function of the reference speed n_m [rpm] for each working condition. The mean values of both $\Delta\Delta P_{PDS,shunt}$ and $\Delta\Delta P_{PDS,probe}$ are evaluated among an observation time T_w equal to 40 min. It can be noticed that the highest $\Delta\Delta P$, corresponding to the maximum power saving, is provided for $T_{em} = T_n$. Furthermore, for each working condition, the trends detected with the two methods are almost comparable between each other, demonstrating that the proposed methodology of measurement can provide the same results even with low-accuracy instrumentation.

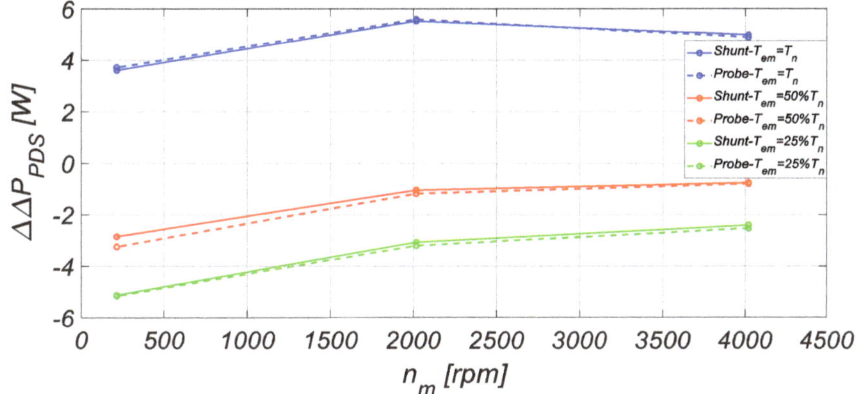

Figure 7. $\Delta\Delta P_{PDS}$ as a function of the reference speed for a variation of d-axis current from $I_d = 0$ A to $I_d = -1$ A and for each of the proposed working condition.

Further analysis has been focused on the computation of the percentage values $\Delta\Delta P_{PDS\%}$ of the quantities $\Delta\Delta P_{PDS,shunt}$ and $\Delta\Delta P_{PDS,probe}$ with respect to the IPMSM rated mechanical power $P_{M,r}$:

$$\Delta\Delta P_{PDS}\% = \frac{\Delta P_{PDS}}{P_{M,r}}100 = \frac{\Delta P_{PDS}}{T_n \omega_n}100 \tag{22}$$

Figure 8 shows the mean values of $\Delta\Delta P_{PDS,shunt}\%$ and $\Delta\Delta P_{PDS,probe}\%$ evaluated through the overall observation time T_w of 40 min.

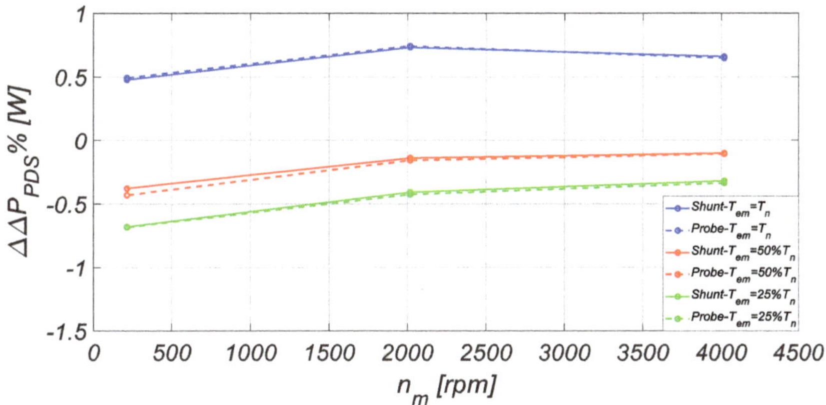

Figure 8. $\Delta\Delta P_{PDS}\%$ as a function of the reference speed for a variation of the d-axis current from $I_d = 0$ A to $I_d = -1$ A and for each of the proposed working condition.

As well as for the previous case, it can be observed that $\Delta\Delta P_{PDS,shunt}\%$ and $\Delta\Delta P_{PDS,probe}\%$ have almost coincident values. Furthermore, the differences between the $\Delta\Delta P_{PDS,shunt}$ and $\Delta\Delta P_{PDS,probe}$ values and between the $\Delta\Delta P_{PDS,shunt}\%$ and $\Delta\Delta P_{PDS,probe}\%$ are determined by means of the following relationships:

$$\Delta\Delta P_{diff} = \Delta\Delta P_{PDS,shunt} - \Delta\Delta P_{PDS,probe} \tag{23}$$

$$\Delta\Delta P_{diff}\% = \Delta\Delta P_{PDS,shunt}\% - \Delta\Delta P_{PDS,probe}\% \tag{24}$$

Table 12 summarizes the numerical results of the previously reported differences. It can be observed that the highest percentage difference between the results provided by the two methods is limited to 0.04%, demonstrating that the $\Delta\Delta P$ index, which was introduced as a valuable parameter in order to compare the performances of the same electric drive under test controlled with different control algorithms, is not affected by uncertainties due to systematic errors. Therefore, the determination of this index can be carried out by adopting sensors, instrumentation and acquisition boards with low accuracy, avoiding the need for high-cost and high-accuracy tools.

Table 12. Numerical results of the differences between the $\Delta\Delta P_{PDS}$ and $\Delta\Delta P_{PDS}\%$.

Working Points	$\Delta\Delta P_{diff}$ [W]	$\Delta\Delta P_{diff}\%$
1	0.062	0.008
2	−0.087	−0.011
3	−0.118	−0.016
4	0.058	0.008
5	0.125	0.016
6	0.301	0.04
7	0.115	0.015
8	0.132	0.017
9	0.052	0.007

Further investigation has been carried out on the $\Delta\Delta P_{PDS}$ measurement uncertainty. More in detail, the uncertainty includes only the uncertainty due to random errors, which is equal to the $\Delta\Delta P_{PDS}$ standard deviation computed with respect to the overall observation time T_w equal to 40 min. This analysis is carried out by considering the expanded uncertainty with a confidence interval of 99%, which can be obtained by multiplying the standard deviation by 2.58. The expanded uncertainty of $\Delta\Delta P_{PDS}$ with a confidence interval of 99%, evaluated with the shunt resistor $u_{shunt}(\Delta\Delta P_{PDS})$ and evaluated with the Fluke current probe $u_{probe}(\Delta\Delta P_{PDS})$ for each motor working condition are reported in Table 13.

Table 13. $\Delta\Delta P_{PDS}$ uncertainty for each working condition and for case 1.

Working Points	$u_{shunt}(\Delta\Delta P_{PDS})$ [W]	$u_{probe}(\Delta\Delta P_{PDS})$ [W]
1	0.42	0.24
2	0.29	0.27
3	0.46	0.26
4	0.24	0.21
5	0.27	0.17
6	0.26	0.27
7	0.2	0.18
8	0.17	0.11
9	0.31	0.22

The detected value of expanded uncertainty of $\Delta\Delta P_{PDS}$ leads to state that the difference between the possible $\Delta\Delta P_{PDS}$ values evaluated with the two measurement systems is quite limited, demonstrating the validity of the proposed measurement procedure.

7. Conclusions

In this paper, an accurate measurement procedure for the control algorithm's impact on the electrical drive's efficiency has been addressed and discussed. For this purpose, an extended experimental investigation has been carried out on an IPMSM electric drive with the use of two different measurement equipment that presents different accuracy and, consequently, different costs. In order to validate the proposed measurement methodology, the behaviors of two different control algorithms have been simulated by changing the magnetization level of the IPMSM under test. This methodology allows estimating the power losses variations $\Delta\Delta P$, which represents a valuable index for the comparison between different PDS control algorithms. A first analysis of the experimental results has shown that a highly accurate and expensive measurement system is needed for the PDS classification according to the standard IEC-61800. However, further analysis of the $\Delta\Delta P$ uncertainty has been carried out. This analysis has shown that the $\Delta\Delta P_{PDS}$ values are coincident for both systems, demonstrating that the $\Delta\Delta P$ parameter results as a valuable index for the characterization of the control system since it can be measured even with low-precision and cheap measurement system.

Author Contributions: Conceptualization, C.N. and C.S.; methodology, C.S.; software, A.O.D.T. and C.N.; validation, G.L., C.N. and C.S.; formal analysis, C.N.; investigation, G.L. and C.N.; resources, R.M. and C.S.; data curation, C.N. and C.S.; writing—original draft preparation, M.C. and C.N.; writing—review and editing, M.C. and R.M.; visualization, M.C., C.N. and R.M.; supervision, A.O.D.T., R.A.M., R.M., C.S. and M.T.; project administration, R.A.M., R.M. and M.T. All authors have read and agreed to the published version of the manuscript.

Funding: This research received no external funding.

Acknowledgments: This work was financially supported by PON R&I 2015–2020 "Propulsione e Sistemi Ibridi per velivoli ad ala fissa e rotante—PROSIB", CUP no:B66C18000290005, by H2020-ECSEL-2017-1-IA-two-stage "first and european sic eightinches pilot line-REACTION", by Prin 2017—Settore/Ambito di intervento: PE7 linea C—Advanced power-trains and -systems for full electric aircrafts, by PON R&I 2014-2020—AIM (Attraction and International Mobility), project AIM1851228-1 and by ARS01_00459-PRJ-0052 ADAS+ "Sviluppo di tecnologie e sistemi avanzati per la sicurezza dell'auto mediante piattaforme ADAS".

Conflicts of Interest: The authors declare no conflict of interest.

References

1. Ferreira, F.J.T.E.; De Almeida, A.T. Overview on energy saving opportunities in electric motor driven systems—Part 1: System efficiency improvement. In Proceedings of the 2016 IEEE/IAS 52nd Industrial and Commercial Power Systems Technical Conference (I&CPS), Detroit, MI, USA, 1–5 May 2016; pp. 1–8.

2. Ferreira, F.J.T.E.; De Almeida, A.T. Overview on energy saving opportunities in electric motor driven systems—Part 2: Regeneration and output power reduction. In Proceedings of the 2016 IEEE/IAS 52nd Industrial and Commercial Power Systems Technical Conference (I&CPS), Detroit, MI, USA, 1–5 May 2016; pp. 1–8.

3. Torrent, M.; Perat, J.I.; Jiménez, J.A. Permanent Magnet Synchronous Motor with Different Rotor Structures for Traction Motor in High Speed Trains. *Energies* **2018**, *11*, 1549. [CrossRef]

4. Łebkowski, A. Design, Analysis of the Location and Materials of Neodymium Magnets on the Torque and Power of In-Wheel External Rotor PMSM for Electric Vehicles. *Energies* **2018**, *11*, 2293. [CrossRef]

5. Kutt, F.; Michna, M.; Kostro, G. Non-Salient Brushless Synchronous Generator Main Exciter Design for More Electric Aircraft. *Energies* **2020**, *13*, 2696. [CrossRef]

6. Zhao, T.; Wu, S.; Cui, S. Multiphase PMSM with Asymmetric Windings for More Electric Aircraft. *IEEE Trans. Transp. Electrif.* **2020**, *6*, 1592–1602. [CrossRef]

7. Hansen, J.F.; Wendt, F. History and State of the Art in Commercial Electric Ship Propulsion, Integrated Power Systems, and Future Trends. *Proc. IEEE* **2015**, *103*, 2229–2242. [CrossRef]

8. Kasha, A.; Lin, R.; Sudhoff, S.; Chalfant, J.; Alsawalhi, J. A comparison of permanent magnet machine topologies for marine propulsion applications. In Proceedings of the 2017 IEEE Electric Ship Technologies Symposium (ESTS), Arlington, VA, USA, 14–17 August 2017; pp. 437–444. [CrossRef]

9. Ye, M.; Shi, T.; Wang, H.; Li, X.; Xia, C. Sensorless-MTPA Control of Permanent Magnet Synchronous Motor Based on an Adaptive Sliding Mode Observer. *Energies* **2019**, *12*, 3773. [CrossRef]

10. Sun, J.; Lin, C.; Xing, J.; Jiang, X. Online MTPA Trajectory Tracking of IPMSM Based on a Novel Torque Control Strategy. *Energies* **2019**, *12*, 3261. [CrossRef]

11. Caruso, M.; Di Tommaso, A.O.; Miceli, R.; Nevoloso, C.; Spataro, C.; Trapanese, M. Maximum Torque Per Ampere Control Strategy for Low-Saliency Ratio IPMSMs. *Int. J. Renew. Energy Res.* **2019**, *9*, 374–383.

12. Cavallaro, C.; Di Tommaso, A.; Miceli, R.; Raciti, A.; Galluzzo, G.R.; Trapanese, M. Analysis a DSP implementation and experimental validation of a loss minimization algorithm applied to permanent magnet synchronous motor drives. In Proceedings of the IECON'03. 29th Annual Conference of the IEEE Industrial Electronics Society (IEEE Cat. No.03CH37468), Roanoke, VA, USA, 2–6 November 2003; Volume 1, pp. 312–317.

13. Caruso, M.; Di Tommaso, A.O.; Genduso, F.; Miceli, R. Experimental investigation on high efficiency real-time control algorithms for IPMSMs. In Proceedings of the 2014 International Conference on Renewable Energy Research and Application (ICRERA), Milwaukee, WI, USA, 19–22 October 2014; pp. 974–979.

14. Caruso, M.; Di Tommaso, A.O.; Miceli, R.; Nevoloso, C.; Spataro, C.; Viola, F. Characterization of the parameters of interior permanent magnet synchronous motors for a loss model algorithm. *Measurement* **2017**, *106*, 196–202. [CrossRef]

15. Antonello, R.; Tinazzi, F.; Zigliotto, M. Energy efficiency measurements in IM: The non-trivial application of the norm IEC 60034-2-3:2013. In Proceedings of the 2015 IEEE Workshop on Electrical Machines Design, Control and Diagnosis (WEMDCD), Torino, Italy, 26–27 March 2015; pp. 248–253.

16. Agamloh, E.B. Power and efficiency measurement of motor-variable frequency drive systems. In Proceedings of the 2015 61st IEEE Pulp and Paper Industry Conference (PPIC), Milwaukee, WI, USA, 14–18 June 2015; pp. 1–8.

17. Stockman, K.; Dereyne, S.; Vanhooydonck, D.; Symens, W.; Lemmens, J.; Deprez, W. Iso efficiency contour measurement results for variable speed drives. In Proceedings of the The XIX International Conference on Electrical Machines—ICEM 2010, Rome, Italy, 6–8 September 2010; pp. 1–6.

18. Rotating Electrical Machines—Part 2-1: Standard Methods for Determining Losses and Efficiency from Test. IEC 60034-2-1. Available online: https://webstore.iec.ch/publication/121 (accessed on 1 September 2020).

19. Rotating Electrical Machines—Part 2-2: Specific methods for Determining Separate Losses of Large Machines from Tests—Supplement to IEC 60034-2-1. IEC 60034-2-2. Available online: https://webstore.iec.ch/publication/122 (accessed on 1 September 2020).

20. Rotating Electrical Machines—Part 2-3: Specific Test Methods for Determining Losses and Efficiency of Converter-Fed AC Motors. IEC 60034-2-3. Available online: https://webstore.iec.ch/publication/30919 (accessed on 1 September 2020).

21. Adjustable Speed Electrical Power Drive Systems—Part 9-1: Ecodesign for Power Drive Systems, Motor Starters, Power Electronics and Their Driven Applications—General Requirements for Setting Energy Efficiency Standards for Power Driven Equipment Using the extended Product Approach (EPA) and Semi Analytic Model (SAM). IEC 61800-9-1. Available online: https://www.iec.ch/dyn/www/f?p=103:38:4308635742220::::FSP_ORG_ID,FSP_APEX_PAGE,FSP_PROJECT_ID:1416,21,22026 (accessed on 1 September 2020).

22. Adjustable Speed Electrical Power Drive Systems—Part 9-2: Ecodesign for Power Drive Systems, Motor Starters, Power Electronics & Their Driven Applications—Energy Efficiency Indicators for Power Drive Systems and Motor Starters. IEC 61800-9-2. Available online: https://webstore.iec.ch/publication/31527 (accessed on 1 September 2020).

23. Bucci, G.; Ciancetta, F.; Fiorucci, E.; Ometto, A. Uncertainty Issues in Direct and Indirect Efficiency Determination for Three-Phase Induction Motors: Remarks about the IEC 60034-2-1 Standard. *IEEE Trans. Instrum. Meas.* **2016**, *65*, 2701–2716. [CrossRef]

24. IEEE Trial-Use Guide for Testing Permanent Magnet Machines. IEEE Std 1812-2014, 2015; pp. 1–56. Available online: https://ieeexplore.ieee.org/document/7047988 (accessed on 1 September 2020).

25. Deusinger, B.; Lehr, M.; Binder, A. Determination of efficiency of permanent magnet synchronous machines from summation of losses. In Proceedings of the 2014 International Symposium on Power Electronics, Electrical Drives, Automation and Motion, Ischia, Italy, 18–20 June 2014; pp. 619–624.

26. Lateb, R.; Da Silva, J. Indirect testing to determine losses of high speed synchronous permanent magnets motors. In Proceedings of the 2014 17th International Conference on Electrical Machines and Systems (ICEMS), Hangzhou, China, 22–25 October 2014; pp. 1520–1526.

27. Aarniovuori, L.; Kolehmainen, J.; Kosonen, A.; Niemela, M.; Pyrhonen, J. Uncertainty in motor efficiency measurements. In Proceedings of the 2014 International Conference on Electrical Machines (ICEM), Berlin, Germany, 2–5 September 2014; pp. 323–329.

28. Yogal, N.; Lehrmann, C.; Henke, M. Determination of the Measurement Uncertainty of Direct and Indirect Efficiency Measurement Methods in Permanent Magnet Synchronous Machines. In Proceedings of the 2018 XIII International Conference on Electrical Machines (ICEM), Alexandroupoli, Greece, 3–6 September 2018; pp. 1149–1156.

29. Dambrauskas, K.; Vanagas, J.; Zimnickas, T.; Kalvaitis, A.; Ažubalis, M. A Method for Efficiency Determination of Permanent Magnet Synchronous Motor. *Energies* **2020**, *13*, 1004. [CrossRef]

30. Mademlis, C.; Margaris, N. Loss minimization in vector-controlled interior permanent-magnet synchronous motor drives. *IEEE Trans. Ind. Electron.* **2002**, *49*, 1344–1347. [CrossRef]

31. Uddin, M.N.; Zou, H.; Azevedo, F. Online Loss-Minimization-Based Adaptive Flux Observer for Direct Torque and Flux Control of PMSM Drive. *IEEE Trans. Ind. Appl.* **2015**, *52*, 425–431. [CrossRef]

32. Hoang, K.D.; Aorith, H.K.A. Online Control of IPMSM Drives for Traction Applications Considering Machine Parameter and Inverter Nonlinearities. *IEEE Trans. Transp. Electrif.* **2015**, *1*, 312–325. [CrossRef]

33. Cataliotti, A.; Cosentino, V.; Di Cara, D.; Lipari, A.; Nuccio, S.; Spataro, C. A PC-Based Wattmeter for Accurate Measurements in Sinusoidal and Distorted Conditions: Setup and Experimental Characterization. *IEEE Trans. Instrum. Meas.* **2011**, *61*, 1426–1434. [CrossRef]

Publisher's Note: MDPI stays neutral with regard to jurisdictional claims in published maps and institutional affiliations.

Article

Study of a Combined Demagnetization and Eccentricity Fault in an AFPM Synchronous Generator

Alexandra C. Barmpatza * and Joya C. Kappatou

Department of Electrical and Computer Engineering, University of Patras, 26504 Patras, Greece;
joya@ece.upatras.gr
* Correspondence: abarmpatza@upatras.gr; Tel.: +30-694-2479943

Received: 22 September 2020; Accepted: 26 October 2020; Published: 27 October 2020

Abstract: This article investigates the combined partial demagnetization and static eccentricity fault in an Axial Flux Permanent Magnet (AFPM) Synchronous Generator. The machine is simulated using 3D FEM, while the EMF spectrum is analyzed in order to export the fault related harmonics using the FFT analysis. Firstly, the partial demagnetization fault, without the coexistence of eccentricity, and both the static angular and axis eccentricity faults, without the coexistence of partial demagnetization, are studied. In the case of eccentricity fault, the phase EMF sum spectrum has also been used as a diagnostic mean, because, when only eccentricity fault exists in the generator (either angular or axis) new harmonics do not appear in the EMF spectrum. Secondly the combination of partial demagnetization fault with static axis and static angular eccentricity is investigated and different comparisons are made when the demagnetization and the eccentricity level changes. The investigation revealed that the combination of eccentricity and demagnetization creates new harmonics in the EMF spectrum. The novelty of the article is that these combined faults are studied for the first time in the international literature, and the phase EMF sum spectrum has not been previously used for eccentricity diagnosis in this machine type.

Keywords: axial flux; demagnetization; finite element analysis; permanent magnet; static eccentricity; synchronous generator

---

1. Introduction

Eccentricity and demagnetization are two critical mechanical faults that can both occur in electrical machines, creating vital problems in the industry. Eccentricity appears when the stator is not placed correctly in relation to the rotor, a phenomenon that can occur during assembly or during machine operation. In other words, eccentricity is the result of manufacturing imperfections like unbalanced mass, poor alignment and excessive tolerances. 60% of the faults that appear in electrical machines are mechanical and 80% of them can create eccentricity [1]. This fault is responsible for unbalanced magnetic forces, vibration, and acoustic noise, creating problems in the machine operation and reducing its lifetime. If the level of eccentricity severity is quite high, stator and rotor can both be scraped, leading to the damage of the generator. Especially, the Axial Flux Permanent Magnet (AFPM) synchronous machines are prone to eccentricity fault because their overall axial length is short and as consequence, the ratio of machine diameter to length is high [2]. In addition, this type of machine contains permanent magnets that can get demagnetized or crack easily. The high temperatures, the structural defects, and the degradation of the coercive force are responsible for this fault. The demagnetization can be partial or total irreversible. The early diagnosis of both faults is a vital need for the interrupted operation of the systems in the industry.

In the international literature, several methods are proposed for demagnetization [3–6] and eccentricity detection [5–7]. The most commonly used methods are the Time Domain, the Frequency

---

Analysis, and the Time Scale Analysis Methods, like Discrete or Continuous Wavelet Transform. The Motor Current Signature Analysis (MCSA) and the Motor Voltage Current Analysis (MVCA) are the most frequently used online methods for fault detection, since there is no need for any additional connections or hardware.

The majority of existing studies investigates these faults in Radial Flux Permanent Magnet (RFPM) synchronous machines. During recent years, the demagnetization and the eccentricity fault have also been studied in the AFPM synchronous machine. More specifically [2,8–21] study the eccentricity fault, while [21–30] investigate the demagnetization fault in AFPM synchronous machines. In [31] a combined eccentricity and demagnetization fault in a double-sided AFPM synchronous machine using a time analytical model is presented.

This study investigates the partial demagnetization fault, the static eccentricity (angular and axis) faults and the combined partial demagnetization and static eccentricity (angular and axis) faults. The Electromotive Force (EMF) spectrum will be used for fault diagnosis purposes and the fault signatures harmonics will be extracted. The machine simulation is performed while using the three-dimensional (3D)-Finite Element Analysis (FEA) that gives more accurate results for this machine type. In all demagnetized cases, one magnet of the generator is partially demagnetized in different percentages, while the generator speed in constant, 375 rpm. Section 2 portrays the basic characteristics of the AFPM synchronous generator, in which the faults are studied. Section 3 provides a validation of the Finite Element Method (FEM) model of the machine, while Section 4 explains the two different types of eccentricity. Section 5 presents the fault signature analysis and Section 6 investigates the partial demagnetization fault in the AFPM synchronous generator without the coexistence of the static eccentricity. Two percentages of partial demagnetization (50% and 80%) are examined. Section 7 studies the static angular and the static axis eccentricity faults. An additional spectrum, the spectrum of the phase EMF sum, has been used for fault diagnosis cases in this specific section. Subsequently, Section 8 studies the combined fault of partial demagnetization in combination with static angular and static axis eccentricity. The fault related harmonics in the EMF spectra are exported and comparisons are made when the level of partial demagnetization changes and the severity of eccentricities remains constant and when the level of partial demagnetization does not change, but the severity of eccentricities increases. Finally, Section 9 is the conclusion section which summarizes the basic assumptions. The novelty of the paper is that these combined faults have not been previously studied in the international literature as well as the phase EMF sum waveform has not been previously used for fault detection in this type of generator under these specific faults.

2. The AFPM Synchronous Generator

The machine, in which the faults are investigated, is a three phase, star connected with neutral, double-sided rotor AFPM synchronous generator [32]. Figure 1 depicts the axial representation of the generator. The generator has 375 rpm nominal speed, 50 Hz nominal frequency, 80 V nominal voltage, 250 W nominal power, 16 poles in each rotor, 12 coils, and 210 turns per coil. The rotor, the magnet, the airgap, and the stator have 12 mm, 10 mm, 3 mm, and 18 mm axial thicknesses, respectively, while the stator external radius is 158mm and internal 60 mm.

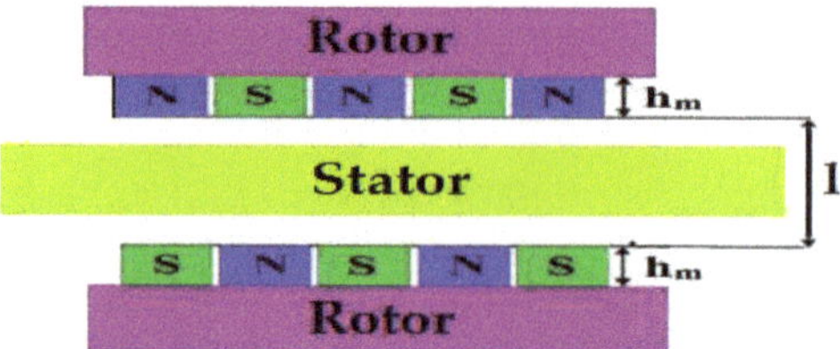

Figure 1. The axial representation of the generator.

Beside the two rotors, there is a coreless stator with concentrated, non-overlapping, and trapezoidal, windings embedded in resin. Figure 2 depicts its layout, where b_{co} is 34.3 mm, b_{ci} is 9 mm, b_{sc} is 16.52 mm, l_c is 63.2 mm, R_i is 58.35 mm, R_o is 121.55 mm, and r is 89.95 mm. Each rotor has 16 permanent magnets of trapezoidal shape made by NdFeB and their layout is presented in Figure 3, where b_{mo} is 47 mm, b_{mi} is 6 mm, R_i is 77.57 mm, R_o is 138.2 mm, and R is 107.885 mm.

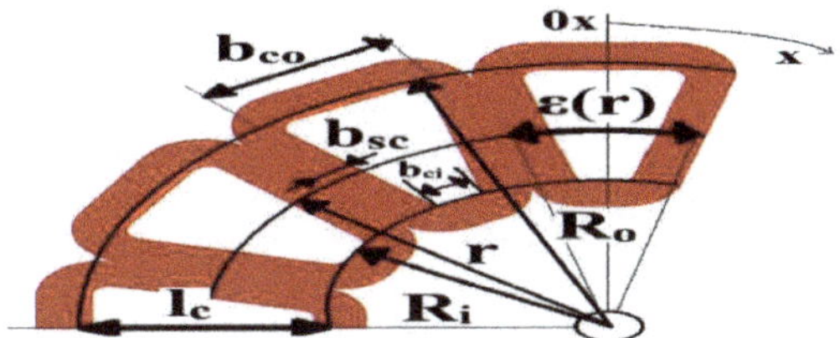

Figure 2. The basic dimension parameters of the generator winding.

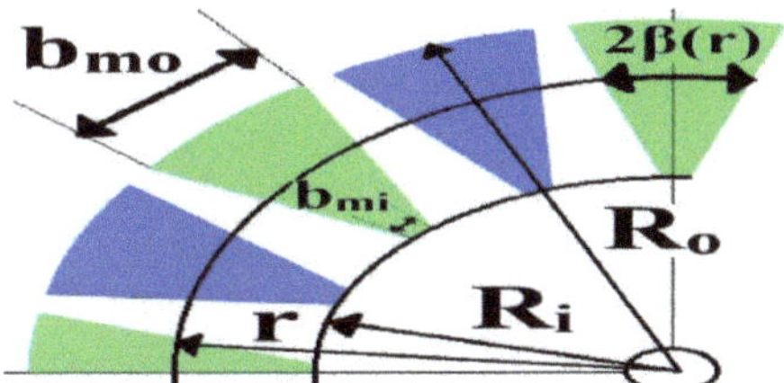

Figure 3. The basic dimension parameters of the permanent magnets of the generator.

In the axial flux permanent magnet synchronous generator, the axial component of the magnetic flux density is divided to the axial component of the magnetic flux density due to the winding Magnetomotive Force (MMF) and the axial component of the magnetic flux density due to the permanent magnets that can be given by (1):

$$B_z = B_{z\_MMF} + B_{z\_PM} \tag{1}$$

where B_z is the axial component of the magnetic flux density in the AFPM, B_{z_MMF} is the axial component of the magnetic flux density due to the winding MMF, and B_{z_PM} is the axial component of the magnetic flux density due to the permanent magnets.

In an AFPM machine with three phases, the axial component of the magnetic flux density due to the MMF can be given by (2):

$$B_{z\_MMF} = \sum_{a=1}^{3} \frac{\mu_0}{1 + 2h_m} F_a \tag{2}$$

where

$$F_a = \sum_{v \in P} i_a \frac{1}{\pi} \frac{w_s}{|v|} \sin\left(|v| \frac{\varepsilon_r}{2}\right) \frac{\sin\left(|v| \frac{a_{sc}(r)}{2}\right)}{|v| \frac{a_{sc}(r)}{2}} e^{jv(x-x_a)} \tag{3}$$

and

$$x_a = (a-1) \frac{2\pi}{3p_s} \tag{4}$$

where F_α is the MMF of the phase α winding, w_s the number of phase winding turns, p_s the number of phase coils, x the location according to the stator, $\varepsilon(r) = \frac{b_c}{r}$, $b_c \approx \frac{b_{c0}+b_{ci}}{2}$, $a_{sc}(r) = \frac{b_{sc}}{r}$, and $P = \{ \dots -3p_s, -2p_s, -p_s, p_s, 2p_s, 3p_s \dots \}$ [33].

The axial component of the magnetic flux density due to the permanent magnets can be given by (5):

$$B_{z\_PM} = \sum_{\varsigma \in Q} \frac{4}{\pi} \frac{B_r}{\varsigma} \frac{h_m}{2h_m + 1} p\sin(\varsigma\beta(r))e^{j\varsigma(x-\varphi)} \tag{5}$$

where

$$\beta(r) = \frac{b_m}{2r} \tag{6}$$

where $b_m \approx \frac{b_{mo}+b_{mi}}{2}$, φ the angle of a rotor position and $Q = \{\dots -5p, -3p, -p, p, 3p, 5p \dots\}$ [33].

The harmonics that will be created in the magnetic flux density spectrum are responsible for the harmonics that will appear in the EMF spectrum.

3. Model Validation

For the validation of the model in the healthy condition, we present the waveforms of the stator current, derived from the simulation and experiment, respectively, when the machine has nominal speed 375 rpm and supplies the nominal resistive load 70 Ohm, as Figure 4 shows. It can be observed that the two waveforms are qualitatively and quantitatively similar and they validate the accuracy of our FEM model. In addition, Figure 5 depicts the waveforms of the stator current when one magnet is totally demagnetized derived from simulation and experimental procedure when the generator has nominal speed and supplies a load 70 Ohm. It can be seen that the FEM waveform also agrees with the experimental waveform. More specifically, we have used the 3D-Opera mesher while our model contains 5128737 elements. The transient electromagnetic analysis with motion has been used. On a PC (Intel i7-4770 with 8 GB RAM) the finite element analysis requires 12 h in order to reach the steady state condition.

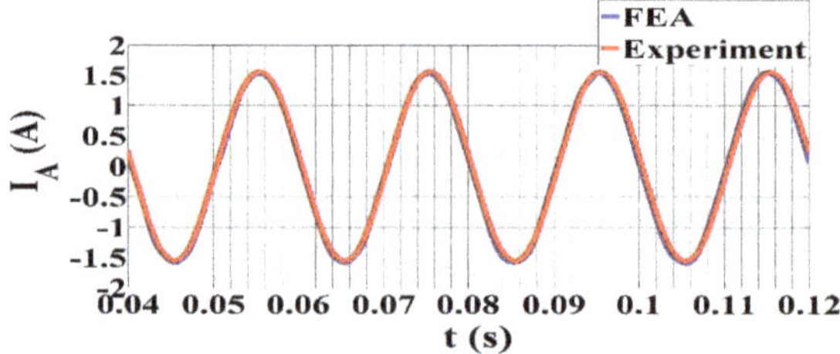

Figure 4. The stator current waveform in the healthy condition when the generator has nominal speed 375 rpm and supplies the nominal load of 70 Ohm (blue line-simulation results, red line-experimental results).

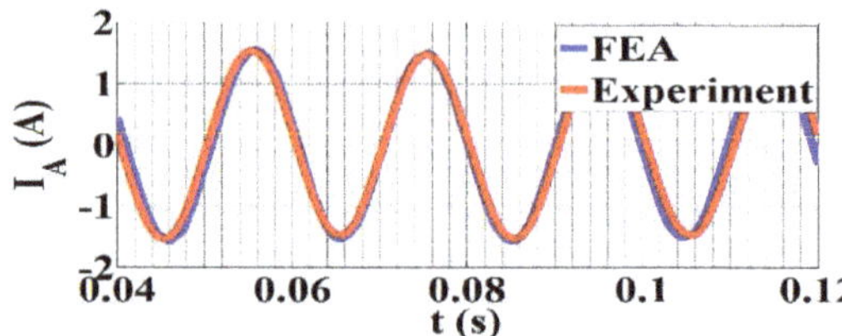

Figure 5. The stator current waveform when one magnet is totally demagnetized, the generator has nominal speed 375 rpm and supplies the nominal load of 70 Ohm (blue line-simulation results, red line-experimental results).

4. Types of Eccentricity

Two types of static eccentricity appear in the literature [15,16]: the static angular and the static axis eccentricity. The first type occurs when the rotor axis coincides with the rotation axis but does not coincide with the stator axis. In this case, the air gap is not uniform, but, during the rotation, the maximum and minimum air gap positions are constant. In other words, the air gap does not change in time. The second type occurs when the stator and rotor are offset from each other in the axis direction. Figure 6 depicts the axial representation of the machine when the two different types of static eccentricity exist in the generator.

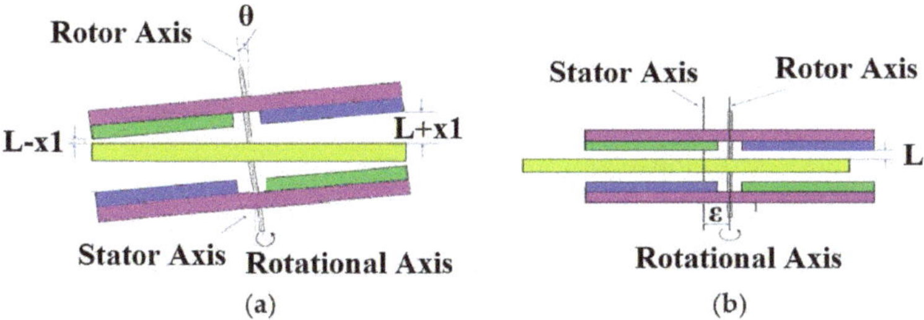

Figure 6. The axial representation of the machine when the generator is: (**a**) static angular eccentric and (**b**) static axis eccentric. Axial representation of the generator.

5. Fault Signature Analysis

From the literature [7,34–36], it is known that the stator current spectrum of a RFPM synchronous machine with demagnetization or eccentricity fault contains new harmonics that are given by Equation (7):

$$f_{demag} = f_s\left(1 \pm \frac{k}{p}\right) \tag{7}$$

where f_{demag} are the frequencies of the fault related harmonics, fs is the fundamental frequency, p the number of machine poles pairs, and k an integer number. Previous studies [29,30] prove that this Equation can predict also the fault related harmonics in the EMF and stator current spectra of an AFPM synchronous machine with totally demagnetized magnets. In this article, it is examined whether this Equation is applicable in the case of the combined fault, in order to interpret the fault related harmonics.

6. Partial Demagnetization

First, the partial demagnetization fault without the coexistence of the eccentricity fault is studied. In all of the investigated cases, one magnet is partially demagnetized in two different percentages (50% and 80% partial demagnetization). Figure 7 depicts the machine 3D-FEA model when one magnet is partially demagnetized, and Figure 8 shows the waveforms of the axial component of the magnetic flux density when the fault exists. The increment of the severity of the demagnetization leads to the decrement of the amplitude of the waveform in the location of the faulty magnet.

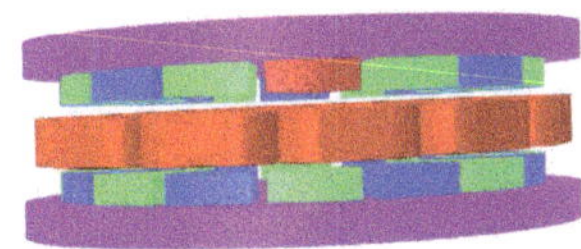

Figure 7. The three-dimensional (3D)-FEA model of the machine when one magnet is partially demagnetized.

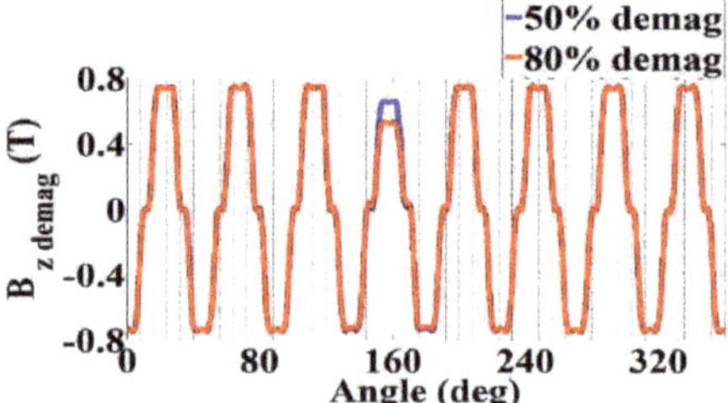

Figure 8. The waveform of the axial component of the magnetic flux density when one magnet is partially demagnetized (blue line—50% partial demagnetization, red line—80% partial demagnetization).

The above waveform can be given by Equation (8) [37]:

$$B_{z\_one\_dem} = B_{z\_tot} - y(t) \qquad (8)$$

where B_{z_tot} is the axial component of the total magnetic flux density in the healthy case and $y(t)$ is the product between the B_{z_tot} and the square wave $x(t)$ divided by V_{dem}, which is the B_{z_tot} amplitude immersion due to TDF, as it can be seen by (9):

$$y(t) = \frac{B_{z\_tot}}{V_{dem}} x(t) \qquad (9)$$

while $x(t)$ can be expressed in Fourier series using (10):

$$x(t) = \frac{1}{2p} + \sum_{k=1}^{\infty} \frac{2}{k\pi} \sin\left(\frac{\pi k}{2p}\right) \cos\left(\frac{2k\pi f_s t}{p}\right) \qquad (10)$$

Substituting (9) and (10) in (8) implies (11):

$$B_{z\_one\_dem} = B_{z\_tot} - \frac{B_{z\_tot}}{2pV_{dem}} - \sum_{k=1}^{\infty} \frac{2B_{z\_tot}}{k\pi V_{dem}} \sin\left(\frac{k\pi}{2p}\right) \cos\left(\frac{2k\pi f_s t}{p}\right) \qquad (11)$$

The harmonics that appear in Equation (11) are responsible for the harmonics that will be created in the EMF spectrum in the faulty condition. Equation (11) is suitable to interpret every percentage of partial demagnetization, because, as can been seen below, when the fault severity changes the kind of the fault related harmonics does not change but their amplitude changes. In Equation (11), it will be modification in V_{dem} when the severity of the fault changes.

Figures 9 and 10 depict the EMF waveforms and the corresponding spectra when one magnet is 50% and 80% partially demagnetized. The increment of the fault severity reduces the amplitude of the EMF waveform. In addition, the fault creates new harmonics in the corresponding spectra, which Table 1 summarizes. The new harmonics are of frequencies 25 Hz, 75 Hz, 100 Hz, 125 Hz, 175 Hz, 200 Hz, and 225 Hz and their amplitudes increase when the severity of demagnetization increases. The fundamental harmonic decreases in amplitude when the level of demagnetization increases. The fault related harmonics are both even and fractional and of the same frequencies, like the case wjere one magnet is totally demagnetized [30].

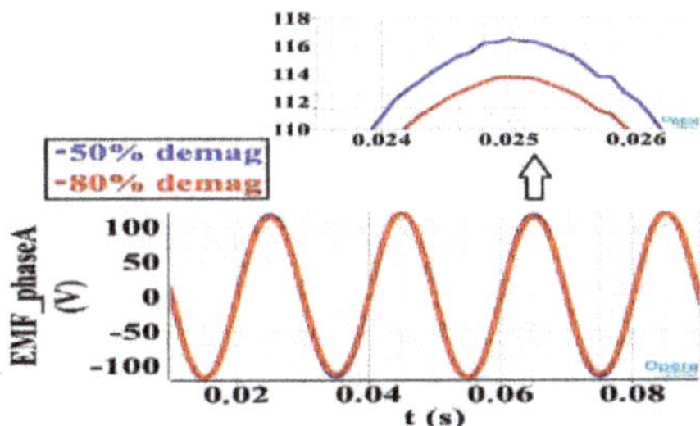

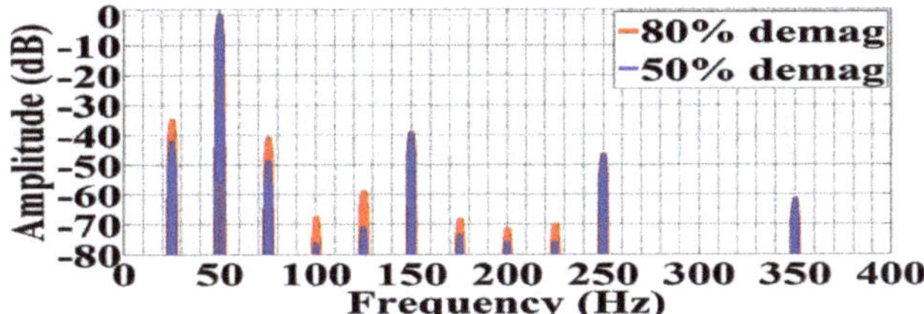

Figure 9. The EMF waveform when one magnet is partially demagnetized (blue line—50% partial demagnetization, red line—80% partial demagnetization).

Figure 10. The EMF spectrum when one magnet is partially demagnetized (blue line—50% partial demagnetization, red line-80% partial demagnetization).

Table 1. Fundamental Harmonic and Fault Related Harmonics in the Spectrum of the EMF When One Magnet is Partially Demagnetized.

Harmonic Order	F (Hz)	Generator with One 50% Partially Demagnetized Magnet		Generator with One 80% Partially Demagnetized Magnet	
		(dB)	(V)	(dB)	(V)
0.5	25	−42.33	0.2455	−35.10	0.5581
1	50	0	32.12	0	31.74
1.5	75	−48.88	0.1156	−41.29	0.2733
2	100	−76.84	0.004686	−67.97	0.01299
2.5	125	−71.04	0.009101	−59.38	0.03411
3.5	175	−73.79	0.006643	−68.48	0.01198
4	200	−76.03	0.00501	−71.77	0.008362
4.5	225	−76.1	0.005209	−70.28	0.00991

7. Static Eccentricity Fault

7.1. Static Angular Eccentricity Fault

In this section, the static angular eccentricity fault will be studied. As it is already proven in [18], the static angular eccentricity fault does not create new harmonics in the EMF and the stator current spectra. This can be justified by the fact that during this fault the airgap in a double-sided machine increases from the one side and decreases from the other size resulting in a constant total airgap. For that reason, the EMF waveform, the stator current waveform, and the corresponding spectra remain approximately unaffected by the fault. However, the spectrum of the phase EMF sum presents variation when static angular eccentricity exists in the generator. Figure 11 depicts the phase EMF sum spectra for two different severities of static angular eccentricity (30% and 40%). Equation (12) describes the phase EMF sum waveform. This signal has a fundamental frequency of 150 Hz ($3f_s$), three times the fundamental frequency of the EMF waveform of each generator phase (f_s). In both cases, the harmonic component of frequency 50Hz is the fault related harmonic that, in the faulty case, its amplitude increases more than the other amplitudes when compared to the corresponding healthy spectrum for spectra normalized to the $3f_s$ frequency (150 Hz). When the level of eccentricity increases, the amplitude of this harmonic component also increases, as can be seen by Table 2. In other words, we can tell that the component of frequency f_s Hz is the most dominant fault related harmonic component in the EMF sum spectrum and it indicates the existence of static angular eccentricity fault. Finally, from Table 2, it can be seen that the absolute value of the $3f_s$ harmonic component (150 Hz) also increases when the eccentricity severity increases.

$$PhaseEMF_{sum} = EMF_{phaseA} + EMF_{phaseB} + EMF_{phaseC} \tag{12}$$

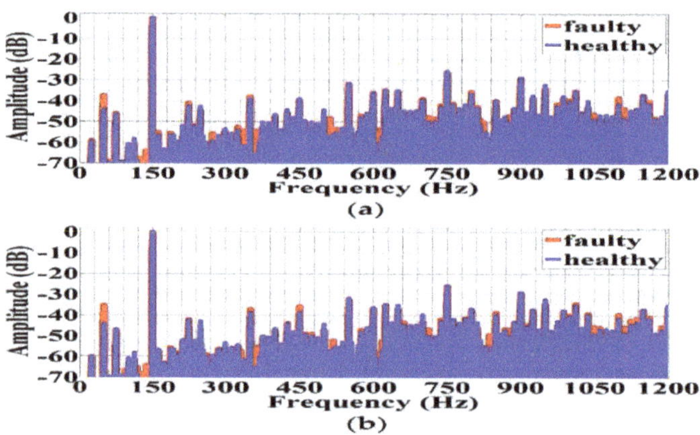

Figure 11. The phase EMF sum spectra of the double-sided rotor generator when a static angular eccentricity exists in the machine: (**a**) 30% fault severity, (**b**) 40% fault severity. (blue line-healthy case, red line-faulty case).

Table 2. The Harmonics of Frequencies 50 Hz and 150 Hz in the Spectrum of the phase EMF sum When Static Angular Eccentricity exists in the Generator.

Harmonic	f (Hz)	Healthy Generator		Generator with 30% Static Angular Eccentricity		Generator with 40% Static Angular Eccentricity	
		(dB)	(V)	(dB)	(V)	(dB)	(V)
f_s	50	−44.33	0.00634	−37.51	0.01392	−35.29	0.01816
$3f_s$	150	0	1.043	0	1.045	0	1.056

7.2. Static Axis Eccentricity Fault

In this section, the static axis eccentricity fault will be studied. Like with the previous case, the static axis eccentricity fault does not create new harmonics in the phase EMF and the stator current spectra. In [18], it is referred that, when the severity of eccentricity increases, the amplitude of the third harmonic of the phase EMF spectrum slightly increases, while the amplitudes of the fifth and seventh harmonics slightly decrease. However, in the phase EMF sum, new harmonic components appear as Figure 12 depicts. As it can be observed, the harmonic component of frequency f_s Hz is a fault related harmonic, like to the case of static angular eccentricity. Consequently, the increment of the amplitude of the harmonic component of 50 Hz indicates static eccentricity fault but we cannot separate the two faults. Finally, Table 3 summarizes the amplitude in dB and in absolute value of the harmonic components of frequencies 50 Hz and 150 Hz.

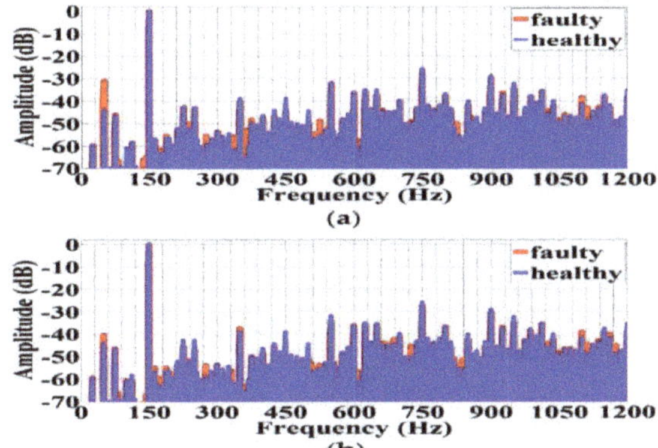

Figure 12. The phase EMF sum spectra of the double-sided rotor generator when a static axis eccentricity exists in the machine: (**a**) 2mm fault severity, (**b**) 3mm fault severity (blue line—healthy case, red line—faulty case).

Table 3. The Harmonics of Frequencies 50 Hz and 150 Hz in the Spectrum of the phase EMF sum When Static Axis Eccentricity exists in the Generator.

Harmonic f (Hz)		Healthy Generator		Generator with 2mm Static Axis Eccentricity (dB)		Generator with 3mm Static Axis Eccentricity (dB)	
		(dB)	(V)	(dB)	(V)	(dB)	(V)
f_s	50	−44.33	0.00634	−31.01	0.03033	−40.47	0.01051
$3f_s$	150	0	1.043	0	1.077	0	1.109

8. The Combined Fault

8.1. The Combined Partial Demagnetization and Static Angular Eccentricity Fault

In this paragraph, the combined partial demagnetization and static angular eccentricity fault is studied. In all cases, one magnet is partially demagnetized. Figure 13 depicts the EMF spectra when the severity of static angular eccentricity remains constant and the level of demagnetization changes, while Figure 14 shows the EMF spectra when the level of demagnetization remains constant and the severity of static angular eccentricity changes. Although the static angular eccentricity does not create new harmonics in this spectrum [18], when it is combined with demagnetization, new harmonic components appear. The new harmonics due to the combined fault agree with Equation (7) for k =−5, −4, −3, −1, 1, 3, 4, 5, 8, 12, 20, 24, 28. The machine odd harmonics (third, fifth, and seventh) appear variation when the combined fault exists. More specifically, the amplitude of the third harmonic decreases when there is combined fault in the machine, while the amplitudes of the fifth and seventh harmonics increase, like the case that only static angular eccentricity exists in the generator [18]. In other words, the variation of the amplitude of these harmonic components is due to static angular eccentricity fault. Tables 4 and 5 summarize the amplitudes of the fault related harmonics derived from Figures 13 and 14 respectively. As it can be seen by Table 4, when the demagnetization level increases and the static angular eccentricity level remains constant, the amplitude of all combined fault related harmonics also increases. However, the absolute value of the fundamental frequency, 50 Hz, slightly decreases. Observing Table 5, we can see that when the severity of the partial demagnetization remains invariable and the level of static angular eccentricity increases the amplitude of all combined fault related harmonics increases too with exception the demagnetization fault related harmonics. In other words, the increment of the eccentricity also creates an increment of the amplitude of the harmonics with frequencies 18.75 Hz, 31.25 Hz, 43.75 Hz, 56.25 Hz, 68.75 Hz, 81.25 Hz. The amplitude of the demagnetization harmonic components (0.5th, 1.5th, 2nd, 2.5th, 3.5th, 4th, and 4.5th) remains approximately constant, as expected when taking into consideration that the level of demagnetization does not change. In addition, the absolute value of the fundamental frequency presents slightly increment, as Table 5 depicts.

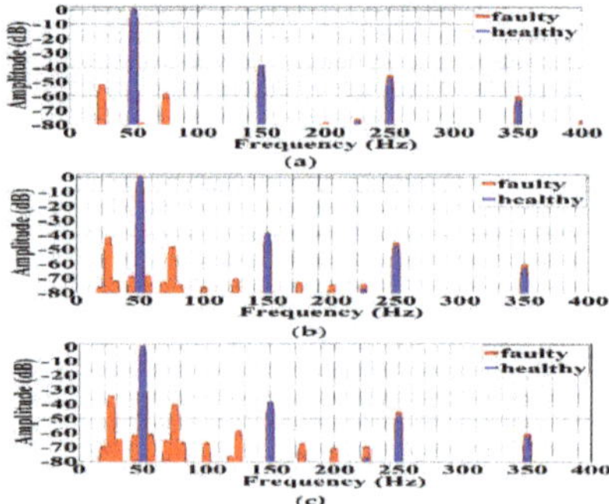

Figure 13. The EMF spectra when a combined fault of partial demagnetization and static angular eccentricity exists in the generator and the level of demagnetization changes: (**a**) 20% demagnetization and 30% static angular eccentricity, (**b**) 50% demagnetization and 30% static angular eccentricity, and (**c**) 80% demagnetization and 30% static angular eccentricity.

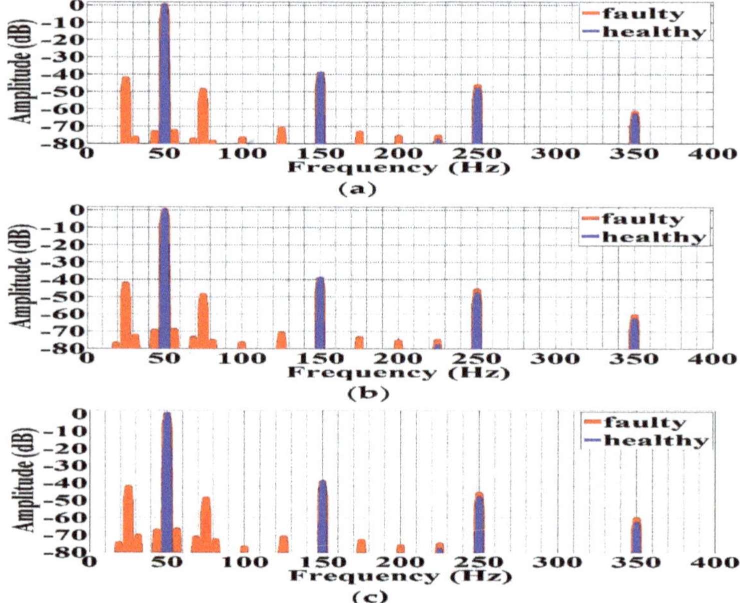

Figure 14. The EMF spectra when a combined fault of partial demagnetization and static angular eccentricity exists in the generator and the level of eccentricity changes: (**a**) 50% demagnetization and 20% static angular eccentricity, (**b**) 50% demagnetization and 30% static angular eccentricity, and (**c**) 50% demagnetization and 40% static angular eccentricity.

Table 4. Fundamental and Fault Related Harmonics in the Spectrum of the EMF for the Combined Partial Demagnetization and Static Angular Eccentricity Fault When Changes the Level of Demagnetization.

Harmonic Order	f (Hz)	Healthy Generator		Generator with One 20% Partially Demagnetized Magnet and 30% Static Angular Eccentricity		Generator with One 50% Partially Demagnetized Magnet and 30% Static Angular Eccentricity		Generator with One 80% Partially Demagnetized Magnet and 30% Static Angular Eccentricity	
		(dB)	(V)	(dB)	(V)	(dB)	(V)	(dB)	(V)
0.375	18.75	-	5.366×10^{-5}	-	0.001382	−77.24	0.004418	−70.1	0.009934
0.5	25	-	0.00219	−52.52	0.07655	−42.35	0.2452	−35.1	0.5586
0.625	31.25	-	2.246×10^{-5}	-	0.002323	−72.76	0.007402	−65.6	0.01669
0.875	43.75	-	8.122×10^{-5}	−79.67	0.003362	−69.71	0.01052	−62.51	0.02379
1	50	0	32.07	0	32.36	0	32.16	0	31.78
1.125	56.25	-	5.785×10^{-5}	−79.35	0.003489	−69.18	0.01117	−61.86	0.02566
1.375	68.75	-	0.0002289	-	0.002056	−73.73	0.006621	−68.75	0.01541
1.5	75	-	0.001492	−59.27	0.03521	−48.87	0.1158	−41.29	0.2739
1.625	81.25	-	0.0001917	-	0.00158	−75.43	0.005441	−67.7	0.0131
2	100	-	0.001133	-	0.001901	−76.74	0.004682	−67.75	0.01302
2.5	125	-	0.0009809	-	0.001735	−71.1	0.008965	−59.4	0.03406
3.5	175	-	0.0006477	-	0.002879	−73.7	0.006641	−68.56	0.01186
4	200	-	0.001959	−79.36	0.003483	−75.64	0.005314	−71.25	0.008699
4.5	225	−78.11	0.003984	−76.9	0.004623	−75.35	0.00549	−70.03	0.01001

Table 5. Fundamental and Fault Related Harmonics in the Spectrum of the EMF for the Combined Partial Demagnetization and Static Angular Eccentricity Fault When Changes the Level of Eccentricity.

Harmonic Order	F (Hz)	Healthy Generator		Generator with One 50% Partially Demagnetized Magnet and 20% Static Angular Eccentricity		Generator with One 50% Partially Demagnetized Magnet and 30% Static Angular Eccentricity		Generator with One 50% Partially Demagnetized Magnet and 40% Static Angular Eccentricity	
		(dB)	(V)	(dB)	(V)	(dB)	(V)	(dB)	(V)
0.375	18.75	-	5.366×10^{-5}	-	0.002912	−77.24	0.004418	−74.76	0.005873
0.5	25	-	0.00219	−42.33	0.2456	−42.35	0.2452	−42.35	0.2458
0.625	31.25	-	2.246×10^{-5}	−76.4	0.004863	−72.76	0.007402	−70.34	0.009796
0.875	43.75	-	8.122×10^{-5}	−73.24	0.006998	−69.71	0.01052	−67.39	0.01375
1	50	0	32.07	0	32.13	0	32.16	0	32.2
1.125	56.25	-	5.785×10^{-5}	−72.59	0.005742	−69.18	0.01117	−66.75	0.01481
1.375	68.75	-	0.0002289	−77.46	0.004302	−73.73	0.006621	−71.39	0.008676
1.5	75	-	0.001492	−48.87	0.1157	−48.87	0.1158	−48.85	0.1163
1.625	81.25	-	0.0001917	−78.57	0.003786	−75.43	0.005441	−72.82	0.007364
2	100	-	0.001133	−76.76	0.004668	−76.74	0.004682	−76.98	0.004557
2.5	125	-	0.0009809	−71.22	0.008831	−71.1	0.008965	−71.18	0.008893
3.5	175	-	0.0006477	−73.55	0.00675	−73.7	0.006641	−73.21	0.007038
4	200	-	0.001959	−75.9	0.00515	−75.64	0.005314	−76.18	0.004998
4.5	225	−78.11	0.003984	−75.53	0.005375	−75.35	0.00549	−75.16	0.005619

8.2. The Combined Partial Demagnetization and Static Axis Eccentricity Fault

This paragraph investigates the combined partial demagnetization with static axis eccentricity fault. Figure 15 presents the EMF spectra for the combined fault when the level of static axis eccentricity remains constant and the severity of demagnetization changes, while Figure 16 depicts the EMF spectra for the combined fault when the level of demagnetization remains constant and the level of static axis eccentricity changes. The corresponding amplitudes of the combined fault related harmonics are depicted in Tables 6 and 7, respectively. The combined fault creates new harmonic components in the EMF spectra, in contrast to the case that only static axis eccentricity exists in the generator and in the EMF spectrum do not appear new harmonic components that are related to the fault, as [18] proves. The combined fault related harmonics are of frequencies 18.75 Hz, 25 Hz, 31.25 Hz, 43.75 Hz, 56.25 Hz, 68.75 Hz, 75 Hz, 81.25 Hz, 100 Hz, 125 Hz, 175 Hz, 200Hz, and 225 Hz. These harmonics agree with Equation (7) for k = −5, −4, −3, −1, 1, 3, 4, 5, 8, 12, 20, 24, and 28, like in the previous section, Section 8.1. In other words, in two combined faults appear the same fault related harmonics. The machine odd harmonics (third, fifth, and seventh) remain approximately constant when the combined fault exists when compared to the healthy case, like to the case that only static axis eccentricity exists in the generator [18]. Observing Table 6, it can be seen that, when the severity of demagnetization increases while the level of static axis eccentricity remains constant, the amplitude of the combined fault related harmonics also increases, like the case where a combined partial demagnetization and static angular eccentricity fault exists in the generator. Although the absolute value of the fundamental frequency presents a small decrement, as Table 6 depicts. Table 7 shows that, when the level of demagnetization remains invariable and the severity of static axis eccentricity increases, the amplitude of all combined fault related harmonics slightly increases apart from the harmonic components with frequencies 25 Hz, 75 Hz, 100 Hz, 125 Hz, 175 Hz, 200 Hz, and 225 Hz. These components are related with demagnetization and for that reason do not change amplitude when the demagnetization level is constant, like Section 8.1. Finally the absolute value of the fundamental frequency slightly decreases when changes the level of the eccentricity and the demagnetization level remains constant. To conclude, it can be observed that when only partial demagnetization exists on the generator, fault related harmonics of frequencies of 25 Hz, 75 Hz, 100 Hz, 125 Hz, 175 Hz, 200 Hz, and 225 Hz appear in the EMF spectrum, when only static eccentricity, either angular or axis, exists do not appear new harmonics in the EMF spectrum, while, in the cases of combined faults, fault related harmonics of frequencies 18.75 Hz, 25 Hz, 31.25 Hz, 43.75 Hz, 56.25 Hz, 68.75 Hz, 75 Hz, 81.25 Hz, 100 Hz, 125 Hz, 175 Hz, 200Hz, and 225 Hz appear. In other words, in the last cases, the demagnetization fault related harmonics and some sideband harmonics appear. The phase EMF sum signal can also be used in the case of combined faults, but the reason that we did not investigate it is because the EMF signal is able to provide the fault related harmonics that are related to combined faults, in contrast to Section 7. The fault identification using the EMF signal and not the phase EMF sum signal makes the detection process simpler and more cost effective, as we should measure one signal every time and not three different signals. However, the study of the phase EMF sum signal in combined faults can be an object of investigation in a future article.

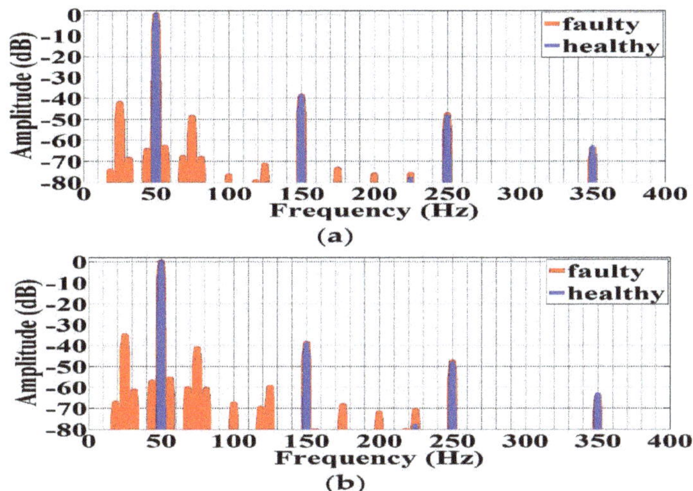

Figure 15. The EMF spectra when a combined fault: (**a**) 50% demagnetization and 2 mm static axis eccentricity, (**b**) 80% demagnetization and 2 mm static axis eccentricity exist in the generator.

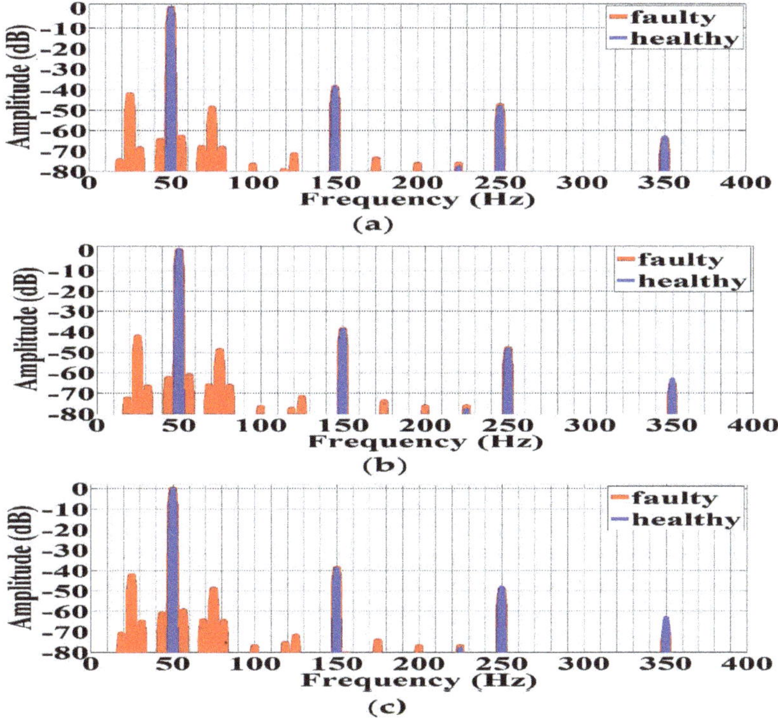

Figure 16. The EMF spectra when a combined fault: (**a**) 50% demagnetization and 2 mm static axis eccentricity, (**b**) 50% demagnetization and 2.5 mm static axis eccentricity, and (**c**) 50% demagnetization and 3 mm static axis eccentricity exist in the generator.

Table 6. Fundamental and Fault Related Harmonics in the Spectrum of the EMF for the Combined Partial. Demagnetization and Static Axis Eccentricity Fault When Changes the Level of Demagnetization.

Harmonic Order	f (Hz)	Healthy Generator		Generator with One 50% Partially Demagnetized Magnet and 2mm Static Axis Eccentricity		Generator with One 80% Partially Demagnetized Magnet and 2mm Static Axis Eccentricity	
		(dB)	(V)	(dB)	(V)	(dB)	(V)
0.375	18.75	-	5.366×10^{-5}	−74.56	0.005965	−67.27	0.01365
0.5	25	-	0.00219	−42.3	0.2448	−35.06	0.5566
0.625	31.25	-	2.246×10^{-5}	−68.75	0.01165	−61.59	0.02624
0.875	43.75	-	8.122×10^{-5}	−64.64	0.01869	−57.39	0.04255
1	50	0	32.07	0	31.89	0	31.52
1.125	56.25	-	5.785×10^{-5}	−63.1	0.02231	−55.78	0.05123
1.375	68.75	-	0.0002289	−67.99	0.01271	−60.57	0.02951
1.5	75	-	0.001492	−48.95	0.1139	−41.36	0.2694
1.625	81.25	-	0.0001917	−68.4	0.01212	−60.73	0.02897
2	100	-	0.001133	−76.79	0.004618	−67.78	0.01287
2.5	125	-	0.0009809	−71.59	0.008395	−59.72	0.03254
3.5	175	-	0.0006477	−73.73	0.006561	−68.59	0.01172
4	200	-	0.001959	−76.38	0.004839	−72.15	0.007781
4.5	225	−78.11	0.003984	−76.22	0.004926	−70.75	0.00914

Table 7. Fundamental and Fault Related Harmonics in the Spectrum of the EMF for the Combined Partial Demagnetization and Static Axis Eccentricity Fault When Changes the Level of Eccentricity.

Harmonic Order	f (Hz)	Healthy Generator		Generator with One 50% Partially Demagnetized Magnet and 2mm Static Axis Eccentricity		Generator with One 50% Partially Demagnetized Magnet and 2.5mm Static Axis Eccentricity		Generator with One 50% Partially Demagnetized Magnet and 3mm Static Axis Eccentricity	
		(dB)	(V)	(dB)	(V)	(dB)	(V)	(dB)	(V)
0.375	18.75	-	5.366×10^{-5}	−74.56	0.005965	−72.63	0.007425	−70.99	0.008928
0.5	25	-	0.00219	−42.3	0.2448	−42.28	0.2445	−42.26	0.2439
0.625	31.25	-	2.246×10^{-5}	−68.75	0.01165	−66.8	0.01454	−65.21	0.01739
0.875	43.75	-	8.122×10^{-5}	−64.64	0.01869	−62.73	0.02323	−61.21	0.02782
1	50	0	32.07	0	31.89	0	31.79	0	31.66
1.125	56.25	-	5.785×10^{-5}	−63.1	0.02231	−61.19	0.02773	−59.6	0.03313
1.375	68.75	-	0.0002289	−67.99	0.01271	−66.02	0.0159	−64.55	0.01875
1.5	75	-	0.001492	−48.95	0.1139	−48.98	0.1131	−49.02	0.1121
1.625	81.25	-	0.0001917	−68.4	0.01212	−66.4	0.01522	−64.79	0.01823
2	100	-	0.001133	−76.79	0.004618	−76.66	0.004671	−77.04	0.00445
2.5	125	-	0.0009809	−71.59	0.008395	−71.92	0.008061	−71.93	0.008013
3.5	175	-	0.0006477	−73.73	0.006561	−74.01	0.006336	−74.28	0.006118
4	200	-	0.001959	−76.38	0.004839	−76.5	0.004758	−77.16	0.004392
4.5	225	−78.11	0.003984	−76.22	0.004926	−76.48	0.004767	−77.1	0.00442

9. Conclusions

This paper investigates the partial demagnetization fault, the static angular and the static axis eccentricity fault, and the combination of partial demagnetization and static angular or static axis eccentricity fault in an AFPM synchronous double-sided rotor generator. The machine has been simulated while using 3D-FEA, while the EMF spectra are studied for fault diagnosis purposes in order to extract the fault related harmonics. In all cases one magnet is partially demagnetized. Firstly, the partial demagnetization fault without the coexistence of eccentricity is studied for two different percentages of partial demagnetization. This fault creates both even and fractional harmonics (with frequencies of 25 Hz, 75 Hz, 100 Hz, 125 Hz, 175 Hz, 200 Hz, and 225 Hz) in the EMF spectra and the harmonic amplitude increases when the severity of partial demagnetization also increases. In addition, the static angular and the static axis eccentricity faults are investigated. Both of the faults do not create new harmonics in the corresponding EMF spectra. However, these faults create new harmonic components in the phase EMF sum spectrum. The phase EMF sum waveform is a signal that has a fundamental frequency of 150 Hz ($3f_s$), three times the fundamental frequency, f_s, of the EMF waveform. When static eccentricity fault exists, either angular or axis, in the corresponding spectra of the phase EMF sum new harmonic components appear and harmonic component of frequency 50 Hz is the fault related harmonic component that, in the faulty case, its amplitude increases more than the amplitudes of other harmonics when compared to the corresponding healthy spectrum. In other words, this harmonic component in the phase EMF sum spectrum can predict eccentricity, but cannot identify whether the static eccentricity is angular or axis. Then the combined fault of partial demagnetization with static angular eccentricity is studied. The analysis proves that this fault creates new harmonics of frequencies 18.75 Hz, 25 Hz, 31.25 Hz, 43.75 Hz, 56.25 Hz, 68.75 Hz, 75 Hz, 81.25 Hz, 100 Hz, 125 Hz, 175 Hz, 200Hz, and 225 Hz in the EMF spectrum when compared to the healthy case. Comparisons are made when the level of demagnetization changes and the level of eccentricity remains constant and when the level of demagnetization remains constant and the level of eccentricity changes. In the first case, the amplitude of all fault related harmonics increases when the severity of demagnetization increases, while, in the second case, when the severity of eccentricity increases, the amplitude of the fault related harmonics increases too apart from the harmonics that also appear when only demagnetization exist in the machine (0.5th, 1.5th, 2nd, 2.5th, 3.5th, 4th, and 4.5th). In addition, the combined partial demagnetization and static axis eccentricity fault is studied. The fault related harmonics are of frequencies 18.75 Hz, 25 Hz, 31.25 Hz, 43.75 Hz, 56.25 Hz, 68.75 Hz, 75 Hz, 81.25 Hz, 100 Hz, 125 Hz, 175 Hz, 200Hz, and 225 Hz. Like previous cases, comparisons are made when the level of demagnetization increases, and the severity of eccentricity remains constant and when the level of demagnetization remains invariable and the level of static axis eccentricity changes. In the first case, all of the combined fault related harmonics increase in amplitude, while, in the second case, there is increment in the amplitude of all harmonic components apart from these that are related to demagnetization fault. A future step of this study is to find a formula that is able to separate the two eccentricity cases, because, as can be seen in both eccentricity faults, fault related harmonics of the same frequencies appear.

Author Contributions: The presented work was carried out through the cooperation of all authors. A.C.B. conducted the research and wrote the paper and J.C.K. supervised the whole study and edited the manuscript. All authors have read and agreed to the published version of the manuscript.

Funding: This research received no external funding.

Conflicts of Interest: The authors declare no conflict of interest.

References

1. Mirimani, S.M.; Vahedi, A.; Marignetti, F.; De Santis, E. Static Eccentricity Fault Detection in Single-Stator–Single-Rotor Axial-Flux Permanent-Magnet Machines. *IEEE Trans. Ind. Appl.* **2012**, *48*, 1838–1845. [CrossRef]
2. Mirimani, S.M.; Vahedi, A.; Marignetti, F.; Di Stefano, R. An Online Method for Static Eccentricity Fault Detection in Axial Flux Machines. *IEEE Trans. Ind. Electron.* **2015**, *62*, 1931–1942. [CrossRef]
3. Faiz, J.; Nejadi-Koti, H. Demagnetization fault indexes in permanent magnet synchronous motors—An Overview. *IEEE Trans. Magnet.* **2016**, *52*. [CrossRef]
4. Faiz, J.; Mazaheri Tehrani, E. Demagnetization modeling and fault diagnosing techniques in permanent magnet machines under stationary and non-stationary conditions an overview. *IEEE Trans. Ind. Appl.* **2016**, *51*, 2772–2785. [CrossRef]

5. Capolino, G.-A.; Antonino-Daviu, J.A.; Riera-Guasp, M. Modern Diagnostics Techniques for Electrical Machines, Power Electronics, and Drives. *IEEE Trans. Ind. Electron.* **2015**, *62*, 1738–1745. [CrossRef]

6. Duan, Y.; Toliyat, H. A Review of Condition Monitoring and Fault Diagnosis for Permanent Magnet Machines. In Proceedings of the 2012 IEEE Power and Energy Society General Meeting, San Diego, CA, USA, 22–26 July 2012.

7. Ebrahimi, B.M.; Faiz, J.; Roshtkhari, M.J. Static-, Dynamic-, and Mixed-Eccentricity Fault Diagnoses in Permanent-Magnet Synchronous Motors. *IEEE Trans. Ind. Electron.* **2009**, *56*, 4727–4739. [CrossRef]

8. Tarek, M.T.B.; Das, S.; Sozer, Y. Comparative Analysis of Static Eccentricity Faults of Double Stator Single Rotor Axial Flux Permanent Magnet Motors. In Proceedings of the 2019 Energy Conversion Congress and Exposition (ECCE), Baltimore, MD, USA, 29 September–3 October 2019.

9. Mirimani, S.M.; Vahedi, A.; Marignetti, F. Effect of Inclined Static Eccentricity Fault in Single Stator-Single Rotor Axial Flux Permanent Magnet Machines. *IEEE Trans. Magnet.* **2012**, *48*, 143–149. [CrossRef]

10. Marignetti, F.; Vahedi, A.; Mirimani, S.M. An Analytical Approach to Eccentricity in Axial Flux Permanent Magnet Synchronous Generators for Wind Turbines. *Electr. Power Componets Syst.* **2015**, *43*, 1039–1050. [CrossRef]

11. Ogidi, O.O.; Barendse, P.S.; Khan, M.A. Detection of Static Eccentricities in Axial-Flux Permanent-Magnet Machines with Concentrated Windings Using Vibration Analysis. *IEEE Trans. Ind. Appl.* **2015**, *51*, 4425–4434. [CrossRef]

12. Ogidi, O.O.; Barendse, P.S.; Khan, M.A. Fault diagnosis and condition monitoring of axial-flux permanent magnet wind generators. *Electr. Power Syst. Res.* **2016**, *136*, 1–7. [CrossRef]

13. Huang, Y.; Guo, B.; Hemeida, A.; Sergeant, P. Analytical Modeling of Static Eccentricities in Axial Flux Permanent-Magnet Machines with Concentrated Windings. *Energies* **2016**, *9*, 892. [CrossRef]

14. Ogidi, O.O.; Barendse, P.S.; Khan, M.A. Influence of Rotor Topologies and Cogging Torque Minimization Techniques in the Detection of Static Eccentricities in Axial-Flux Permanent-Magnet Machine. *IEEE Trans. Ind. Appl.* **2017**, *53*, 161–170. [CrossRef]

15. Thiele, M.; Heins, G. Computationally Efficient Method for Identifying Manufacturing Induced Rotor and Stator Misalignment in Permanent Magnet Brushless Machines. *IEEE Trans. Ind. Appl.* **2016**, *48*, 3033–3040. [CrossRef]

16. Guo, B.; Huang, Y.; Peng, F.; Guo, Y.; Zhu, J. Analytical Modeling of Manufacturing Imperfections in Double-Rotor Axial Flux PM Machines: Effects on Back EMF. *IEEE Trans. Magnet.* **2017**, *53*. [CrossRef]

17. Barmpatza, A.C.; Kappatou, J.C.; Skarmoutsos, G.A. Investigation of Static Angular and Axis Misalignment in an AFPM Generator. In Proceedings of the 2019 IEEE Workshop on Electrical Machines Design, Control and Diagnosis (WEMDCD), Athens, Greece, 22–23 April 2019.

18. Barmpatza, A.C.; Kappatou, J.C. A Study of Static Angular and Axis Eccentricity in a Double-Sided Rotor AFPM Generator using 3D-FEM. In Proceedings of the 2019 IEEE International Symposium on Diagnostics for Electrical Machines, Power Electronics and Drives (SDEMPED), Toulouse, France, 27–30 August 2019.

19. Di Gerlando, A.; Foglia, G.M.; Iacchetti, M.F.; Perini, R. Effects of Manufacturing Imperfections in Concentrated Coil Axial Flux PM Machines: Evaluation and Tests. *IEEE Trans. Ind. Electron.* **2014**, *61*, 5012–5024. [CrossRef]

20. Guo, B.; Huang, Y.; Peng, F.; Dong, J.; Li, Y. Analytical Modelling of Misalignment in Axial Flux Permanent Magnet Machine. *IEEE Trans. Ind. Electron* **2020**, *67*, 4433–4443. [CrossRef]

21. Ajily, E.; Ardebili, M.; Abbaszadeh, K. Magnet defect and rotor eccentricity modeling in axial-flux permanent-magnet machines via 3-D Field Reconstruction Method. *IEEE Trans. Energy Convers.* **2016**, *31*, 486–495. [CrossRef]

22. Guo, B.; Huang, Y.; Peng, F.; Dong, J. General Analytical Modeling for Magnet Demagnetization in Surface Mounted Permanent Magnet Machines. *IEEE Trans. Ind. Electron.* **2019**, *66*, 5830–5838. [CrossRef]

23. Verkroost, L.; De Bisschop, J.; Vansompel, H.; De Belie, F.; Sergeant, P. Active Demagnetization Fault Compensation for Axial Flux Permanent Magnet Synchronous Machines Using an Analytical Inverse Model. *IEEE Trans. Energy Convers.* **2020**, *35*, 591–599. [CrossRef]

24. De Bisschop, J.; Abou-Elyazied Abdallh, A.; Sergeant, P.; Dupré, L. Analysis and selection of harmonics sensitive to demagnetisation faults intended for condition monitoring of double rotor axial flux permanent magnet synchronous machines. *IET Electr. Power Appl.* **2018**, *12*, 486–493. [CrossRef]

25. De Bisschop, J.; Abdallh, A.; Sergeant, P.; Dupré, L. Identification of demagnetization faults in axial flux permanent magnet synchronous machines using an inverse problem coupled with an analytical model. *IEEE Trans. Magnet.* **2014**, *50*. [CrossRef]

26. Saavedra, H.; Riba, J.-R.; Romeral, L. Magnet shape influence on the performance of AFPMM with demagnetization. In Proceedings of the 2013 Annual Conference of the IEEE Industrial Electronics Society (IECON), Vienna, Austria, 10–13 November 2013.

27. Bahador, N.; Darabi, A.; Hasanabadi, H. Demagnetization analysis of axial flux permanent magnet motor under three phase short circuit fault. In Proceedings of the 2013 Annual International Power Electronics, Drive Systems and Technologies Conference, Tehran, Iran, 13–14 February 2013.

28. De Bisschop, J.; Vansompel, H.; Sergeant, P.; Dupré, L. Demagnetization Fault Detection in Axial Flux PM Machines by Using Sensing Coils and an Analytical Model. *IEEE Trans. Magnet.* **2017**, *53*. [CrossRef]

29. Barmpatza, A.C.; Kappatou, J.C. Demagnetization Faults Detection in an Axial Flux Permanent Magnet Synchronous Generator. In Proceedings of the 2017 IEEE International Symposium on Diagnostics for Electrical Machines, Power Electronics and Drives (SDEMPED), Tinos, Greece, 29 August–1 September 2017.

30. Barmpatza, A.C.; Kappatou, J.C. Study of the Demagnetization Fault in an AFPM Machine in Relation with the Magnet Location. In Proceedings of the 2018 International Conference on Electrical Machines (ICEM), Alexandroupoli, Greece, 3–6 September 2018.

31. De Bisschop, J.; Sergeant, P.; Hemeida, A.; Vansompel, H.; Dupré, L. Analytical Model for Combined Study of Magnet Demagnetization and Eccentricity Defects in Axial Flux Permanent Magnet Synchronous Machines. *IEEE Trans. Magnet.* **2017**, *53*. [CrossRef]

32. Kappatou, J.C.; Zalokostas, J.D.; Spyratos, D.A. 3-D FEM Analysis, Prototyping and Tests of an Axial Flux Permanent-Magnet Wind Generator. *Energies* **2017**, *10*, 1269. [CrossRef]

33. Radwan-Praglowska, N.; Wegiel, T.; Borkowski, D. Parameters Identification of Coreless Axial Flux Permanent Magnet Generator. *Arch. Electr. Eng.* **2018**, *62*, 391–402. [CrossRef]

34. Urresty, J.C.; Riba, J.-R.; Romeral, L. A Back-emf based method to detect magnet failures in PMSMs. *IEEE Trans. Magnet.* **2013**, *49*, 591–598. [CrossRef]

35. Urresty, J.C.; Riba, J.-R.; Romeral, L. Influence of the stator windings configuration in the currents and Zero-Sequence voltage harmonics in permanent magnet synchronous motors with demagnetization faults. *IEEE Trans. Magnet.* **2013**, *49*, 4885–4893. [CrossRef]

36. Goktas, T.; Zafarani, M.; Akin, B. Discernment of broken magnet and static eccentricity faults in permanent magnet synchronous motors. *IEEE Trans. Energy Convers.* **2016**, *31*, 578–587. [CrossRef]

37. Barmpatza, A.C.; Kappatou, J.C. Study of the Total Demagnetization Fault of an AFPM Wind Generator. *IEEE Trans. Energy Convers.* **2020**. [CrossRef]

Publisher's Note: MDPI stays neutral with regard to jurisdictional claims in published maps and institutional affiliations.

Article

Parameter Estimation of Inter-Laminar Fault-Region in Laminated Sheets Through Inverse Approach

Osaruyi Osemwinyen [1,*], Ahmed Hemeida [2], Floran Martin [1], Anouar Belahcen [1] and Antero Arkkio [1]

[1] Department of Electrical Engineering and Automation, Aalto University, 02150 Espoo, Finland; floran.martin@aalto.fi (F.M.); anouar.belahcen@aalto.fi (A.B.); antero.arkkio@aalto.fi (A.A.)

[2] Electrical Power and Engineering Department, Cairo University, Cairo 12613, Egypt; A.hemeida@cu.edu.eg

* Correspondence: osaruyi.osemwinyen@aalto.fi

Received: 1 June 2020; Accepted: 19 June 2020; Published: 23 June 2020

Abstract: Estimating the additional power losses caused by an inter-laminar short circuit in electromagnetic devices using thermal measurements depends on many parameters such as thermal conductivity, heat capacity, convective heat coefficient, and size of the fault points. This paper presents a method for estimating these parameters using experimental measurement and a numerical model. The surface temperature rise due to inter-laminar short circuit fault was obtained using an infrared camera. Based on the initial temperature rise method, the least square non-linear approximation technique was used to determine the best fitting parameters of the fault region from the numerical model. To validate the results obtained, the fault region temperature rise and the total loss of the experimental sample were compared with the numerical model using the obtained parameters for different current supply conditions. The study shows that surface temperature distribution can be used to estimate the inter-laminar short circuit fault parameters and localized losses.

Keywords: fault size; inter-laminar fault; localized losses; thermographic measurement; thermal-electric coupling

---

1. Introduction

The manufacturing process of electrical machines that includes punching/cutting, core assembly, and welding induces mechanical and thermal stresses to the electrical steel sheet. The induced stresses cause degradation of the magnetic properties of the laminated steel sheet around the cutting edge and the welded region [1–5], which has a direct impact on the core losses of the final manufactured electrical machine. Furthermore, the mechanical stress during the punching process inevitably causes burr on the cut edge of the steel, when the sheets are stacked and welded together to form the machine core, can lead to the formation of an inter-laminar short circuit fault between laminations. Since the core is subjected to a time-varying magnetic field during operation, circulating eddy currents are induced around the short circuit region. These currents create an additional localized power loss around the short circuit region and if this electrical shorting covers several laminations, high currents can circulate leading to a significant increase in the power loss and excessive localized heating. In the absence of adequate cooling, this phenomenon might lead to an insulation breakdown of the machine laminations. Thus, it causes the potential of a complete machine failure.

In order to improve the efficiency of the manufacturing process and reduce the losses in the machine, it is important to study and accurately measure the power loss associated with the inter-laminar short circuit fault in electrical machines. Researchers have used analytical, numerical and experimental methods to study the inter-laminar faults and investigate their effects on the core loss of electrical machines in recent years. Wang and Zhang in [6] analyzed the effect of weld bead radius on eddy current distribution in laminated electrical steels and they concluded that the eddy current losses increase as the weld radius increases. The impact of inter-laminar short circuit and fault positions on eddy current losses was also studied using experimental measurements in [7,8]. The results showed that the fault position and the number of shorted laminations significantly affected the total core losses. However, the position, size, and number of short circuit laminations in the core of electrical machines are generally stochastic in nature, this makes modeling and measuring of losses associated with inter-laminar short circuit faults in electrical machines difficult.

Another approach used in [9–12] to determine power loss in electrical machines, is the initial rate of the temperature rise method. This approach is based on the principle that losses generated in the core of an electrical machine contribute directly to the temperature rise of the core. Hence, by measuring the temperature rise during the instant of heat generation, the core loss distribution can be directly determined. However, to obtain accurate results, the temperature rise must be measured within 5–10 s of the power switch on [10]. In [9], a thermographic imaging technique was used based on the initial temperature rise to directly measure the core losses in electrical steel sheets. Their analysis showed that the material properties also affect the accuracy of the results obtained.

In this paper, a contactless method for estimating the localized losses due to inter-laminar short circuit faults is developed based on inverse approximation techniques. The approach used in this research work is based on both experimental measurements and numerical modeling. First, the initial temperature rise of the hotspot formed by inter-laminar contact is measured using a thermal camera. The detailed explanation of the experimental sample and measurement setup is given in Section 2. To estimate the parameters of the fault region, a 3D finite element model of a coupled electric field and heat transfer is developed using COMSOL Multi-physics in Section 3. Finally, an inverse approach is used to obtain the fault size, thermal conductivity, electrical conductivity, and convective coefficient of the faulty region using the least square non-linear method to minimize the error between the measured and simulated surface temperature rise. Analysis of the results obtained shows that the simulated hotspot temperature rise from the numerical model using the estimated parameters is in agreement with the experimental measurements for different current supply conditions.

2. Experimental Measurement

In this section, the experimental measurement setup used in the study of inter-laminar short circuit fault is described together with the thermal camera measuring procedure.

2.1. Measurement Setup

A stack of two (26 × 40 mm) non-oriented laminated electrical sheets were used for the experimental study. The sheet thickness is 0.5 mm and the material grade is M450-50A and it is semi-processed. The sheets are cut to be burr free. Hence, there will be no galvanic contact between them when stacked together. To model inter-laminar faults between the laminations, laser beam spot welding is used to create a force contact on one end of the stacked laminations. The assembly of the 2-sheets is done carefully and without additional forces, so that the only place where the insulation could be damaged is in the proximity of the welding. Figure 1 shows the sample used in the experimental study with a magnified image of the fault region when observed from the top using a magnifying camera IMPERX GEV-B1922C-SC000.

To perform the experimental measurements, the sample is energized with a DC current supplied through copper wires attached to the other end of the sample. The temperature rise distribution of the fault region is recorded during the current switch on, using an infrared thermal camera. The sample total loss was determined from the measured voltage and input current at the supply spot on the sample during the switch on period. The thermal measurements are described in the next subsection.

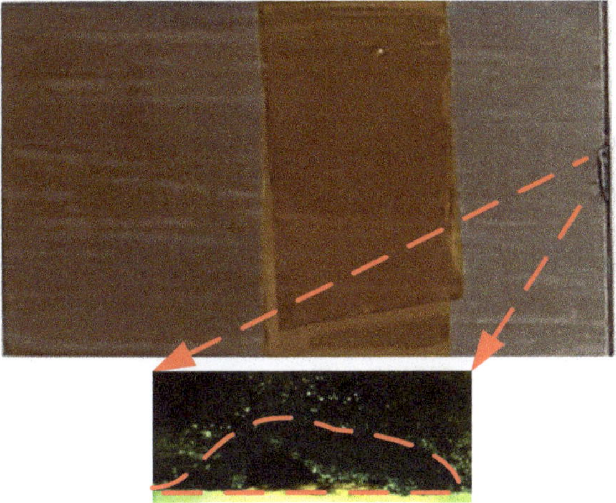

Figure 1. Experimental sample with the magnified image of the fault region.

2.2. Thermal Measurement

The thermal measurement setup used for the experimental study is shown in Figure 2. The measured sample is placed inside a black cardboard box to reduce the reflective temperature effect from surrounding objects. An infrared camera is used to measure the surface temperature distribution of the fault region at the instant of the power switch. The infrared camera specifications used for the measurements are given in Table 1.

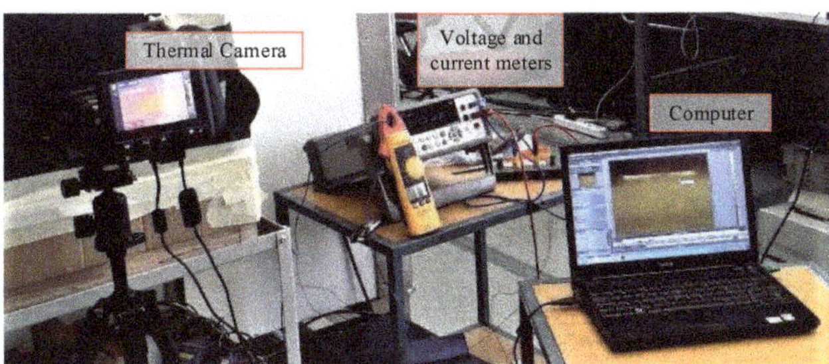

Figure 2. Thermal-electric measurement setup for investigation of inter-laminar short circuit fault.

Table 1. FLIR T640 thermal camera specifications.

Frame rate	30 Hz
Temperature range	233.15–2273 K
Accuracy	$\pm$2% of reading
Field of view	32 $\times$ 24 mm
Detector type	uncooled microbolometer (640 $\times$ 480 pixels)

3. Thermal-Electric Finite Element (FE) Model

A 3D time-dependent coupled electric and heat transfer simulation was carried out using COMSOL. To simplify the geometry of the fault region in the modeled sample, a semi-cylindrical shape was used to represent the fault region as shown in the model geometry description in Figure 3 and the physics used in the model is derived below.

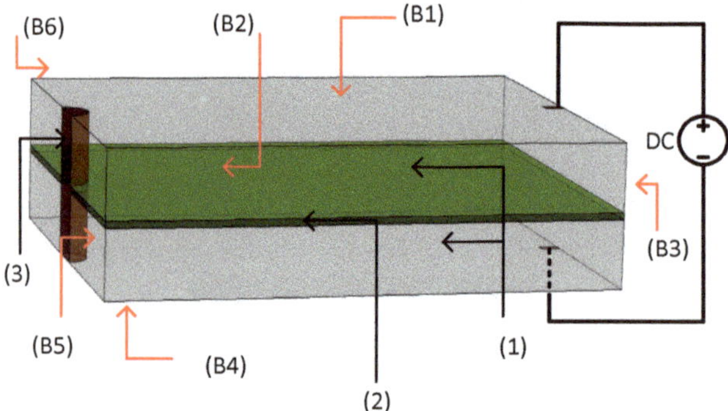

Figure 3. Geometry description: Lamination (**1**), insulation (**2**), fault region (**3**), thermal boundary condition (**B1–B6**), and DC supply.

When DC current is applied to the sample, steady current starts to flow inside the lamination and through the short circuit path, as shown in Figure 4.

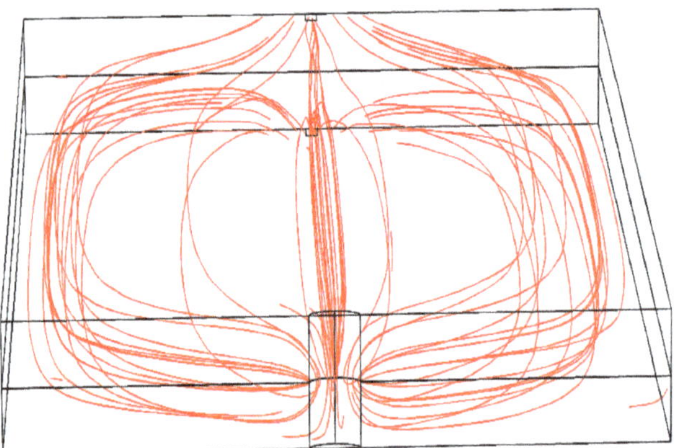

Figure 4. Current flowing inside the lamination and through the short circuit region.

The resistive loss distribution p_{gen} in (W/m^3) generated inside the lamination and in the fault region is computed using (1) and coupled with the heat transfer study as the source.

$$p_{\text{gen}} = \mathbf{I}^2 \left(\frac{R_{\text{weld}}}{v_{\text{weld}}} + \frac{R}{v} \right) = \mathbf{J}^2 \left(\frac{1}{\sigma_{\text{weld}}} + \frac{1}{\sigma} \right) \tag{1}$$

where, $\mathbf{I}$ is the applied current, R_{weld} is the inter-laminar short circuit fault resistance, R, is the resistance of the remaining part of the sample, σ_{weld} is the electrical conductivity of the fault region, σ is the electrical conductivity of the sample, v_{weld} is the volume of the fault region, v is the volume of the remaining part of the sample, and $\mathbf{J}$ is the current density. The coupled solution for the temperature distribution is derived from the 3D general heat equation expressed as shown in (2), where p_{gen} is the heat source obtained from the electric field solution in (1).

$$\rho C_p \frac{\partial T}{\partial t} + \nabla \cdot \mathbf{q} = p_{gen} \tag{2}$$

$$\mathbf{q} = -k\nabla T \tag{3}$$

$$\mathbf{n} \cdot \mathbf{q} = h_{conv}\left(T - T_{ext}\right) \tag{4}$$

where, ρ, C_p, k, T, T_{ext}, h_{conv}, $\mathbf{n}$, and $\mathbf{q}$, are the mass density, specific heat capacity, thermal conductivity, temperature, the initial temperature of the sample, convective heat coefficient, the surface normal to the surrounding, and heat flux of the material respectively. Hence, to solve the heat Equation (2) the boundary condition and initial conditions of the model sample must be specified. Convective heat flux defined by (4) is assigned to all open boundaries (B1–B6), as shown in Figure 3. The coupled model is simulated using a 1 mm fault size and known material properties for 10 s. The temperature distribution of the simulated model is shown in Figure 5.

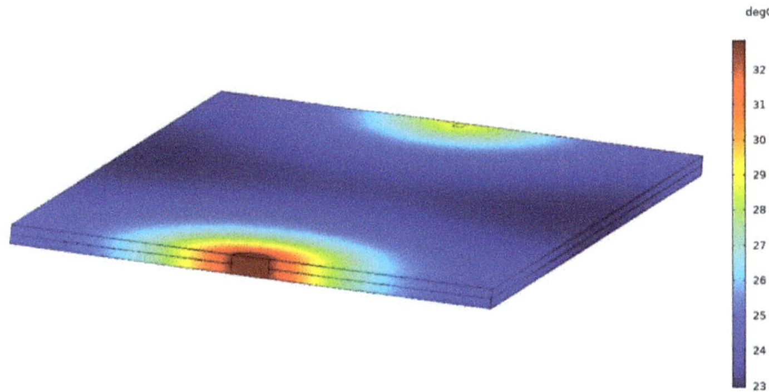

Figure 5. Simulated temperature distribution of the sample at 10 s.

Further analysis of the temperature rise in the fault zone shows that parameters such as thermal conductivity, electrical conductivity, fault size, and convective heat transfer to the surrounding affect the temperature distribution around that region. Therefore, to determine the actual parameters of the experimental sample, an inverse model is developed to fit the simulated temperature rise from the fault region to the measured temperature rise.

4. The Inverse Approach

The target of the inverse problem is to predict the size of the fault region and also to determine the material properties such as convective heat coefficient, thermal conductivity and electric conductivity of the faulty region, which are essential for accurate estimation of the localized loss through temperature measurements. The methodology used in this research is presented in the model algorithm flowchart in Figure 6. The model takes as an input the average surface temperature rise on the fault area and the measured voltage from the forward FE model and experimental sample. Using the least square non-linear approximation method, the error between the measured and simulated inputs is minimized iteratively over time until the best fit parameters for the fault region are obtained.

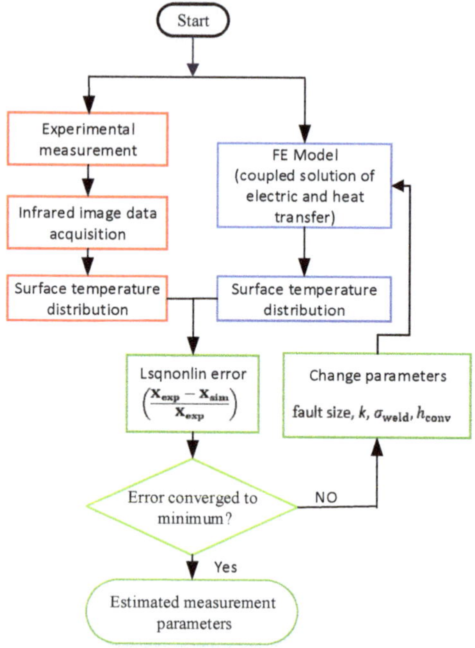

Figure 6. Inverse model algorithm flowchart.

5. Results and Discussion

In this study, two samples of different fault widths were measured. The faults were imposed on the samples by using different laser beam weld widths. The measured sample surface was painted black to improve the accuracy of the measured temperature rise. The infrared camera is calibrated for the measurements, as shown in Table 2. A DC current of 20 A is first supplied to each of the experimental samples at ambient temperature conditions.

Table 2. Infrared camera calibration.

Emissivity	0.96
Measurement distance	0.09 m
Atmospheric temperature	20 °C
Relative humidity	50

The voltage drop over the sample is measured using a voltmeter and the surface temperature distribution on the fault region is obtained with an infrared camera. The thermal image of the measured sample surface at 8 s after power was switched on is shown in Figure 7. From the image, it was observed that current streaming through the fault region creates high loss density, which results in the formation of a hotspot around the short circuit. Since we are interested in the hotspot region (area enclosed in the white dash box), the rise in temperature of the fault region is obtained and used in the inverse algorithm.

In the parameter search, to reduce the number of iterations, a sensitivity test was carried out and to see the effect of each parameter on the measured voltage and temperature using initial known material properties on the sample. Based on the sensitivity, lower and upper bounds were fixed for each parameter in the final fitting process The best fit parameters for each of the model samples obtained from the approximation are shown in Table 3 below.

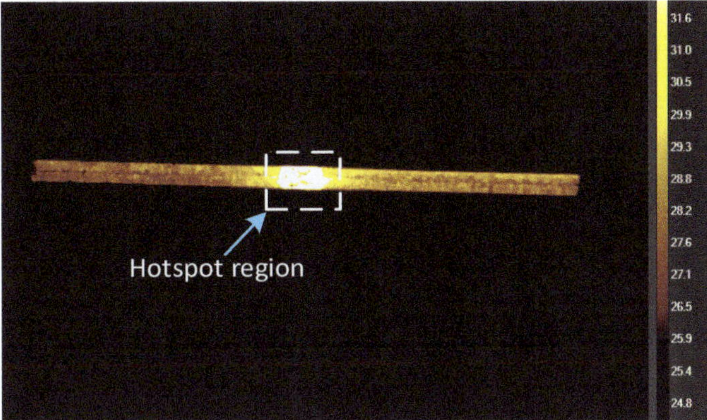

Figure 7. Surface temperature distribution from infrared camera at 8 s.

Table 3. Estimated parameters of the fault region obtained from the Lsqnonlin approximation.

Parameters	Sample 1	Sample 2
Fault Width [mm]	2.2	3.0
Thermal conductivity, k [W/mK]	27.7	27.7
Convective heat coefficient, h_{conv} [W/m²K]	15	15
Electric conductivity, σ_{weld} [S/m]	2.994×10^5	2.994×10^5

To demonstrate the accuracy of the results obtained from the inverse approximation in Table 3, the size of the predicted fault region was compared to the size of the experimental sample fault area observed with a magnifying camera with a standard measurement scale for each modeled sample. Figure 8 shows the top view image of the experimental sample observed with a magnifying camera and the predicted fault region used in the inverse optimization solution. Comparing the scale measurements of the experimental fault regions represented by the red contour lines as shown in Figures 8a,b with the predicted fault shown in Figure 8c,d, it is observed that the fault width along the surface of the sheet is predicted correctly with a less than seven percent error. However, less accuracy was obtained in predicting the fault depth, but usually for inter-laminar faults, this depth is less than the thickness of the laminated sheet.

Furthermore, two other measurements are performed at 10 and 30 A current supplies to validate the parameters obtained from the least-square non-linear optimization. The voltage drop over the sample, and the surface temperature rise of the fault region were obtained for each supply case and compared with the results from the simulated model. First, the total resistive losses in the measured sample were computed and compared to the total loss from the simulated model for each of the studied samples, as shown in Table 4.

The results obtained show that the predicted total resistive loss closely follows the measured loss, with a maximum deviation of 8.4 percent. The deviation is assumed to be caused by non-uniformity of the actual fault compared to the modeled fault and the fact that the convection coefficient slightly depends on the temperature difference between the sample and the ambient, i.e., it is a slightly non-linear term.

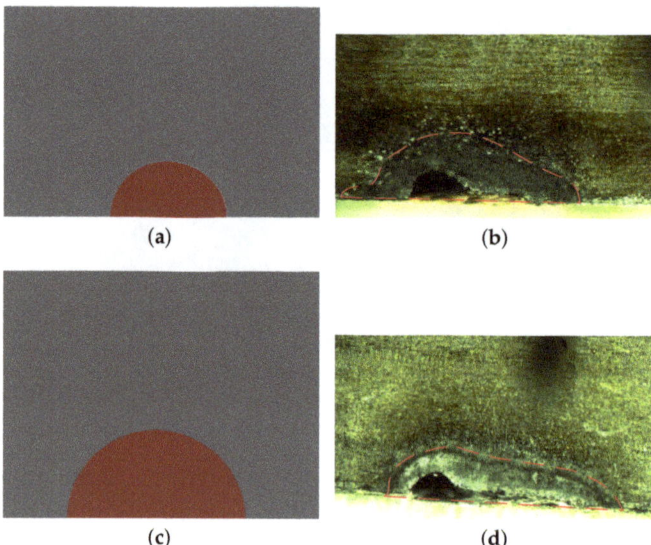

Figure 8. Comparing predicted fault with experimental sample. (**a**) Model sample with 2.2 mm fault width. (**b**) Measured sample with 2.06 mm fault width. (**c**) Model sample with 3.0 mm fault width. (**d**) Measured sample with 2.82 mm fault width.

Table 4. Total restive loss comparison for the studied samples. (**a**) Sample 1 with fault width of 2.2 mm (numbers in bold represent the test measurement used to obtain the fault parameters). (**b**) Sample 2 with fault width of 3.0 mm.

(a)			
Current [A]	Measured Loss [W]	Predicted Loss [W]	Difference [%]
10	0.39	0.41	5.12
20	**1.6**	**1.7**	**6.2**
30	3.57	3.87	8.4
(b)			
Current [A]	Measured Loss [W]	Predicted Loss [W]	Difference [%]
10	0.37	0.39	5.4
20	**1.51**	**1.56**	**3.3**
30	3.42	3.72	8.77

Finally, the measured surface temperature rise from the fault region is compared with the predicted temperature rise for each current supply case in Figure 9. The results show that for each supply case, the predicted temperature rise closely follows the measured temperature rise for the two studied samples.

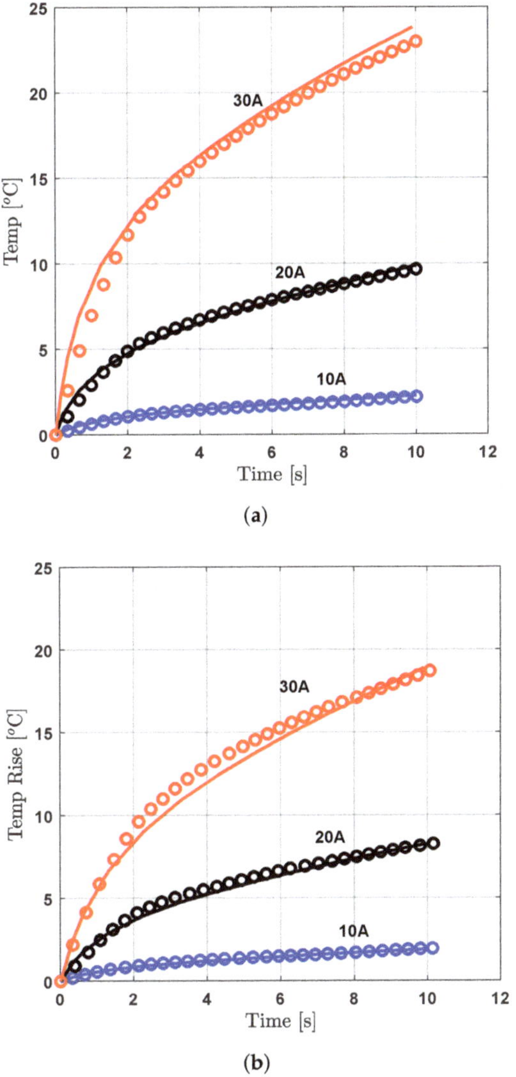

Figure 9. Measured (o) and predicted (-) temperature rise comparison for different current supplies. (**a**) Sample 1 with fault width of 2.2 mm. (**b**) Sample 2 with fault width of 3.0 mm.

6. Conclusions

In this paper, a method for estimating the sizes and localized losses due to inter-laminar short faults in laminated steel sheets has been developed based on the surface temperature measurement and numerical modeling. This method has been applied to determine the fault parameters on two stacks of electrical steel sheets with an artificial weld contact to represent inter-laminar short circuit fault. Experimental measurement of surface temperature rise using an infrared camera is conducted on two samples with different fault sizes under a DC power supply condition. An inverse modeling technique is used to obtain the best fitting parameter of the fault region from the coupled numerical model. The parameters obtained from the inverse approximation are validated by comparing the measured temperature rise of the fault region and total loss with the numerical model for different DC current supply conditions. The analysis of the results obtained showed that the material properties of the fault region are closer to prior known material properties of the sample used and the fault size depends on the number of laminations and the length of fault. However, these parameters affect the temperature rise of the fault region. Therefore, the technique developed for obtaining the fault parameters can be generalized for studying inter-laminar short faults in electrical machines. Also, the visual image of the hotspot region can be used for fast detection of inter-laminar short circuit faults in the core of electrical machines.

Author Contributions: Conceptualization, O.O.; funding acquisition, A.A.; investigation, O.O.; methodology, O.O. and A.H.; supervision, A.B.; writing—original draft, O.O.; writing—review and editing, A.H., F.M. and A.B. All authors have read and agreed to the published version of the manuscript.

Funding: The research leading to these results has received funding from the European Research Council under the European Unions Seventh Framework Programme (FP7/2007-2013) ERC grant agreement no 339380.

Conflicts of Interest: The authors declare no conflict of interest.

References

1. Harstick, H.M.S.; Ritter, M.; Riehemann, W. Influence of Punching and Tool Wear on the Magnetic Properties of Nonoriented Electrical Steel. *IEEE Trans. Magn.* **2014**, *50*, 1–4. [CrossRef]
2. Siebert, R.; Schneider, J.; Beyer, E. Laser Cutting and Mechanical Cutting of Electrical Steels and its Effect on the Magnetic Properties. *IEEE Trans. Magn.* **2014**, *50*, 1–4. [CrossRef]
3. Kuo, S.; Lee, W.; Lin, S.; Lu, C. The Influence of Cutting Edge Deformations on Magnetic Performance Degradation of Electrical Steel. *IEEE Trans. Ind. Appl.* **2015**, *51*, 4357–4363. [CrossRef]
4. Schoppa, A.; chneider, J.; Wuppermann, C.D. Influence of the manufacturing process on the magnetic properties of non-oriented electrical steels. *J. Magn. Magn. Mater.* **2000**, *215–216*, 74–78. [CrossRef]
5. Dems, M.; Komeza, K.; Kubiak, W.; Szulakowski, J. Impact of Core Sheet Cutting Method on Parameters of Induction Motors. *Energies* **2020**, *13*, 1960, doi:10.3390/en13081960. [CrossRef]
6. Wang, H.; Zhang, Y. Modeling of Eddy-Current Losses of Welded Laminated Electrical Steels. *IEEE Trans. Ind. Electron.* **2017**, *64*, 2992–3000. [CrossRef]
7. Eldieb, A.; Anayi, F. Evaluation of Loss Generated by Edge Burrs in Electrical Steels. *IEEE Trans. Magn.* **2016**, *52*, 1–4. [CrossRef]
8. Hamzehbahmani, H.; Anderson, P.; Jenkins, K.; Lindenmo, M. Experimental study on inter-laminar short-circuit faults at random positions in laminated magnetic cores. *IET Electric Power Appl.* **2016**, *10*, 604–613. [CrossRef]
9. Shimoji, H.; Borkowski, B.E.; Todaka, T.; Enokizono, M. Measurement of Core-Loss Distribution Using Thermography. *IEEE Trans. Magn.* **2011**, *47*, 4372–4375. [CrossRef]
10. Gilbert, A.J. A method of measuring loss distribution in electrical machines. *Proc. IEE Part A Power Eng.* **1961**, *108*, 239–244. [CrossRef]
11. Hamzehbahmani, H.; Moses, A.J.; Anayi, F.J. Opportunities and Precautions in Measurement of Power Loss in Electrical Steel Laminations Using the Initial Rate of Rise of Temperature Method. *IEEE Trans. Magn.* **2013**, *49*, 1264–1273. [CrossRef]
12. Bousbaine, A. A thermometric approach to the determination of iron losses in single phase induction motors. *IEEE Trans. Energy Convers.* **1999**, *14*, 277–283. [CrossRef]

Review

Advances in Power Quality Analysis Techniques for Electrical Machines and Drives: A Review

Artvin-Darien Gonzalez-Abreu [1], Roque-Alfredo Osornio-Rios [1,*], Arturo-Yosimar Jaen-Cuellar [1], Miguel Delgado-Prieto [2], Jose-Alfonso Antonino-Daviu [3] and Athanasios Karlis [4]

[1] HSPdigital CA-Mecatronica, Engineering Faculty, Autonomous University of Queretaro, San Juan del Rio 76807, Mexico; agonzalez63@alumnos.uaq.mx (A.-D.G.-A.); ayjaen@hspdigital.org (A.-Y.J.-C.)

[2] MCIA Research Center Department of Electronic Engineering, Technical University of Catalonia (UPC), 08034 Barcelona, Spain; miguel.delgado@mcia.upc.edu

[3] Instituto Tecnológico de la Energía, Universitat Politècnica de València (UPV), Camino de Vera s/n, 46022 Valencia, Spain; joanda@die.upv.es

[4] Department of Electrical & Computer Engineering, Democritus University of Thrace, 69100 Komotini, Greece; akarlis@ee.duth.gr

* Correspondence: raosornio@hspdigital.org

Abstract: The electric machines are the elements most used at an industry level, and they represent the major power consumption of the productive processes. Particularly speaking, among all electric machines, the motors and their drives play a key role since they literally allow the motion interchange in the industrial processes; it could be said that they are the medullar column for moving the rest of the mechanical parts. Hence, their proper operation must be guaranteed in order to raise, as much as possible, their efficiency, and, as consequence, bring out the economic benefits. This review presents a general overview of the reported works that address the efficiency topic in motors and drives and in the power quality of the electric grid. This study speaks about the relationship existing between the motors and drives that induces electric disturbances into the grid, affecting its power quality, and also how these power disturbances present in the electrical network adversely affect, in turn, the motors and drives. In addition, the reported techniques that tackle the detection, classification, and mitigations of power quality disturbances are discussed. Additionally, several works are reviewed in order to present the panorama that show the evolution and advances in the techniques and tendencies in both senses: motors and drives affecting the power source quality and the power quality disturbances affecting the efficiency of motors and drives. A discussion of trends in techniques and future work about power quality analysis from the motors and drives efficiency viewpoint is provided. Finally, some prompts are made about alternative methods that could help in overcome the gaps until now detected in the reported approaches referring to the detection, classification and mitigation of power disturbances with views toward the improvement of the efficiency of motors and drives.

Keywords: electrical drives; electrical machines; energy efficiency; energy-saving; induction motor; power quality

Citation: Gonzalez-Abreu, A.-D.; Osornio-Rios, R.-A.; Jaen-Cuellar, A.-Y.; Delgado-Prieto, M.; Antonino-Daviu, J.-A.; Karlis, A. Advances in Power Quality Analysis Techniques for Electrical Machines and Drives: A Review. *Energies* **2022**, *15*, 1909. https://doi.org/10.3390/en15051909

Academic Editors: Saeed Golestan and Ryszard Palka

Received: 18 February 2022
Accepted: 1 March 2022
Published: 5 March 2022

Publisher's Note: MDPI stays neutral with regard to jurisdictional claims in published maps and institutional affiliations.

1. Introduction

The energy conversion through electrical and electromechanical machines allows for performing a wide variety of man activities that were considered complex to be carried out by themselves. These devices are installed widespread around the world and, according to several authors in the literature, they consume between 60% and 80% of the total energy in the industrial sector [1,2]. Most of the machines used in the industrial processes are the electric motors, which transform the electrical energy nature, whether continuous or alternating, into a mechanical one, also known as kinetic energy generation, ensuring the movement on an output shaft. The electric motors are coupled to a mechanical ensemble for generating motion, rotational or linear, in purposes such as: pushing, heating, pumping, transporting, among others. In order to carry out the aforementioned actions, a necessary element that has been integrated with the electrical machines, the drive, is required, which is the system used for controlling the motion of the electric motors. The purpose of a drive is to adjust the output parameters of the motor, such as the speed, through variations in voltage or frequency [3]. Thus, the electric drive is the linkage between the mechanical and the electrical engineering. A typical drive system is assembled with an electric motor and a sophisticated controller unit that manipulates the rotation of the motor shaft. This control can be carried out quickly with the help of hardware and software.

Despite of being the more recurrent equipment for controlling the industrial machines, both the electric motor and its drive cause adverse effects to the electrical grid by inducing power disturbances to it [4]. For instance, a motor startup could generate voltage disturbances such as sags, swells, and flickers in weak power systems. In addition to this, the drives induce harmonic and inter-harmonic content during the motor feeding when the frequency is variated [5,6]. In counterpart, in this regard it must be highlighted that a poor power quality, in turn, affects the normal operation of the motors and drives, causing equipment malfunctioning, failures, or even worse, irreparable damage [7]. Whenever a machine transforms energy from one form to another, and this combined with power quality disturbances in the electric grid yields an unavoidable loss in the equipment, it is normally manifest as an increase in the temperature and an efficiency reduction [8,9]. Therefore, since the electrical machines use a significant part of the total electric power generated worldwide and its performance impact directly in the productivity costs, any improvement in its operation and control that increases its efficiency will have a meaningful impact [10–13].

Due to the abovementioned points, the power quality monitoring represents an essential aspect to consider in today's electrical environments or power grids. As a matter of fact, the critical aspect to be considered is the relationship between electrical motors and drives with the power quality. Indeed, several methodologies have been developed for detecting faults and identifying, classifying, mitigating, and suppressing power quality disturbances [14]. The important points related to the employ of such techniques address: (i) the analysis of the effects produced by the power quality disturbances (PQDs) on electrical devices o machinery, (ii) the parameters involved with the disturbance generation in the electrical grid, and (iii) the proper action to be taken once the electrical phenomenon has occurred. Therefore, it is important to conduct an exhaustive review of the reported works in two main aspects; those studies that focus on techniques developed and applied to detect, classify, and mitigate electrical events or power disturbances, and those investigations that attend what has been carried out regarding how poor power quality affect the electric motors and drives and reduces their efficiency.

Regarding to the existing electric machine technologies, a generalized classification can be made according to [3,15]. This cataloguing concerns to the form of the power supply and applies for both electrical and electromechanical machines such as motors, drives, transformers, etc. Two main branches can be considered being direct current (DC) and alternating current (AC) electric machines, and from these other subcategories are derived. In one hand, the case of direct current machines consists on DC generators and DC motors. On the other hand, for the case of alternating current machines there exist synchronous and asynchronous electromechanical devices. In a similar way, as in direct current, the synchronous machines are divided into AC generators and AC motors. Meanwhile, asynchronous technologies involve only induction machines. Apart from this, other categorization is made from the standpoint of performance losses, in this the electrical machines may be divided into two groups: those with rotary parts (motors, generators), and those with static parts (transformers, reactors). Under this point of view, the electrical and mechanical losses are produced in rotating machines, whereas only electrical losses are produced in stationary machines. Finally, another classification can be made by takin into account if the machine uses single-phase or three-phase alternating current (AC) supply [16]. In addition to that, it is worth noticing that the motor drives can be classified according to the prime mover they handle, such as electric motors, diesel or petrol engines, gas or steam turbines, steam engines, and hydraulic motors [17].

From the aforementioned classifications, the asynchronous or induction motor together with its electrical drive are the most widely used in industry, the reason for which they are going to be selected for analysis in this review. The main advantage of the induction motor is that it eliminates all sliding contacts, resulting in an exceedingly simple and rugged construction. Moreover, the rapid development of new induction machines and the emerging of power drive technologies in the past few decades, ranging from a few watts to many megawatts [17], enables them to be used in many fields involving conversion processes [18], whether in the generation, transmission, or electrical energy consumption [19]. Therefore, the electrical motors and drives are used in industrial, commercial, and domestic applications such as transportation systems [20,21], rolling mills [22], paper machines [23], textile mills [24], machine tools [25], pumps [26], robots [27], fans [28], and vehicle propulsion [29], among others [30].

In relation to the applications of motors and drives some examples are described next. Some of the most recent studies on electrical machines are focused on the new applications for industries equipment supplying and that can be beneficial for the environmental issues by using more efficiently motors and drives an combining them with emerging technologies [16,31–33]. For instance, the development of electric vehicles by improving their motors for driving, transportation, and mobility applications [34]. Regarding the electrical drives, many studies have been conducted in areas such as high-speed rotating mechanical machinery [35]. Concerning to the power generation topic, the efforts look for developing electric machines to be the element that allows a clean and efficient generation of energy [36]. Currently, two important aspects are currently being addressed: the best usage in the conversion of energy by electrical machines and at the same time heed that the use of these devices does not introduce anomalies to the electrical network. These considerations are being sought from regulatory points of view. An example of the above mentioned is the power factor regulation by using capacitive or inductive elements depending on the case. This power factor is penalized by electrical regulatory agencies.

Some international organizations such as NEMA, ANSI, or IEEE define the standards and fix the tolerances for the operational parameters for electrical machines [37]. These standards specify power, speed, voltage, and operating frequency ranges in order to guarantee that the power source is as pure as possible, which is known as Power Quality (PQ) [38]. Nowadays, the tendency for electrical machines is to be more efficient, to require less maintenance, to have high power density, robustness, and applicability in different areas [39,40]. Some investigations present the central energy efficiency-related regulations, the most applicable efficiency increasing technical solutions, and the possibility of replacing the most widely used squirrel cage induction machines with more efficient variants. However, the industrial power supply is typically contaminated due to all the loads connected to the grid, and also their non-linear behavior because of its integrated elements that inject power quality disturbances (PQD) such as noise, sags, swells, interruptions, flicker, harmonics and inter-harmonic content affecting the PQ [41]. In the end, these PQ affectations are also reflected back on the electrical machines by decreasing their efficiencies, provoking malfunctioning and damage to their components [42]. Power electronics are an important part of any power conversion system. Notwithstanding, these devices have a non-linear behavior and generate PQ issues that must be addressed [43]. All in all, monitoring PQ is not an easy task because measurements devices are expensive, and it is financially impractical to monitor every segment of a power network [44]. Additionally, another aspect to consider is that power signals are seldom stationary and the nonstationary nature of waveforms could corrupt the spectrum analysis results [45]. Among the main parameters of an efficient power supply system are its reliability and its quality; moreover, it is aimed to have the possible shortness time after a failure. The monitoring systems of the power grid have areas for improvement [46]. Energy saving is taken into account by institutions, companies, and industry, promoting the best use of electrical machines [47].

In the case of domestic commercial applications in buildings and residential installations, several works have studied how they impact mainly in the energy consumption, energy saving, and energy management. For example, there are studies dedicated to analyzing and estimate the consumption of energy in constructions, residentials and publics, due to the common commercial equipment. The topic of real-time monitoring for energy saving is tackled in [48]. In other work, a study of power consumption was carried out with the aim to reach costs savings, by developing a community structure based on smart homes in electric network systems [49]. By its part in [50], artificial intelligence is used to estimate the energy consumption profile in commercial buildings in order to contract adequate energy plans with public services companies considering the load projections. Finally, in [51] a scheme for the accurate assessment of the electrical energy demand of modern medical equipment operated in laboratories is presented, and it is found that only a few plug load groups mainly contributed to the total energy consumption. Although the domestic commercial equipment also impacts the energy efficiency, this work will focus only on the industrial equipment, specifically electric motors and their drives.

This work presents an overview of the advances in the methodologies applied to the power quality analysis for detecting, identifying and classifying power disturbances that affects the operation of motors and drives, but also how the motors and drives generate adverse effects to the grid. This relation between the power grid and the industrial machines also impacts their own efficiencies and this field is an area of opportunity. A detailed discussion of the methodologies that are the trends in these topics and those approaches is also provided that, by its own characteristics, must be considered to be explored since it represents potential solutions capable to provide accurate results with high reliability, overcoming the drawbacks of the conventional reported techniques. The remainder of the review is organized as follows. In Section 2, the efficiency concept and how it is calculated in both aspects, for the electrical machines and for the electric power, is discussed, providing a quick overview of power quality phenomena and its existing relation with the electrical machines' efficiency reduction. Section 3 sets out the techniques for identifying, detecting, and classifying PQDs following state-of-the-art methodologies and provides a general overview of how this type of study is being carried out and which techniques are currently in trend. Section 4 affords the techniques applied in electrical machines to detect, mitigate, or manage the condition when an electrical phenomenon is presented. Section 5 furnishes a discussion of the techniques presented in the review and the alternative approaches that could be explored in this same context. Finally, in Section 6, the conclusions drawn for this review are presented.

2. Electrical Machines and Energy Efficiency

In general, the term of efficiency is very important when using electrical machines, motors and drives, as well as in the analysis of power quality, since they have a close relation between them. In brief, to this framework this review addresses two types of efficiency: the performance of the electric machine, and that defined by the electrical power supply. Generally speaking, the efficiency of an electrical machine is its capacity to convert the electrical active power into mechanical power. Therefore, the above sentence can be defined, technically speaking, as the ratio of the power output to the power input expressed in percentage terms [52]. Thus, it is necessary to know the values of the mechanical and the electrical active power for determining the efficiency of an electrical machine [53]. On one hand, the relation of parameters for calculating the electrical active power in a three-phase motor, $\sqrt{3}$, is through the voltage, V, the current intensity, I, and the power factor, $cos(\phi)$. Where ϕ is the phase angle between V and I. On the other hand, the mechanical power is obtained with the relation of torque, T_s, and the angular velocity, ω_m. The Table 1 summarizes these parameters relations to calculate the efficiency, η, in an electrical machine, this table was created based on the equations presented in [54].

Table 1. Relationships to calculate electrical power, mechanical power, and efficiency for three-phase motors.

Parameter	Relationship
Electric Active Power	$P_{elec} = \sqrt{3} \cdot V \cdot I \cdot cos(\phi)$
Mechanical Power	$P_{mec} = T_s \omega_m$
Efficiency	$\eta = \frac{P_{mec}}{P_{elec}} \cdot 100\%$

From table it is observed that efficiency states a relation between the electrical parameters and the design criteria of a machine, hence a bad or inadequately design, or failures on its construction, could affect to the electrical power grid [55]. As previously mentioned, the construction, the electrical components, the operation, and the auxiliary elements, to keep the operation of an induction motor through its drive, induce to the power grid electrical disturbances [38,41]. As examples, some typical causes of induced anomalies in the power grid are the non-linear characteristics of loads, sudden switching of loads to the grid, transformers connected in asymmetrical banks, the significant presence of single-phase loads [56], motors current peaks demand, frequency variations by the drives, the usage of static starters and power converters [57], changes of the impedance caused by variations in the capacitive and inductive components feed with AC voltage, equipment failures [58]. In this sense, the type of an electric machine, motor with its drive, predominantly determines its efficiency characteristics and the affectations caused to the grid [59]. Thereby, any improvement on these, or in their configuration topologies, helps to keep a low energy consumption and to rise their efficiency [60]. All these aspects need to be considered, since according to the Department of Energy (DOE) data from USA the industrial motors consume one billion kilowatt-hours of energy each year, approximately the 50% of the world's energy usage [61]. In consequence, regulations in developed countries are moving towards higher efficiency machine classes tending to reduce greenhouse gas emissions and efficient energy usage [62]. For instance, the Table 2 presents the efficiency levels of electric machines according to the standards under NEMA and IEC organizations. The class IE stands for "International Efficiency", and the IEC 60034-30 standard describes it [63].

Table 2. Efficiency classes and levels for electrical machines.

Efficiency Levels	Classes	
	IEC (International)	NEMA (USA)
Standard	IE1	-
High	IE2	Energy Efficient EPACT
Premium	IE3	Premium
Super-Premium	IE4	Super-Premium
Ultra-Premium	IE5	Ultra-Premium

Along the years, the electrical machines are generally mass-produced, meeting specific design and efficiency requirements. Additionally, one of the current objectives of many countries, companies, and industries is to adopt an energy efficiency higher than IE4 class to reflect that they are within the framework of the new global regulations concerning better environmental practices. The latest motors models, as minimum, must be classified IE3 class as stated in these international regulations. Today, high-efficient electrical machines are a new and mandatory trend in motors manufacturing in Europe and the United States of America. The efforts in upgrading the motors have resulted in excellent solutions to environmental problems [64].

On the other hand, it is important to highlight that a chain exists in the process for generating, transporting, converting and distributing the electric energy to the final users. Such a chain has several links and steps in which the energy efficiency is affected. For each link in the chain, the issue of the power quality must be considered, since problems, losses, or affectations may occur. In the framework of this research, the last link is the industrial machine; hence, the efficiency of the power grid impact in the efficiency of the machine. In this sense, the efficiency of the electric power can be considered as the pure sine waveform and its maximum exploitation for feeding electrical equipment [65]. To this respect, a deficient power supply such as drops in voltage, or leakage current, affects the proper operation of a machine, reducing its efficiency, producing malfunctioning, reducing its lifespan expectancy, or even causing irreparable damage to the equipment [66,67]. Therefore, the power quality (PQ), in this context, can be defined as an adequate power supply to electrical equipment and devices for their proper operation. According to the international standards such as IEEE, IEC [68], the power supply voltage must be following established references and limits in terms of amplitude and frequency. Any deviation from these parameters is considered an electrical disturbance or power quality disturbance (PQD) [69]. The international standards define some of these disturbances as amplitude changes referred to as sags, swells, or interruptions. The standards also define frequency change disturbances such as harmonic or inter-harmonic content, and other disturbances associated with minor changes in voltage such as oscillatory transients, fluctuations, and notching. In Table 3 are summarized the different kinds of disturbances, their category, and principal causes and effects for each of them, according to [70]. The flicker term is the effect produced by the voltage fluctuations as indicated in IEEE 1159 [71].

Table 3. PQDs and their causes and effects.

Type of Disturbance	Categories	Causes	Effects
Transients [72]	Impulsive	Lightning strikes, transformer energization, capacitor switching	Power system resonance
	Oscillatory	Line, capacitor or load switching	System resonance
Short duration voltage variation [41]	Sag	Motor starting, single line to ground faults	Protection malfunction, loss of production
	Swell	Capacitor switching, large load switching, faults	Protection malfunction, stress on computers and home appliances
	Interruption	Temporary faults	Loss of production, malfunction of fire alarms
Long duration voltage variation [41]	Sustained interruption	Faults	Loss of production
	Undervoltage	Switching on loads, capacitor de-energization	Increased losses, heating
	Overvoltage	Switching offloads, capacitor energization	Damage to household appliances
Power imbalance [73]		Single-phase load, single phasing	Heating of motors
Waveform distortion [74]	D.C. offset	Geomagnetic disturbance, rectification	Saturation in transformers
	Harmonics	ASDs, nonlinear loads	Increased losses, poor power factor
	Interharmonics	ASDs, nonlinear loads	Acoustic noise in power equipment
	Notching	Power Electronic converters	Damage to capacitive components
	Noise	Arc furnaces, arc lamps, power converters	Capacitor overloading, disturbances to appliances
Voltage fluctuations [75]		Load changes	Protection malfunction, light intensity changes
Power frequency variation [76]		Faults, disturbances in isolated customer-owned systems, and islanding operations	Damage to generator and turbine shafts.

It is common to find PQ problems in industrial electrical systems, such as voltage deviation, unbalance, and harmonics. These issues may adversely affect the operation of induction motors and the electrical drives connected to the grid [77]. In general, the effects of an electrical source with a poor power source and contaminated with PQD on the installed induction motors of industrial processes can be detailed such as: the voltage sag is concerned with affectations in torque, power, speed, and stalling; the harmonic and inter-harmonic content is associated with losses and torque reduction; the voltage unbalance cause extra losses of iron and copper, thus leading to increments of the temperature in the machines and vibrations; the short interruptions generate mechanical shock and possible stall; the impulse surges are related to isolation damage; the overvoltage is related with expected lifespan-shortening; and the undervoltage is concerned with overheating and low speed [78]. Needless to say, the electrical machines depend entirely on being supplied with adequate electrical power to function correctly. Consequently, the electrical network must satisfy the minimum requirements considered for a suitable utilization of the energy. In order to improve the energy performance indicators at industry, it is essential to know the operating status of electrical machines such as motors and drives [79].

Analyses related to the energy-efficient operation of induction motors show that PQDs also affect isolated systems such as marine systems or ships. It is necessary to have the motors' good energy-efficient operation [80]. PQDs can trigger protective devices immediately to trip off motors. However, motors can ride through most of the voltage sags because sag durations are commonly short [81]. Some standards do not consider the effect of the simultaneous disturbances on the electrical machinery. Since several years ago, there has been an increase in protecting the electrical equipment in the industry [82]. Sometimes a non-invasive sensor is considered to monitor the condition of electrical machines [83]. Among the parameters to be monitored in electric motors is the power factor. PQ monitoring is often avoided as a measure for enhancing energy efficiency [80].

3. Techniques for Power Quality Detection, Identification, and Mitigation

It is very important to highlight that the industrial processes require to have power networks with a Power Quality (PQ) as good as possible, since the equipment connected to the grid is very sensible and can easily be affected, as described above, in such a way that the final repercussions are reflected as economic losses and environmental problems. In this sense, the PQ analysis becomes a fundamental study in order to develop methodologies capable of detecting, identifying and mitigating the PQDs present in the power grids. As aforementioned, several works exist that have addressed the study of PQDs from different viewpoints. For example, the studies for detecting power disturbances mainly focus on the development of techniques capable of find out the presence of anomalies in electric signals no matter the nature of the disturbance. In another example, recent works have tackled the detection and identification of the anomalies in the electric signals by classifying them as a particular disturbance from those presented in Table 3. Additionally, there are few studies that really handle the mitigation or minimization of the effects of PQDs on the equipment connected to the power grid. Typically, their solution to this industrial problem is very general, by applying strategies of loads balancing or capacitors banks, but these solutions only work for some disturbances. All the studies are important, and in the following paragraphs they are discussed according to the issue that they address.

Regarding to the identification problem of electric disturbances, the detection techniques that have been developed are very important in order to enhance the quality of a power system [84]. In the first decades of analysis, traditional approaches have been probabilistic-based over signals in the time domain, assuming that the disturbances do not affect the analytical process [85]. Later, for the energy quality monitoring, the process was a fault diagnosis where the electrical signal is processed through different techniques, usually implicating some transformation. Among the most common are those techniques such as Fourier transform and its variants such as fast Fourier transform (FFT), the short time Fourier transform (STFT), the discrete Fourier transform (DFT) [86–88], the discrete wavelet transform (DWT) [89–92], the Hilbert–Huang transform [93–95], the S-transform [96], among others. After decomposing, or transforming, the analyzed signal, an extraction of indicators for making the disturbances detection is performed, the most typical is to use statistical ones in the time domain, the frequency domain, and the time–frequency domain. Recently, due to the difficulty for detecting various and more complex disturbances that may appear in the electrical network, techniques with the ability to handle and process large volumes of data and find relations among the types of disturbances have been considered. For instance, classical machine learning techniques like support vector machines (SVM) [97–99], artificial neural networks (ANN) [85,100–102], deep learning (DL) [103,104], and other machine-learning techniques. Notwithstanding, several studies identify a combination of power disturbances described in the standards [105]. Such patterns could be considered novelty results, and their study has been proposed as an important prospective in the field of electric power disturbances detection.

In reference to the problem for classifying PQDs are presented the following works. The nature of a PQD present in the electric network generates profiles (or patterns) with high complexity on the loads, also connected to the grid, characterized during the operation by non-periodicities and disparities in the combinations of the disturbances observed by the meassuring system [106]. Therefore, power disturbances detection and classification with such profiles are still topics of interest because reported approaches are not robust enough for treating them, having drawbacks and limitations, since they only tackle the disturbaces by separate, or simple combinations [91]. To overcome these drawbacks, the artificial intelligence techniques, the heuristic techniques, and deep learning are being used every time more frequently. The reason is very simple, these techniques are more suitable for treating problems where the prior knowledge of the system is not required, a big amount of data need to be processed, high accuracy is required, data with non-linear behavior, between other advantages [107–109]. Several works in the state of the art that address the tasks of detecting and clasifying power disturbances mention that methodologies based on data-driven could be considered to provide excellent results for the PQ analysis [110]. As a whole, the data-driven procedure consists of three steps: feature extraction, feature reduction, and classification (Figure 1).

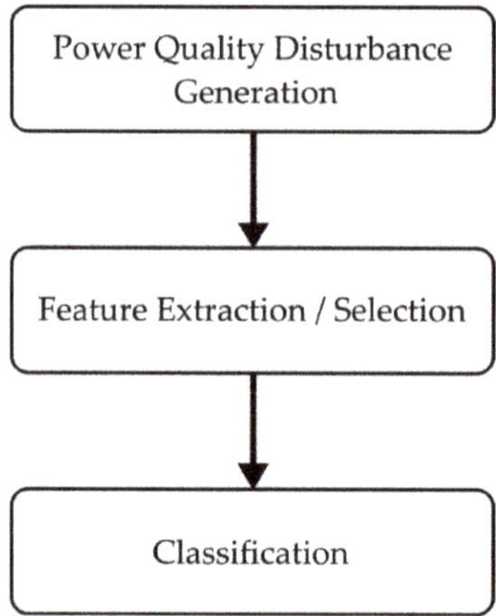

Figure 1. The typical process for Power Quality failure detection methodology [110].

Some examples of the works reported for detecting and classifying electric power disturbances are described in next. The work developed in [98] describes a scheme in which the input signal is first decomposed through the variational mode decomposition (VMD), then the recurrence quantification analysis (RQA) for defining the frequency and duration of the disturbances is performed. This method achieves, by means of data-driven, an adequate parameterization of the present disturbances. Otherwise, in [90] a modified method based on symmetrical components in the time domain for detecting and classifying various PQDs is presented. That implies that a single-phase PQ disturbance and other two ideal phases generated by means of a phase-locked loop (PLL) processed to determine the symmetrical components. Consequently, by triggering points it is possible to detect PQDs in the disturbance phase through the negative sequence component. The detected PQDs have been straightforwardly categorized from the profiles of the waveforms by means of the addition of the sequence components, positive and negative. Then, simulated and real-time results are presented for a wide variety of PQDs to show the effectiveness for detecting and classifying of the proposed method. Other studies such as [111] investigates the efficiency of a methodology for classifying electric disturbances when the manner to extrac the signal features is varied through different classical processing approaches on several data subsets. Although the results obtained are good, a limitation in this strategy is the high amount of resources required to compute the optimal features, since, precisely, several techniques are implemented. On another topic, the work presented in [112] describes a methodology for de PQDs classification; this study uses a higher order of cumulants as feature parameters and the classification approach is based in a quadratic approximation. Here, the signal processing tools are mandatory for obtaining feature vectors from the voltage or current waveform data. Novel, or non-typical, approaches are also performed such as in [113] whose method is out of the typical approaches found in the literature about the processing through sparse signal decomposition on an overcomplete hybrid dictionary, and then the classification stage is performed by a decision tree algorithm. In another example, the work of [114] develops a new method for automatically detecting and classifying electric disturbances by means of Kalman filter (KF). Here, the KF is applied as series of equations for computing the state of a signal measured. The disadvantage is that it is necessary to make a selection of the parameters and verify that the state space model is not incorrect. For microgrids in the photovoltaic (PV) generation there are also a worry about detection of power disturbances generated by the grid inconsistencies. Thus, the work present in [115] presents a variational mode decomposition and empirical wavelet transform used as solution for monitoring and identifying electric disturbances in a distributed generation microgrid. With the advent of Industry 4.0, the aspects involving the condition monitoring of electrical machines have evolved. In consequence, new trends and techniques for signal processing such as artificial intelligence, handling of large volumes of data, and performance improvements are becoming more common, and they have been adapted by more and more users [116]. Recent reviews demonstrate the current literature and tendencies in development and research to aim for the proper detection, identification, and classification of the PQDs [117]. These reviews specifically remark on the works related to digital signal processing (DSP) and machine learning [116]. The recent approaches show their capability to process large data amounts and several signal patterns on the PQ monitoring area that are the current trends. The firsts works that related the use of neural networks with PQDs detection and classification is that presented in [85], where a radial basis function neural network is implemented for the classification of the 20 kinds of disturbances. This scheme is compared with others approaches involving the use of feed forward multilayer network, probabilistic neural network and the generalized regressive neural network. Other works such as in [118] spend their effort in improving the feature extraction, feature calculation and feature selection stages in a common framework of identification and classification of PQDs. This work presents an optimization framework for the optimal selection of features from the different signal domains based on ant-colony optimization. In other case, it is presented in [119] a new approximation for classifying PQDs, firstly, a transformation of the signal from a representation of 1 dimension is carried out into a representation of 2 dimensions for extracting useful indicators. Finally, several approaches for classifyign the disturbances are executed to see wich perform better, between them the machine learning (ML) like k-nearest neighbour (kNN), multilayer perceptrons, and the SVM. In order to validate the aproximation, the PQDs employed are combined defining up to 2 or 3 disturbances at the same time. However, other approaches that address the combination of PQDs conclude that the best way to treat this situation is by data fusion [13]. The reduction of the numerical indicators is very important in the approaches based on data-driven in order to avoid redundancy of information [16]. That means, not useful information must be discarded, or ignored, in order to improve the characterization task of patterns in the analyzed data, for example, the methodologies described in [111,120,121] consider for the reduction task the following techniques: the kNN, the principal component analyzis (PCA), and the sequential forward selection (SFS). Nonetheless, when handling with a large quantity of patterns, as usually recent methodologies for PQ monitoring do, their efficiency is quite restricted [18]. For that reason, the DL approaches were taken into account with more frequency in industry for handling with data sets of high dimensionality and complex pattern behavior [122]. The use of DL provides robustness and efficiency in the classification, recognition, and processing of, images, speech, and video, respectively, but also, recently, in managing of energy [33]. Some good examples of approaches that process data with a high level of complexity are the convolutional neural network (CNN), the recurrent neural network (ReNN), and the autoencoder (AE) technique. Although some of these approaches have been used to test their capabilities for monitoring signals in the power grid, the classification task of PQDs still needs exploration [123]. Even though the achieved performances are good enough, the absence of an standard and simple process to adjust and tune such techniques still represents a drawback that does not allow considering applications in real industrial environments [124]. Meanwhile, the investigation developed in [125] explores the potential of deep learning schemes for classifying PQDs by calculating statistical indicators from four main components through a variant of the PCA and making the disturbances categorization by menas of a CNN. The approach classifies multiple power disturbances in two main classes, reaching accurate results for simulated data. In [104], a novel method based on deep learning is proposed for identifying and classifying PQDs in three main stages: feature extraction from the power system, adaptive pattern recognition by means of AE, and, finally, disturbances classification by NN. Continuing with data-driven strategies, the SVM are becoming important approaches for characterizing multiple patterns that would help to give support to the classification taks. The approach reported in [126] uses a variation of wavelet transform called tunable-Q to efficiently extract features from the signal tuning the Q-factor, and then the disturbances are classified by dual multi-class support vector machines. On the other hand, in [127] a cross wavelet is used, aided by Fischer linear discriminant analysis (LDA), and for the classification of disturbances it uses a Linear SVM. The study referenced in [120] presents a method to classify PQD based on wavelet energy change and the Support Vector Machine. Another scheme that uses a modified version of SVM and variation of wavelet transform is the work presented in [128], which uses empirical wavelet transform arguing is suitability for nonstationary kind of signals such as those presented in electrical disturbances. The method extracts six features that are input to the SVM method for the classification stage. In relation to the space-transform techniques, in [129] several statistical indicators are taken into account to be computed by means of the S-Transform, then the power dirturbances are characterized by appliying an analysis of multi-resolution over such indicators.. The method presented in [88] proposes PQDs recognition by applying the modified S-transform (MST) combined with the parallel stacked sparse autoencoder (PSSAE). Here, the MST uses a Kaiser window in order to concentrate the energy in the matrix of time-frequency and, together with the Fourier transform spectrum, the extraction of features is automatically carried out in order to input them to two sub-models in PSSAE. Moreover, there are performed the reduction of dimensionality and the visual analysis of the features, thus, the classification of the PQDs is finally made with the softmax. Discussing another technique, the approach of [130] uses the S-transform to extract the significant features of the electrical signals, which are the inputs to different machine learning models. This work considers the combination of single disturbances. In the end proposes a hybrid scheme for the classification supported by the single models evaluated at first. A variation of S-transform called double-resolution

S-transform is used in [97] to extract denominated effective features from the signals. Then, the disturbances classification of the signals is made by directed acyclic graph support vector machines (DAG-SVMs). The variations of the typical methods used in this article are supported in the robustness of the techniques and the fact to reduce the computational burden to implement in real-time applications. Involving aspects as the complexity of the signal processing in [86] an optimal multi-resolution fast S-transform is adopted to compress the information obtained from the features extracted and then with a rotation forest made the evaluation of 17 types of PQDs. As can be seen, a base transform is adopted. Depending on the application of the hypothesis to test, it means evaluation in line, monitoring offline, application in embedding systems, different techniques o adaptations of the techniques are considered. Moreover, Mahela in [129] proposed the detection and classification through the S-transform and Fuzzy C-Means. This approach tests their results through simulation signals by software. In other approaches, Sahani in [93] performs a novel signal segmentation method and a new scheme to carry out the classification stage of PQDs based specifically on the use of reduced sample Hilbert–Huang transform combined with class-specific weighted random vector functional link network. These authors based this approach on the implementation in a field programmable gate array (FPGA) environment to then test and validate at online monitoring and see the advantages of their proposal.

Regarding to the mitigation of the effects generated by the PQDs, this field requires more researching, since the reported works are few and they are focused on strategies based on capacitors banks or loads balancing. The effective identification and classification of PQDs is significant for controlling the pollution in the power grid previously to any corrective action. In this matter, the power filtering is an effective way to reduce the effects generated by the PQDs in the electric grids, for example, by using inductive active filters [131]. In [95], the improvement of the microgrid technology is presented, whose applications have increased and gained attention. Nevertheless, distributed generations with intermittency, loads with nonlinearity, and various electrical and electronic devices cause PQ problems in the microgrid, particularly in islanding configurations. A precise and fast method for detecting power disturbances is essential because it is the premise for the PQ control. The proposed approach presented in [58] develops a methodology capable of estimating the expected magnitude for voltage sags in order to provide information of the motor starters applied for ship electrical power. In reference to power imbalance, in [132], a shedding for managing time-optimal loads is presented. In general, this is allowed by using a post event overload mitigation tool that enhances the efficiency of the system by prioritizing the mitigations and ensures the time-dependent network security. Additionally, periodic disturbance mitigation techniques exist based on controllers [133], such works consider as periodic disturbance the harmonic content and by measuring the disturbance and by applying a resonant scheme in the feed forward control or model predictive control the disturbance is mitigated. The power quality is also analyzed in the microgrid systems and here the supra-harmonic (SH) content is also the interest topic. The mitigation strategies for SH are based in the use of dynamic voltage restorers (DVR) for handling voltage sags and swells, but with the limitation of keeping the harmonic content. However, some strategies combine the static synchronous compensator (STATCOM) with static VAR compensator (SVC) for reducing the harmonic effects [134]. Finally, Table 4 summarizes the reported works in the literature and the issue addressed in the PQD analysis.

Table 4. Comparison of PQDs studies in the literature.

Ref.	PQD Issue Addressed	Detection Technique	Classification Technique	Mitigation Technique	Number of PQD Handled	Accuracy Reported
[90]	Detection	SC-PLL	-	-	8	-
[114]	Detection	KF	-	-	14	98.8–100%
[88]	Detection and classification	MST	PSSAE	-	12	99.46%
[98]	Detection and classification	VMD-RQA	SVM	–	7	99.03%
[111]	Detection and classification	FT, STFT, HHT, ST, DWT	ANN, SVM, DT, KNN	-	16	99.31–100%
[112]	Detection and classification	HOC	QC	-	2	98–100%
[113]	Detection and classification	EMD	SVM	-	4	98%
[115]	Detection and classification	VMD-EWT	RKRR	-	12	99%
[85]	Detection and classification	DWT	RBFNN	-	20	96.3%
[118]	Detection and classification	1DST	DT	–	14	99.93%
[119]	Detection and classification	2DRT-MOGWO	KNN	-	18	99.26%
[122]	Detection and classification	DL	CNN	-	16	98.13–99.96%
[123]	Detection and classification	PSR	CNN	-	10	99.8%
[128]	Detection and classification	EWT	SVM	–	15	95.56%
[86]	Detection and classification	ST	DT	–	16	99.47%
[125]	Detection and classification	PCA	CNN	–	11	99.92%
[104]	Detection and classification	FFT, EMD, SAE	SMNN	-	17	98.06%
[119]	Detection and classification	2DRT	KNN	–	17	99.26%
[126]	Detection and classification	TQWT	MSVM	-	14	96.42–98.78%
[135]	Detection and classification	HOS	NT	-	19	97.8%
[131]	Mitigation	HPF	-	IAF	2	-
[95]	Mitigation	HHT	ANN	SVG	4	-
[132]	Mitigation	-	-	PEOM	2	-
[133]	Mitigation	KF	-	RSC, MPC	4	-
[134]	Mitigation	BPF-FFT	-	DVR, STATCOM+SVC	6	-

4. Techniques for Power Quality Related to Electrical Machines and Electrical Drives

As described in previous sections, the PQ is an important topic to be analyzed for industrial equipment connected to the grid because they could be adversely affected yielding important economic losses. Therefore, this section describes and analyses, through the discussion of several works, how a poor power quality affects the main equipment used at industry level, particularly speaking about motors and drives. It is worth mentioning that typically the electric grid is polluted with anomalies such as those described above as PQDs, which not only cause malfunctioning, failures, or damage to the motors and drives, but also reduces its efficiency.

Next, a discussion has begun based on those works that handle the affectations on motors and drives caused by PQDs related with changes in the amplitude of the power source such as voltage sags, swells, interruptions, and unbalance. For instance, regarding discussing or studying the effects of sag disturbances in induction motors, the work presented in [36] calculates the motor performance by analyzing the electromagnetic properties under symmetrical voltage sag conditions. Then, by using an adjustable speed drive (ASD), the energy consumption of the motor is reduced. These elements could be configured in so many forms and can be used for motors of medium voltage applications; however, this solution does not present a good performance regarding Power Quality. Additionally, regarding voltage sag propagation, the work presented in [66] develops an analytical tool able to describe the influence of the sag disturbances over a group of induction motors, but also describe the influence of the motors on the voltage sags characteristics. That means, it is explained how a motor and its drive affects the power grid, but also is explained how the contaminated signal from the grid impacts in the motor operation. In [81] the sag disturbance is analyzed in induction motors by presenting a method that determines the maximum allowed time for the motor connection, whereas a sag occurs in the power line. Due to sag events, a reduction in the electromagnetic torque is produced in the induction motors, tending to a deceleration effect. On the other hand, the work described in [136] presents a control scheme for minimizing the impact of the starting of an induction motor, on the network, by using a voltage feedback-based reactive power support from the existing distributed generator units. It is well known that the electronic equipment, the process control systems, among others, are susceptible to this kind of disturbance. A specific case has been observed when induction motors decelerate due to a short circuit occurrence in the power supply, naturally, the motor will accelerate after source conditions restoring demanding a high current value from the supply causing a postfault voltage sag [137]. On the other hand, the work presented in [138] develops a methodology for mitigating voltage sags during starting of three-phase induction motors. In that study, a neighboring voltage supporting distributed generation (VSDG) reduces the starting peak current and quickly restores the power source to typical values. The study presented in [93] analyses the transient characteristics of induction motors under the influence of sag disturbances using a multi-slice field-circuit-motion integrating time-stepping finite element method. Additionally, in [98], the propagation produced by the induction motors in sags disturbances is analyzed. In [97], the first part of a study is presented where the interaction of induction motors against voltage sags disturbances is presented. Additionally, in [98] the effects produced due to short interruptions and voltage sags are investigated. Here, an analysis of protection devices indicates how to maintain the proper operation of the electrical machines.

By the other side, some examples of works are next described regarding to the voltage unbalance issue. In this line, works such as in [139] present a strategy that is developed to mitigate the voltage unbalance that occurs when energizing induction motors to support the restoration of the grid after this event. To do this, a VSDG injects reactive power into the grid once it is calculated through optimal feedback control of distributed generators. The authors argue that distributed generators are capable of improving the power quality by providing ancillary services as reactive power injection, voltage unbalance compensation, and harmonic filtering. In other works, the wavelet transform is used, such as in [140], to deal with nonstationary signals and where a model is proposed to handle overvoltages caused by pulse-width modulation in voltage source inverters. These disturbances are often presented due to the response from the motor to the inverter pulse voltages. On other topic, a method to estimate the shaft power of an induction motor operating under voltage unbalance and with harmonic content is presented in [56]. Additionally, an algorithm of search in conjunction with an equivalent circuit are developed as the corresponding solution. Another example is the study described in [141] that analyzes induction motors connected to unbalanced three-phase voltages in the steady-state through an index called "the complex voltage unbalance factor". The study carried out in [100] presents a methodology based on thermal effects to monitor an induction motor under unbalanced disturbance conditions. By having thermal profiles, it can be determined when a motor is under the effects of this electrical disturbance. Additionally, the research reported in [67] proposes a new power quality index to determine two kinds of PQDs, voltage unbalance and harmonic content, typically presented in the power supply. The introduction of this new index aims to show the thermal effects of the disturbances into the induction motors simultaneously. In [79], the authors asset the specific effect of the positive sequence of voltage on derating three-phase induction motors under voltage unbalance. This power disturbance could present in motors an overheating, decreasing in efficiency, and reduction in the output torque. In order to mitigate these adverse effects, the motor must be kept in an optimal operational state.

There are works that address the harmonic content issue, such as the work presented in [61], where a new configuration is presented for induction motors. Here, the typical operation of the electrical motor drives is held as the configuration named vector control mode, because this configuration offers a similar performance as in the case of dc motors. However, the use of these drives results in the harmonic injection to the current line affecting the power quality. Different active or passive wave shaping techniques are used to mitigate the harmonic content effects [30]. In this sense, the IEEE 141-1993, the IEC 60034-26, and the NEMA MGI-2003 establish derating factors for induction motors under unbalanced conditions of voltage and harmonic content. Additionally, the work presented in [142] investigates the harmful effects of the harmonic content as a power disturbance, which means changes in frequency above and under the fundamental frequency waveform, e.g., subharmonic and interharmonic content. It is known that these disturbances cause power losses, rotational speed changes, electromagnetic torque variations and windings temperature risings. Additionally, in that study it is reported that vibrations are also generated in induction motors due to the harmonic content caused by the nonlinear loads.

Regarding the works focused on affectations by the PQDs in the motor efficiency, some works have addressed this analysis such as in [9], where it is shown how the derating factor established by the standards for motors with higher efficiency is insufficient for being applied in medium efficiency motors. This work compares the derating factor from different motor classes to maintain the losses at the rated values according to the standards. Similarly, in [8] a comparison was carried out between the motor classes IE2, IE3 and IE4 under two different PQDs, unbalanced voltage and harmonic content, where the research focus was mainly on showing the life expectancy of the motors. The study presents several factors to be considered, in a very comprehensive manner, to properly select the motors for a better operation and reliability, nevertheless, the study is mainly dedicated to three-phase squirrel-cage induction motors (SCIMs). Several former works are reported in [143], where the studied effects in electrical machines by PQDs are the harmonic content and the voltage unbalance. This study presents the arrival of the adjustable speed drives (ASD), as a new enthusiasm for this topic back to the beginning of the 2000s. Additionally, it is performed the economic analysis and provide recommendations for mitigating the harmonic content affectations. Also, in this research is proposed an adequate instrument for assessing and monitoring the motors based on a coefficient related to the energy performance, since this coefficient can indicate the equipment efficiency, or if there exist excessive losses. Last, but not least, another power disturbance that adversely affect the proper operation of induction motors and drives is the flicker. In relation to these phenomenon effects in the induction motors, a research is described in [144] where it is asserted that the studies involving induction motors in the transfer and attenuation of fluctuations need to be modeled in a better way. It has been reported that loads of induction motors contribute to the attenuation of this phenomenon. Table 5 summarizes the works that address the PQD issue in motors and drives and the different approaches proposed.

Table 5. Literature dealing with PQDs in electrical machines and drives.

Reference	Electrical Machine	PQD	Method	Year
[36]	Induction motors	Voltage sags	System modelling	2008
[66]	Induction motors	Voltage sags	Analytical tool for sag description	2008
[81]	Induction motors	Voltage sags	Analysis of critical clearance time of symmetrical voltage sags	2014
[136]	Induction motors	Voltage sags	Voltage supporting distributed generation	2019
[137]	Induction motors	Voltage sags	Voltage supporting distributed generation	1995
[138]	Induction motors	Voltage sags	Coordinated control for distribution feeders	2018
[139]	Induction motors	Voltage sags	Coordinated optimal feedback control for distributed generators	2020
[140]	Induction drives	Voltage sags	Wavelet modelling of motor drives	2004
[93]	Induction motors	Voltage sags	Reduced-Sample Hilbert–Huang transform	2019
[97]	Induction motors	Voltage sags	S-Transform with double resolution and SVM	2016
[98]	Induction motors	Voltage sags	Qualitative-quantitative hybrid approach	2020
[56]	Induction motors	Voltage unbalance	Estimation of shaft power	2016
[141]	Induction motors	Voltage unbalance	Analysis on the angles of complex voltage unbalance, Index CVUF	2001
[100]	Induction motors	Voltage unbalance	Discrete wavelet transform, mathematical morphology and speed variation drive	2018
[67]	Induction motors	Harmonic content	New power quality index	2010
[61]	Induction motors	Harmonic content	Adjustable speed drive with a multiphase staggering modular transformer	2019
[30]	Induction motors	Harmonic content	Pulse Multiplication in AC–DC Converters	2006
[79]	Induction motors	Harmonic content	Analysis of positive sequence voltage on derating 3-phase induction motors	2013
[142]	Induction motors	Harmonics, subharmonics and interharmonics	Vibration Analysis	2019
[8]	Induction motors	Voltage unbalance and harmonic content	Comparison between classes efficiency with driver metrics	2015

5. Analysis of Techniques Trends

This section addresses two main topics: the first one identifies and summarizes the present problematics and the tendencies towards solutions that have been applied regarding electric rotating machines and drives and their relationship with power quality; the second one includes possible approaches as solutions to the niches of opportunity detected in the related analysis to the detection, classification, and mitigation of the effects of the power disturbances.

5.1. Overview on the Proposed Solutions regarding Power Quality Issues on Motors and Its Drives

The quality of an electric network, at the industrial facilities, is reduced by the influence of disturbances that normally appear by different factors, internal and external. For example, the induced disturbances to the power signals are due to the loads represented by electrical equipment switched to the network, their electric and electronic components, and their non-linear behavior. As a counterpart, the industrial equipment is affected, in turn, by a poor power quality provoking malfunctioning, reducing its lifespan expectancy, causing irreparable damage, and reducing it efficiency. From the analyzed works, an evolution is observed in the manner that every problem is tackled; for instance, the first study-developed methodologies focused on the detection of power disturbances without considering the anomaly nature. Such approaches mainly used space transformations (FFT, DWT) in the time domain, or the frequency domain, or the time–frequency domain in order to posteriorly make a manual analysis.

Later, further works evolving not only for detection of power disturbances tasks, but also for identifying and classifying them. To this respect, several techniques were used to extract what is known as features from the measured signals, which are values computed from statistical, electrical, mechanical parameters, etc. An interesting topic is the manner in which the classification task was carried out, e.g., by integrating the artificial intelligence (AI) techniques such as the artificial neural networks. Later other AI approaches were employed to identify and classify PQDs, such as the fuzzy logic, many variants of the NN, SVM, DT, among others. Some important drawbacks are presented in the PQ diagnosis, in recent years, such as the big amount of data generated by the high sampling frequency of the data acquisition systems, the high complexity in the hardware of new devises, the appearance of several PQDs combined. To overcome these limitations the heuristic techniques and the machine learning were integrated, which are capable of handling high amount of data, treating with non-linearities, providing results with high accuracy, working without previous knowledge of the problem, etc. Recently, for raising the accuracy and the reliability of the results, the techniques for features extraction aim for the generation of high-dimensional indicators matrices, with the aim to have as many data as possible to obtain valuable information of several combined disturbances. Posteriorly, the redundant information is eliminated (only the useful information is kept) by applying techniques such as LDA or PCA. The advantages of the reduction techniques are the simplified representations of the data which become useful outputs for conventional classification techniques. The mitigation of PQDs is a problem that has been addresses by few works, since they focus their efforts on strategies based on capacitors banks or loads balancing, for that reason, it is observed that mitigation of PQDs is an area of opportunity.

Finally, from the works that address the PQDs analysis to the particular application in motors and drives, it can be mentioned that recent approaches consider only the effects of voltage sags, generated by motor starters. Additionally, the use of electric drives also induces to the power system harmonic content (harmonics, interharmonic, and subhamonics). The voltage unbalance is other typical problem affecting the induction motors operation and its drives, since they are switched to the grid generating asymmetric loads in the lines. The methodologies that tackle these problems, as observed in Table 5, are based on controlled systems by distribution generators, systems modelling, development of tools for describing the disturbances characteristics, and adjustable speed drives, among others.

5.2. Techniques That Could Be Possible Potential Solutions to the Existing Problems

As described in previous subsection, the effects of the PQDs and its combinations on the efficiency of motors and its drives are not full studied yet. Additionally, there is a niche of opportunity related to the study about the effects of electric motors and its drives over the electrical grid. The electric drives and the motors, especially the first ones, could affect the power quality and to produce undesirable effects that may not be considered, yet by these standards, some recent methodologies call this analysis a novelty detection. The PQ analysis could be still with views toward solutions about the identification, classification and mitigation issues but considering the use of alternative methodologies capable to overcome the drawbacks that reported methodologies cannot, not only considering some isolated disturbances such as voltage sags, voltage unbalance or harmonic content, but by considering other varieties of PQDs, their combinations and their mitigations. Therefore, a well-structured approach that combines the best of such alternative approaches, with a general procedure capable of treating a large amount of data and to provide high accurate and reliable results could be helpful in this area. With any technique used, an important aspect to be considered is that the information obtained by the approach must be used in the development of strategies for the mitigation of power disturbances.

Regarding the alternative methodologies that still are not considered, in the next lines the novelty detection approaches are discussed as possible potential solutions to the field of PQ and motors and its drives. The detection of problems, such as electric disturbances in the grid and faults conditions into the induction motors, can be tackled through novelty detection (ND) [82]. The purpose of ND is to observe a system behavior and to decide whether an observation belongs to the same distribution of the existing observations, or if it must be considered different. In the framework of the PQ, the observations during the normal operation of the power grid, or of the motor and its drive, could be considered the reference (typical distribution), and any deviation from this behavior is susceptible to be considered as atypical. Between the different schemes to apply ND, (i) probabilistic techniques, (ii) distance-based techniques, (iii) reconstruction techniques, and (iv) domain-based techniques exist.

The probabilistic techniques include the Gaussian Mixture Models [145], the Extreme Value Theory [146], the State-Space Models [147], the Kernel Density Estimators [148], and the Negative Selection [149]. These techniques estimate the value of density from the normal class, and assume that areas of low density in the training set indicate a low probability to contain normal objects. A drawback of these methods is the limited performance when the training set is too small. Thus, when the dimensionality of data space growths, all data points extend to a bigger volume. Therefore, the signals measured in the electric grid and the physical magnitudes captured from the motor and its drive such as current, voltage, vibrations, temperature, etc., could be employed to perform the probabilistic analysis. This way, the applicability of these techniques could be explored in the identification and classification of power disturbances in the grid, as well as the faults conditions in motors and its drives. The analysis from the probabilistic viewpoint would be helpful to define classes with densities variations according to the anomalies detected (power disturbance or fault condition). Additionally, the probabilistic approach could provide a new indicator index associated with the efficiency reduction caused by the power disturbances in the grid, or the fault conditions into a motor.

On the other hand, the distance-based techniques include the k-Nearest Neighbor [150] and the Clustering k-Means [151]. These methods assume tightly grouping, as clusters, for normal data, but different data are located far respect to their nearest neighbors. Additionally, adequate distance metrics are defined to establish the similitude between two points, even within spaces with high dimensionality. There are some drawbacks when using these techniques, for example, they just identify global points, and their flexibility is not enough for detecting local novelty when the data sets present arbitrary shapes and diverse densities. Additionally, the computing of distance between data points represents high cost of the computational resources, mainly in data sets of high dimensionality; as a consequence, these techniques lack scalability. Finally, the approaches based in grouping of data suffer because they must select an appropriate cluster width and they are sensible to the dimensionality variation. Similar to the probabilistic methods, several physical magnitudes from the motors, or from the power network, can be used to extract features that define such clusters. Therefore, these approaches are also sensible to be used for detecting and classifying electric disturbances, since each disturbance contains different characteristic that allow them to be grouped by a distance among them, the same scheme could be defined to the faut conditions detection in motors and its drives.

For the case of the reconstruction techniques, they include variants of the NN, Auto Associative Networks, Radial Basis Function, Self-Organizing Maps, Sparse Autoencoder, and Subspace Methods [152,153]. These methods imply to use a normal data set for training a regression model. As result, when the trained model process atypical data the difference (reconstruction error) between the regression objective and the real value observed yields to a novelty detection. However, the main drawbacks are, for instance, the requirement of an optimized quantity of parameters for defining the structure of the model, and the direct relation of the performance to these model parameters. Additionally, the networks that use reconstructive models with variable size on its structure suffer because it is necessary to select an effective training method that allows to incorporate new units to the existing model structure. In this same line, the approaches based on the subspace must select correctly the values of the parameters that control the mapping to a subspace of lower dimension. In this particular case, for instance, variants of the neural networks could be very helpful to train regression models that describe power disturbances in the grid, or fault conditions into a motor. Thus, the reconstruction techniques could be applied for the quantification and classification tasks of abnormalities in the grid, or faults in the motor. A novel application of these approaches could be explored for mitigating the effects of power disturbances, for example, by defining a model that generates the opposite behavior to the disturbance to counteract their effects.

By its part, the domain-based techniques include the Support Vector Data Description [154] and the One Class Support Vector Machine [155]. These methods describe a domain that have normal data, also define the limits that round the normal class and that follows the distribution of the data, but they do not provide an explicit distribution of the regions with high density. The usefulness of these techniques is observed mainly in the classification task of abnormal conditions in the power network and motors and its drives, since such conditions are represented by classes according to specific distributions of data. Additionally, a complement can be made through feature extraction and dimensionality reduction through LDA and PCA for all the novelty detection techniques.

In relation to the heuristic approaches, techniques such as the Genetic Algorithms (GA) [156], the Evolutionary Programming (EP) [157], the Particle Swarm Optimization (PSO) [158], and the Expert Systems (ES) [159]. These techniques can handle problems in which the previous knowledge is not necessary, they are good for looking values in searching spaces with non-linearities, non-convexities, and with high dimensionality. Additionally, they are of simple concept and have easiness of implementation. As they were originally designed for optimization problems, they can be adjusted for a wide variety of situations where critical values need to be found. Therefore, the heuristic schemes could be considered as opportune in solving the drawbacks present in the novelty detection techniques. For example, in selecting the parameters needed by the novelty approaches to work with high performance, or in founding the adequate dimension of the clusters. Additionally, the application of heuristic approaches is not limited to provide support for a medullar algorithm, they can also be used in parallel, or as the medullar algorithm, such as in a reconstruction model. A perfect example of this is the GA, which have characteristics that enable it to accurately estimate several parameters (multi-optimization search) of a parameterized model [160]. In this same line the heuristic approaches could have applicability in the quantification of fault conditions in motors, as long as a generalized model of the conditions can be defined. Additionally, these techniques could be explored in the mitigation task of power disturbances by optimizing a model that generates the opposite behavior to the disturbance to counteract its effects.

6. Conclusions

This review presents the discussion of several works in the state of the art referring to the following aspects: the efficiency of electric machines (motors and drives); the power quality; the relationship between the power quality and electric machines affecting the efficiency; the techniques for power quality disturbances detection, classification, and mitigation; and the techniques for PQD analysis in motors and drives. The discussion of the works related to electrical machines and energy efficiency allows to conclude that there exists a mutual relation between motors and drives with the power quality. For example, the efficacy of the PQ of a power source is reduced by the disturbances induced by electrical equipment connected to the grid, but also, once the grid is contaminated with electric disturbances, they reduce the performance of motors and drives. Is worth to highlight that in the literature, several works have been developed with the purpose to detect, classify, and mitigate the affectations generated by the PQDs. There are several methodologies; the firsts of them were designed only for the detection task, and they were based mainly in space-transform techniques. Later, the integration of artificial intelligence techniques arrived; this brings out the opportunity for performing the classification of the PQDs. The most recent strategies combine the aforementioned techniques to define well-structured approaches for feature extraction, dimensionality reduction and classification. Alternative methodologies such as novelty detection and heuristic techniques have also been addressed, making a discussion about their characteristics which make them potential solutions to give accurate and reliable results to problems where the reported methodologies cannot. For example, by performing the detection, identification and classification of power disturbances not considered yet by the standards, or other types of disturbances different from those tackled by the reported works. Additionally, in the fault conditions monitoring, in motors and its drives, these approaches can be explored for detecting and classifying several faults or their combinations. By the other side, the heuristic schemes can be adopted to give support to the novelty detection methodologies, by selecting the parameters that play a key role in the performance of such methodologies. Additionally, the heuristic approaches could be used for estimating the values of parameterized models that describe the power disturbances and the fault conditions, or their combinations, respectively. Mitigation of PQDs is still an area of opportunity, since few works have handled this issue but only for limited power disturbances. The alternative methodologies proposed in this review could be opportune options for proposing strategies to meet this goal. For instance, novelty detection can provide accurate information about the anomalies in the grid, or in a motor, in order to develop mitigation strategies. One example about mitigation of power disturbances could be developed through the use of heuristic techniques, by defining a parameterized model capable of generating the opposite signal that mitigates (attenuating or minimizing) the effects of the disturbances (or their combinations). Finally, the studies of PQD affecting the efficiency of motors and drives the analysis considered until now limits to some disturbances such as voltage sags, voltage unbalance, and harmonic content. Here, the well-structured approaches could be useful to this matter. Regulatory agencies are introducing energy efficiency requirements and the electric machine must meet these restrictions. Therefore, it is important that the new lines of investigation look towards solutions to mitigate the PQDs in order to rise the electric machines efficiency that in consequence will increase the power grid efficiency.

Author Contributions: A.-D.G.-A. performed the investigation of the majority of the revised works in the state of the art and wrote most of the paper; A.-Y.J.-C. carried out the research of several complementary works added to the paper from the literature, also revised and corrected most of the paper sections; R.-A.O.-R. conceived and developed the idea of this review, and proofread most of the paper; M.D.-P. made the discussion of several papers included in this review and wrote some sections of the paper; J.-A.A.-D. conceived and developed the idea of this research, performed papers analysis, and wrote some of the paper sections; A.K. conceived several ideas for this research and performed the finals proofreads of the manuscript. All authors have read and agreed to the published version of the manuscript.

Funding: This research received no external funding.

Institutional Review Board Statement: Not applicable.

Informed Consent Statement: Not applicable.

Data Availability Statement: No new data were created or analyzed in this study. Data sharing is not applicable to this article.

Conflicts of Interest: The authors declare no conflict of interest.

Glossary

FT	Fourier transform
ST	S-Transform
SVM	Support vector machines
QC	Quadratic classifier
RQA	Recurrence quantification analysis
KNN	K-nearest neighbor
RBFNN	Radial basis function neural network
DL	Deep learning
PSR	Phase space reconstruction
KF	Kalman filter
FFT	Fast Fourier transform

TQWT	Tunable-Q wavelet transform
IAF	Inductive active filtering
PEOM	Post event overload mitigation
BPF	Band-pass filter
STFT	Short time Fourier transform
DWT	Discrete wavelet transform
DT	Decision trees
SC	Symetric components
EMD	Empirical mode decomposition
EWT	Empirical wavelet transform
1DST	1-dinemsional S-transform
MOGWO	Multi-objective grey wolf optimizer
PCA	Principal component analysis
PSSAE	Parallel stacked sparse autoencoder
SMNN	SoftMax neural network
MSVM	Multiclass support vector machine
SVG	Static VAR generator
RSC	Resonant controller
STATCOM	Static synchronous compensator
HHT	Hilbert–Huang transform
ANN	Artificial neural networks
HOC	Higher-Order cumulants
PLL	Phase locked loop
VMD	Variable mode decomposition
RKRR	Reduced kernel ridge regression
2DRT	2-dimensional Riesz transform
CNN	Convolutional neural network
NT	Nutro tree
MST	Modified S-transform
SAE	Sparse autoencoder
HPF	High-pass filter
SVC	Static VAR compensator
MPC	Model predictive controller

References

1. Ghosh, P.K.; Sadhu, P.K.; Basak, R.; Sanyal, A. Energy Efficient Design of Three Phase Induction Motor by Water Cycle Algorithm. *Ain Shams Eng. J.* **2020**, *11*, 1139–1147. [CrossRef]
2. Boteler, R.; Malinowski, J. Review of Upcoming Changes to Global Motor Efficiency Regulations. In Proceedings of the Conference Record of 2009 Annual Pulp and Paper Industry Technical Conference, Birmingham, AL, USA, 21–26 June 2009; pp. 26–30. [CrossRef]
3. Moro Franchi, C. *Electrical Machine Drives Fundamental Basics and Practice*; CRC Press: Boca Raton, FL, USA, 2019; ISBN 9781138099395.
4. De Capua, C.; Landi, C. Quality Assessment of Electrical Drives with Strongly Deformed Supply Waveform. *Meas. J. Int. Meas. Confed.* **2001**, *30*, 269–278. [CrossRef]
5. LeDoux, K.; Visser, P.W.; Hulin, J.D.; Nguyen, H. Starting Large Synchronous Motors in Weak Power Systems. *IEEE Trans. Ind. Appl.* **2015**, *51*, 2676–2682. [CrossRef]
6. Giannoutsos, S.V.; Manias, S.N. A Systematic Power-Quality Assessment and Harmonic Filter Design Methodology for Variable-Frequency Drive Application in Marine Vessels. *IEEE Trans. Ind. Appl.* **2015**, *51*, 1909–1919. [CrossRef]
7. Alimi, O.A.; Ouahada, K.; Abu-Mahfouz, A.M. A Review of Machine Learning Approaches to Power System Security and Stability. *IEEE Access* **2020**, *8*, 113512–113531. [CrossRef]
8. Ferreira, F.J.T.E.; Baoming, G.; De Almeida, A.T. Reliability and Operation of High-Efficiency Induction Motors. In Proceedings of the 2015 IEEE/IAS 51st Industrial & Commercial Power Systems Technical Conference (I&CPS), Calgary, AB, Canada, 5–8 May 2015; pp. 1–13. [CrossRef]
9. Donolo, P.D.; Pezzani, C.M.; Bossio, G.R.; De Angelo, C.H.; Donolo, M.A. Derating of Induction Motors Due to Power Quality Issues Considering the Motor Efficiency Class. *IEEE Trans. Ind. Appl.* **2020**, *56*, 961–969. [CrossRef]
10. Alberti, L.; Troncon, D. Design of Electric Motors and Power Drive Systems According to Efficiency Standards. *IEEE Trans. Ind. Electron.* **2021**, *68*, 9287–9296. [CrossRef]

11. Fernandez-Cavero, V.; Morinigo-Sotelo, D.; Duque-Perez, O.; Pons-Llinares, J. A Comparison of Techniques for Fault Detection in Inverter-Fed Induction Motors in Transient Regime. *IEEE Access* **2017**, *5*, 8048–8063. [CrossRef]

12. Donolo, P.D.; Chiacchiera, E.; Pezzani, C.M.; Lifschitz, A.S.; De Angelo, C. Economic Barriers to the Application of Energy Efficient Motors in Industry. *IEEE Latin Am. Trans.* **2020**, *18*, 1817–1825. [CrossRef]

13. Hubbi, W.; Goldberg, O. A New Method for Quality Monitoring of Induction Motors. *IEEE Trans. Energy Convers.* **1993**, *8*, 726–731. [CrossRef]

14. Bíscaro, A.A.P.; Pereira, R.A.F.; Kezunovic, M.; Mantovani, J.R.S. Integrated Fault Location and Power-Quality Analysis in Electric Power Distribution Systems. *IEEE Trans. Power Deliv.* **2016**, *31*, 428–436. [CrossRef]

15. Bakshi, U.A.; Bakshi, M.V. *Electrical Machines—I*, 1st ed.; Technical Publications: Pune, India, 2020.

16. Mojlish, S.; Erdogan, N.; Levine, D.; Davoudi, A. Review of Hardware Platforms for Real-Time Simulation of Electric Machines. *IEEE Trans. Transp. Electrif.* **2017**, *3*, 130–146. [CrossRef]

17. Leonhard, W. *Control of Electrical Drives*, 3rd ed.; Springer: Berlin/Heidelberg, Germany, 2001; ISBN 9783642626098.

18. Liu, C.; Chau, K.T.; Lee, C.H.T.; Song, Z. A Critical Review of Advanced Electric Machines and Control Strategies for Electric Vehicles. *Proc. IEEE* **2021**, *109*, 1004–1028. [CrossRef]

19. Liu, P.; Wang, Z.; Wei, S.; Bo, Y.; Pu, S. Recent Developments of Modulation and Control for High-Power Current-Source-Converters Fed Electric Machine Systems. *CES Trans. Electr. Mach. Syst.* **2020**, *4*, 215–226. [CrossRef]

20. Cao, R.; Lu, M.; Jiang, N.; Cheng, M. Comparison between Linear Induction Motor and Linear Flux-Switching Permanent-Magnet Motor for Railway Transportation. *IEEE Trans. Ind. Electron.* **2019**, *66*, 9394–9405. [CrossRef]

21. Abdollahi, S.E.; Mirzayee, M.; Mirsalim, M. Design and Analysis of a Double-Sided Linear Induction Motor for Transportation. *IEEE Trans. Magn.* **2015**, *51*, 1–7. [CrossRef]

22. Safaeian, M.; Jalilvand, A.; Taheri, A. A MRAS Based Model Predictive Control for Multi-Leg Based Multi-Drive System Used in Hot Rolling Mill Applications. *IEEE Access* **2020**, *8*, 215493–215504. [CrossRef]

23. Ramírez, G.; Valenzuela, M.A.; Pittman, S.; Lorenz, R.D. Modeling and Evaluation of Paper Machine Coater Sections Part 1: 1-Coater Section and Tension Setpoints. *IEEE Trans. Ind. Appl.* **2019**, *55*, 2144–2154. [CrossRef]

24. Bhavana., M.B.; Hegde, V. Energy conservation using variable frequency drives of a humidification plant in a textile mill. In Proceedings of the 2015 International Conference on Power and Advanced Control Engineering (ICPACE), Bengaluru, India, 12–14 August 2015; pp. 90–94. [CrossRef]

25. Lin, F.J.; Shieh, H.J.; Shieh, P.H.; Shen, P.H. An adaptive recurrent-neural-network motion controller for X-Y table in CNC Machine. *IEEE Trans. Syst. Man Cybern. Part B* **2006**, *36*, 286–299. [CrossRef]

26. Gedıkpinar, M. Design and Implementation of a Self-Starting Permanent Magnet Hysteresis Synchronous Motor for Pump Applications. *IEEE Access* **2019**, *7*, 186211–186216. [CrossRef]

27. Torres, F.J.; Ramírez-Paredes, J.P.; García-Murillo, M.A.; Martínez-Ramírez, I.; Capilla-González, G.; Ramírez, V.A. A Tracking Control of a Flexible-Robot Including the Dynamics of the Induction Motor as Actuator. *IEEE Access* **2021**, *9*, 82373–82379. [CrossRef]

28. Sharma, U.; Singh, B. Design and Development of Energy Efficient Single Phase Induction Motor for Ceiling Fan Using Taguchi's Orthogonal Arrays. *IEEE Trans. Ind. Appl.* **2021**, *57*, 3562–3572. [CrossRef]

29. Jurkovic, S.; Rahman, K.M.; Morgante, J.C.; Savagian, P.J. Induction Machine Design and Analysis for General Motors e-Assist Electrification Technology. *IEEE Trans. Ind. Appl.* **2015**, *51*, 631–639. [CrossRef]

30. Singh, B.; Bhuvaneswari, G.; Garg, V. Power-Quality Improvements in Vector-Controlled Induction Motor Drive Employing Pulse Multiplication in AC–DC Converters. *IEEE Trans. Power Deliv.* **2006**, *21*, 1578–1586. [CrossRef]

31. Nair, R.; Narayanan, G. Emulation of Wind Turbine System Using Vector Controlled Induction Motor Drive. *IEEE Trans. Ind. Appl.* **2020**, *56*, 4124–4133. [CrossRef]

32. Shang, Z.; Gao, D.; Jiang, Z.; Lu, Y. Towards Less Energy Intensive Heavy-Duty Machine Tools: Power Consumption Characteristics and Energy-Saving Strategies. *Energy* **2019**, *178*, 263–276. [CrossRef]

33. Bayindir, R.; Demirbas, S.; Irmak, E.; Cetinkaya, U.; Ova, A.; Yesil, M. Effects of Renewable Energy Sources on the Power System. In Proceedings of the 2016 IEEE International Power Electronics and Motion Control Conference (PEMC), Varna, Bulgaria, 25–28 September 2016; pp. 388–393. [CrossRef]

34. Shao, L.; Karci, A.E.H.; Tavernini, D.; Sorniotti, A.; Cheng, M. Design Approaches and Control Strategies for Energy-Efficient Electric Machines for Electric Vehicles—A Review. *IEEE Access* **2020**, *8*, 116900–116913. [CrossRef]

35. Gerada, D.; Mebarki, A.; Brown, N.L.; Gerada, C.; Cavagnino, A.; Boglietti, A. High-Speed Electrical Machines: Technologies, Trends, and Developments. *IEEE Trans. Ind. Electron.* **2014**, *61*, 2946–2959. [CrossRef]

36. Tleis, N.D. *Power Systems Modelling and Fault Analysis—Theory and Practice*, 2nd ed.; Academic Press (Elsevier): London Wall, UK, 2008; ISBN 9780750680745.

37. Tarasiuk, T.; Jayasinghe, S.G.; Gorniak, M.; Pilat, A.; Shagar, V.; Liu, W.; Guerrero, J.M. Review of Power Quality Issues in Maritime Microgrids. *IEEE Access* **2021**, *9*, 81798–81817. [CrossRef]

38. Zhang, X.-P.; Yan, Z. Energy Quality: A Definition. *IEEE Open Access J. Power Energy* **2020**, *7*, 430–440. [CrossRef]

39. Wang, J.; Sun, K.; Wu, H.; Zhu, J.; Xing, Y.; Li, Y. Hybrid Connected Unified Power Quality Conditioner Integrating Distributed Generation with Reduced Power Capacity and Enhanced Conversion Efficiency. *IEEE Trans. Ind. Electron.* **2021**, *68*, 12340–12352. [CrossRef]

40. Madonna, V.; Giangrande, P.; Migliazza, G.; Buticchi, G.; Galea, M. A Time-Saving Approach for the Thermal Lifetime Evaluation of Low-Voltage Electrical Machines. *IEEE Trans. Ind. Electron.* **2020**, *67*, 9195–9205. [CrossRef]
41. Jaen-Cuellar, A.Y.; Morales-Velazquez, L.; Romero-Troncoso, R.D.; Moriñigo-Sotelo, D.; Osornio-Rios, R.A. Micro-genetic algorithms for detecting and classifying electric power disturbances. *Neural. Comput. Appl.* **2017**, *28*, 379–392. [CrossRef]
42. Tabora, J.M.; De Lima Tostes, M.E.; Bezerra, U.H.; De Matos, E.O.; Filho, C.L.P.; Soares, T.M.; Rodrigues, C.E.M. Assessing Energy Efficiency and Power Quality Impacts Due to High-Efficiency Motors Operating Under Nonideal Energy Supply. *IEEE Access* **2021**, *9*, 121871–121882. [CrossRef]
43. Kumar, D.; Zare, F. A Comprehensive Review of Maritime Microgrids: System Architectures, Energy Efficiency, Power Quality, and Regulations. *IEEE Access* **2019**, *7*, 67249–67277. [CrossRef]
44. Ali, S.; Wu, K.; Weston, K.; Marinakis, D. A Machine Learning Approach to Meter Placement for Power Quality Estimation in Smart Grid. *IEEE Trans. Smart Grid* **2016**, *7*, 1552–1561. [CrossRef]
45. Li, C.; Xu, W.; Tayjasanant, T. Interharmonics: Basic Concepts and Techniques for Their Detection and Measurement. *Electr. Power Syst. Res.* **2003**, *66*, 39–48. [CrossRef]
46. Bolshev, V.; Vinogradov, A.; Jasinski, M.; Sikorski, T.; Leonowicz, Z.; Gono, R. Monitoring the Number and Duration of Power Outages and Voltage Deviations at Both Sides of Switching Devices. *IEEE Access* **2020**, *8*, 137174–137184. [CrossRef]
47. Albertelli, P. Energy Saving Opportunities in Direct Drive Machine Tool Spindles. *J. Clean. Prod.* **2017**, *165*, 855–873. [CrossRef]
48. Alahmad, M.; Wheeler, P.; Schwer, A.; Eiden, J.; Brumbaugh, A. A comparative study of three feedback devices for residential real-time energy monitoring. *IEEE Trans. Ind. Electron.* **2012**, *59*, 2002–2013. [CrossRef]
49. Cherkassky, V.; Chowdhury, S.; Landenberg, V.; Tewari, S.; Bursch, P. Prediction of electric power consumption for commercial buildings. In Proceedings of the International Joint Conference on Neural Networks, San Jose, CA, USA, 31 July–5 August 2011. [CrossRef]
50. Anees, A.; Chen, Y.-P. True real time pricing and combined power scheduling of electric appliances in residential energy management system. *Appl. Energy* **2016**, *165*, 592–600. [CrossRef]
51. Christiansen, N.; Kaltschmit, M.; Dzukowski, F.; Isensee, F. Electricity consumption of medical plug loads in hospital laboratories: Identification, evaluation, prediction and verification. *Energy Build.* **2015**, *107*, 392–406. [CrossRef]
52. Wang, X.; Yong, J.; Xu, W.; Freitas, W. Practical Power Quality Charts for Motor Starting Assessment. *IEEE Trans. Power Deliv.* **2011**, *26*, 799–808. [CrossRef]
53. Zhang, D.; Liu, T. Effects of Voltage Sag on the Performance of Induction Motor Based on a New Transient Sequence Component Method. *CES Trans. Electr. Mach. Syst.* **2019**, *3*, 316–324. [CrossRef]
54. Agamloh, E.B.; Nagorny, A.S. An Overview of Efficiency and Loss Characterization of Fractional Horsepower Motors. *IEEE Trans. Ind. Electron.* **2018**, *60*, 3072–3080. [CrossRef]
55. Santos, V.S.; Felipe, P.R.V.; Sarduy, J.R.G.; Lemozy, N.A.; Jurado, A.; Quispe, E.C. Procedure for Determining Induction Motor Efficiency Working under Distorted Grid Voltages. *IEEE Trans. Energy Convers.* **2015**, *30*, 331–339. [CrossRef]
56. Sousa, V.; Viego, P.R.; Gómez, J.R.; Quispe, E.C.; Balbis, M. Shaft Power Estimation in Induction Motor Operating under Unbalanced and Harmonics Voltages. *IEEE Lat. Am. Trans.* **2016**, *14*, 2309–2315. [CrossRef]
57. Committee, D.; Power, I.; Society, E. *IEEE 519 Recommended Practice and Requirements for Harmonic Control in Electric Power Systems IEEE Power and Energy Society*; IEEE: Piscataway, NJ, USA, 2014; Volume 2014.
58. Su, C.L.; Chen, C.J.; Lee, C.C. Fast Evaluation Methods for Voltage Sags in Ship Electrical Power Systems. *IEEE Trans. Ind. Appl.* **2013**, *49*, 233–241. [CrossRef]
59. Refoufi, L.; Pillay, P. Harmonic Analysis of Slip Energy Recovery Induction Motor Drives. *IEEE Trans. Energy Convers.* **1994**, *9*, 665–672. [CrossRef]
60. Singh, B.; Bhuvaneswari, G.; Garg, V.; Gairola, S. Pulse Multiplication in AC-DC Converters for Harmonic Mitigation in Vector-Controlled Induction Motor Drives. *IEEE Trans. Energy Convers.* **2006**, *21*, 342–352. [CrossRef]
61. Singh, B.; Kant, P. A 40-Pulse Multiphase Staggering Modular Transformer with Power Quality Improvement in Multilevel Inverter Fed Medium-Voltage Induction Motor Drives. *IEEE Trans. Ind. Appl.* **2019**, *55*, 7822–7832. [CrossRef]
62. De Almeida, A.T.; Ferreira, F.J.T.E.; Baoming, G. Beyond Induction Motors-Technology Trends to Move up Efficiency. *IEEE Trans. Ind. Appl.* **2014**, *50*, 2103–2114. [CrossRef]
63. Tabora, J.M.; Tostes, M.E.; De Matos, E.O.; Bezerra, U.H.; Soares, T.M.; De Albuquerque, B.S. Assessing Voltage Unbalance Conditions in IE2, IE3 and IE4 Classes Induction Motors. *IEEE Access* **2020**, *8*, 186725–186739. [CrossRef]
64. Xie, F.; Liang, K.; Wu, W.; Hong, W.; Qiu, C. Multiple Harmonic Suppression Method for Induction Motor Based on Hybrid Morphological Filters. *IEEE Access* **2019**, *7*, 151618–151627. [CrossRef]
65. Steimer, P.K. Enabled by high power electronics—Energy efficiency, renewables and smart grids. In Proceedings of the 2010 International Power Electronics Conference—ECCE ASIA, Sapporo, Japan, 21–24 June 2010; pp. 11–15. [CrossRef]
66. Milanovic, J.V.; Aung, M.T.; Vegunta, S.C. The Influence of Induction Motors on Voltage Sag Propagation—Part I: Accounting for the Change in Sag Characteristics. *IEEE Trans. Power Deliv.* **2008**, *23*, 1063–1071. [CrossRef]
67. Duarte, S.X.; Kagan, N. A Power-Quality Index to Assess the Impact of Voltage Harmonic Distortions and Unbalance to Three-Phase Induction Motors. *IEEE Trans. Power Deliv.* **2010**, *25*, 1846–1854. [CrossRef]
68. *European Standard, EN 50160*; Voltage Characteristics of Electricity Supplied by Public Electricity Networks. CENELEC: Brussels, Belgium, 2011.

69. Bollen, M.H.J.; Gu, I.Y.-H. *Signal Processing of Power Quality Disturbances*; John Wiley & Sons, Inc.: Hoboken, NJ, USA, 2006; ISBN 9780471931317.

70. Singh, B.; Chandra, A.; Al-Haddad, K. *Power Quality: Problems and Mitigation Techniques*, 1st ed.; John Wiley and Sons Ltd.: Chichester, UK, 2015; ISBN 9781118922057.

71. *IEEE Standard 1159-2009 (Revision IEEE Std 1159-2009)*; IEEE Recommended Practice for Monitoring Electric Power Quality. IEEE Power and Energy Society: New York, NY, USA, 2019; Volume 2019, ISBN 9780738159393.

72. Ma, Y.; Li, Q.; Chen, H.; Li, H.; Lei, Y. Voltage Transient Disturbance Detection Based on the RMS Values of Segmented Differential Waveforms. *IEEE Access* **2021**, *9*, 144514–144529. [CrossRef]

73. Kahingala, T.D.; Perera, S.; Jayatunga, U.; Agalgaonkar, A.P. Network-Wide Influence of a STATCOM Configured for Voltage Unbalance Mitigation. *IEEE Trans. Power Deliv.* **2020**, *35*, 1602–16053. [CrossRef]

74. Rodriguez-Guerrero, M.A.; Jaen-Cuellar, A.Y.; Carranza-Lopez-Padilla, R.D.; Osornio-Rios, R.A.; Herrera-Ruiz, G.; Romero-Troncoso, R.d.J. Hybrid Approach Based on GA and PSO for Parameter Estimation of a Full Power Quality Disturbance Parameterized Model. *IEEE Trans. Ind. Inform.* **2018**, *14*, 1016–1028. [CrossRef]

75. Saadat, A.; Hooshmand, R.-A.; Tadayon, M. Flicker Propagation Pricing in Power Systems Using a New Short-Circuit-Based Method for Determining the Flicker Transfer Coefficient. *IEEE Trans. Instrum. Meas.* **2021**, *70*, 1–9. [CrossRef]

76. Wang, Z.; Wang, Y.; Xu, L. Parameter Estimation of Hybrid Linear Frequency Modulation-Sinusoidal Frequency Modulation Signal. *IEEE Signal Process. Lett.* **2017**, *24*, 1238–1241. [CrossRef]

77. Zhang, D.; Liu, T.; He, C.; Wu, T. A New 2-D Multi-Slice Time-Stepping Finite Element Method and Its Application in Analyzing the Transient Characteristics of Induction Motors under Symmetrical Sag Conditions. *IEEE Access* **2018**, *6*, 47036–47046. [CrossRef]

78. Gomez, J.C.; Morcos, M.M.; Reineri, C.A.; Campetelli, G.N. Behavior of Induction Motor Due to Voltage Sags and Short Interruptions. *IEEE Trans. Power Deliv.* **2002**, *17*, 434–440. [CrossRef]

79. Quispe, E.C.; Lopez-Fernandez, X.M.; Mendes, A.M.S.; Marques Cardoso, A.J.; Palacios, J.A. Influence of the Positive Sequence Voltage on the Derating of Three-Phase Induction Motors under Voltage Unbalance. In Proceedings of the 2013 International Electric Machines & Drives Conference, Chicago, IL, USA, 12–15 May 2013; pp. 100–105. [CrossRef]

80. Gnacinski, P.; Tarasiuk, T.; Mindykowski, J.; Peplinski, M.; Gorniak, M.; Hallmann, D.; Pillat, A. Power Quality and Energy-Efficient Operation of Marine Induction Motors. *IEEE Access* **2020**, *8*, 152193–152203. [CrossRef]

81. Wang, Z.; Zhu, K.; Wang, X. An Analytical Method to Calculate Critical Clearance Time of Symmetrical Voltage Sags for Induction Motors. *Dianwang Jishu/Power Syst. Technol.* **2014**, *38*, 509–514. [CrossRef]

82. Ukil, A.; Bloch, R.; Andenna, A. Estimation of Induction Motor Operating Power Factor from Measured Current and Manufacturer Data. *IEEE Trans. Energy Convers.* **2011**, *26*, 699–706. [CrossRef]

83. Sampath Kumar, P.; Xie, L.; Soong, B.H.; Lee, M.Y. Feasibility for Utilizing IEEE 802.15.4 Compliant Radios inside Rotating Electrical Machines for Wireless Condition Monitoring Applications. *IEEE Sens. J.* **2018**, *18*, 4293–4302. [CrossRef]

84. Dugan, R.C.; McGranaghan, M.F.; Santoso, S.; Beaty, H.W. *Electrical Power Systems Quality*, 3rd ed.; McGraw-Hill: New York, NY, USA, 2012; ISBN 9780071761550.

85. Kanirajan, P.; Suresh Kumar, V. Power Quality Disturbance Detection and Classification Using Wavelet and RBFNN. *Appl. Soft Comput. J.* **2015**, *35*, 470–481. [CrossRef]

86. Huang, N.; Wang, D.; Lin, L.; Cai, G.; Huang, G.; Du, J.; Zheng, J. Power Quality Disturbances Classification Using Rotation Forest and Multi-resolution Fast S-transform with Data Compression in Time Domain. *IET Gener. Transm. Distrib.* **2019**, *13*, 5091–5101. [CrossRef]

87. Kapisch, E.B.; Filho, L.M.A.; Silva, L.R.M.; Duque, C.A. Novelty Detection in Power Quality Signals with Surrogates: A Time-Frequency Technique. In Proceedings of the 2020 International Conference on Systems, Signals and Image Processing (IWSSIP), Niteroi, Brazil, 1–3 July 2020; pp. 373–378. [CrossRef]

88. Qiu, W.; Tang, Q.; Liu, J.; Teng, Z.; Yao, W. Power Quality Disturbances Recognition Using Modified S Transform and Parallel Stack Sparse Auto-Encoder. *Electr. Power Syst. Res.* **2019**, *174*, 105876. [CrossRef]

89. Costa, B.; De Souza, B.A.; Brito, D. Detection and Classification of Transient Disturbances in Power Systems. *IEEJ Trans. Power Energy* **2010**, *130*, 910–916. [CrossRef]

90. Kumar, R.; Singh, B.; Shahani, D.T. Symmetrical Components-Based Modified Technique for Power-Quality Disturbances Detection and Classification. *IEEE Trans. Ind. Appl.* **2016**, *52*, 3443–3450. [CrossRef]

91. Wang, J.; Xu, Z.; Che, Y. Power Quality Disturbance Classification Based on DWT and Multilayer Perceptron Extreme Learning Machine. *Appl. Sci.* **2019**, *9*, 2315. [CrossRef]

92. Olshausen, B.A.; Field, D.J. Sparse Coding with an Overcomplete Basis Set: A Strategy Employed by V1? *Vision Res.* **1997**, *37*, 3311–3325. [CrossRef]

93. Sahani, M.; Dash, P.K. FPGA-Based Online Power Quality Disturbances Monitoring Using Reduced-Sample HHT and Class-Specific Weighted RVFLN. *IEEE Trans. Ind. Inform.* **2019**, *15*, 4614–4623. [CrossRef]

94. Sahani, M.; Dash, P.K. Automatic Power Quality Events Recognition Based on Hilbert Huang Transform and Weighted Bidirectional Extreme Learning Machine. *IEEE Trans. Ind. Inform.* **2018**, *14*, 3849–3858. [CrossRef]

95. LI, P.; GAO, J.; XU, D.; WANG, C.; YANG, X. Hilbert-Huang Transform with Adaptive Waveform Matching Extension and Its Application in Power Quality Disturbance Detection for Microgrid. *J. Mod. Power Syst. Clean Energy* **2016**, *4*, 19–27. [CrossRef]

96. Zhong, T.; Zhang, S.; Cai, G.; Li, Y.; Yang, B.; Chen, Y. Power Quality Disturbance Recognition Based on Multiresolution S-Transform and Decision Tree. *IEEE Access* **2019**, *7*, 88380–88392. [CrossRef]

97. Li, J.; Teng, Z.; Tang, Q.; Song, J. Detection and Classification of Power Quality Disturbances Using Double Resolution S-Transform and DAG-SVMs. *IEEE Trans. Instrum. Meas.* **2016**, *65*, 2302–2312. [CrossRef]

98. Cortes-Robles, O.; Barocio, E.; Segundo, J.; Guillen, D.; Olivares-Galvan, J.C. A Qualitative-Quantitative Hybrid Approach for Power Quality Disturbance Monitoring on Microgrid Systems. *Measurement* **2020**, *154*, 107453. [CrossRef]

99. Ray, P.K.; Mohanty, S.R.; Kishor, N. Classification of Power Quality Disturbances Due to Environmental Characteristics in Distributed Generation System. *IEEE Trans. Sustain. Energy* **2013**, *4*, 302–313. [CrossRef]

100. Lopez-Ramirez, M.; Cabal-Yepez, E.; Ledesma-Carrillo, L.; Miranda-Vidales, H.; Rodriguez-Donate, C.; Lizarraga-Morales, R. FPGA-Based Online PQD Detection and Classification through DWT, Mathematical Morphology and SVD. *Energies* **2018**, *11*, 769. [CrossRef]

101. Liu, H.; Hussain, F.; Shen, Y.; Arif, S.; Nazir, A.; Abubakar, M. Complex Power Quality Disturbances Classification via Curvelet Transform and Deep Learning. *Electr. Power Syst. Res.* **2018**, *163*, 1–9. [CrossRef]

102. Cortes-Robles, O.; Barocio, E.; Obushevs, A.; Korba, P.; Sevilla, F.R.S. Fast-Training Feedforward Neural Network for Multi-Scale Power Quality Monitoring in Power Systems with Distributed Generation Sources. *Meas. J. Int. Meas. Confed.* **2021**, *170*, 108690. [CrossRef]

103. Wang, S.; Chen, H. A Novel Deep Learning Method for the Classification of Power Quality Disturbances Using Deep Convolutional Neural Network. *Appl. Energy* **2019**, *235*, 1126–1140. [CrossRef]

104. Gonzalez-Abreu, A.-D.; Delgado-Prieto, M.; Osornio-Rios, R.-A.; Saucedo-Dorantes, J.-J.; Romero-Troncoso, R.-J. A Novel Deep Learning-Based Diagnosis Method Applied to Power Quality Disturbances. *Energies* **2021**, *14*, 2839. [CrossRef]

105. Zhang, L.; Bollen, M.H.J. Characteristic of Voltage Dips (Sags) in Power Systems. In Proceedings of the 8th International Conference on Harmonics and Quality of Power. Proceedings (Cat. No.98EX227), Athens, Greece, 14–16 October 1998; Vloume 1, pp. 555–560. [CrossRef]

106. Khetarpal, P.; Tripathi, M.M. A Critical and Comprehensive Review on Power Quality Disturbance Detection and Classification. *Sustain. Comput. Inform. Syst.* **2020**, *28*, 100417. [CrossRef]

107. Yazdani-Asrami, M.; Taghipour-Gorjikolaie, M.; Song, W.; Zhang, M.; Yuan, W. Prediction of Nonsinusoidal AC Loss of Superconducting Tapes Using Artificial Intelligence-Based Models. *IEEE Access* **2020**, *8*, 207287–207297. [CrossRef]

108. Elvira-Ortiz, D.A.; Jaen-Cuellar, A.Y.; Morinigo-Sotelo, D.; Morales-Velazquez, L.; Osornio-Rios, R.A.; Romero-Troncoso, R.d.J. Genetic Algorithm Methodology for the Estimation of Generated Power and Harmonic Content in Photovoltaic Generation. *Appl. Sci.* **2020**, *10*, 542. [CrossRef]

109. De Oliveira, R.A.; Ravindran, V.; Rönnberg, S.K.; Bollen, M.H.J. Deep Learning Method with Manual Post-Processing for Identification of Spectral Patterns of Waveform Distortion in PV Installations. *IEEE Trans. Smart Grid* **2021**, *12*, 5444–5456. [CrossRef]

110. Minh Khoa, N.; Van Dai, L. Detection and Classification of Power Quality Disturbances in Power System Using Modified-Combination between the Stockwell Transform and Decision Tree Methods. *Energies* **2020**, *13*, 3623. [CrossRef]

111. Jamali, S.; Farsa, A.R.; Ghaffarzadeh, N. Identification of Optimal Features for Fast and Accurate Classification of Power Quality Disturbances. *Measurement* **2018**, *116*, 565–574. [CrossRef]

112. Gerek, Ö.N.; Ece, D.G. Power-Quality Event Analysis Using Higher Order Cumulants and Quadratic Classifiers. *IEEE Trans. Power Deliv.* **2006**, *21*, 883–889. [CrossRef]

113. Manikandan, M.S.; Samantaray, S.R. Detection and Classification of Power Quality Disturbances Using Signal Sparse Signal Decomposition on Hybrid Dictionares. *IEEE Trans. Instrum. Meas.* **2015**, *64*, 27–38. [CrossRef]

114. Abdelsalam, A.A.; Abdelaziz, A.Y.; Kamh, M.Z. A Generalized Approach for Power Quality Disturbances Recognition Based on Kalman Filter. *IEEE Access* **2021**, *9*, 93614–93628. [CrossRef]

115. Chakravorti, T.; Nayak, N.R.; Bisoi, R.; Dash, P.K.; Tripathy, L. A New Robust Kernel Ridge Regression Classifier for Islanding and Power Quality Disturbances in a Multi Distributed Generation Based Microgrid. *Renew. Energy Focus* **2019**, *28*, 78–99. [CrossRef]

116. Mishra, M. Power Quality Disturbance Detection and Classification Using Signal Processing and Soft Computing Techniques: A Comprehensive Review. *Int. Trans. Electr. Energy Syst.* **2019**, *29*, e12008. [CrossRef]

117. Liu, J.; Song, H.; Sun, H.; Zhao, H. High-Precision Identification of Power Quality Disturbances Under Strong Noise Environment Based on FastICA and Random Forest. *IEEE Trans. Ind. Inform.* **2021**, *17*, 377–387. [CrossRef]

118. Singh, U.; Singh, S.N. A New Optimal Feature Selection Scheme for Classification of Power Quality Disturbances Based on Ant Colony Framework. *Appl. Soft Comput.* **2019**, *74*, 216–225. [CrossRef]

119. Karasu, S.; Saraç, Z. Classification of Power Quality Disturbances by 2D-Riesz Transform, Multi-Objective Grey Wolf Optimizer and Machine Learning Methods. *Digit. Signal Process.* **2020**, *101*, 102711. [CrossRef]

120. Kumar, P.K.A.; Vijayalakshmi, V.J.; Karpagam, J.; Hemapriya, C.K. Classification of Power Quality Events Using Support Vector Machine and S-Transform. In Proceedings of the 2016 2nd International Conference on Contemporary Computing and Informatics (IC3I), Greater Noida, India, 14–17 December 2016; IEEE: Piscataway, NJ, USA, 2016; Volume 7, pp. 279–284. [CrossRef]

121. Zhu, R.; Gong, X.; Hu, S.; Wang, Y. Power Quality Disturbances Classification via Fully-Convolutional Siamese Network and k-Nearest Neighbor. *Energies* **2019**, *12*, 4732. [CrossRef]

122. Khodayar, M.; Liu, G.; Wang, J.; Khodayar, M.E. Deep learning in power systems research: A review. *CSEE J. Power Energy Syst.* **2021**, *7*, 209–220. [CrossRef]

123. Cai, K.; Hu, T.; Cao, W.; Li, G. Classifying Power Quality Disturbances Based on Phase Space Reconstruction and a Convolutional Neural Network. *Appl. Sci.* **2019**, *9*, 3681. [CrossRef]

124. Liu, H.; Hussain, F.; Yue, S.; Yildirim, O.; Yawar, S.J. Classification of Multiple Power Quality Events via Compressed Deep Learning. *Int. Trans. Electr. Energy Syst.* **2019**, *29*, e12010. [CrossRef]

125. Shen, Y.; Abubakar, M.; Liu, H.; Hussain, F. Power Quality Disturbance Monitoring and Classification Based on Improved PCA and Convolution Neural Network for Wind-Grid Distribution Systems. *Energies* **2019**, *12*, 1280. [CrossRef]

126. Thirumala, K.; Prasad, M.S.; Jain, T.; Umarikar, A.C. Tunable-Q Wavelet Transform and Dual Multiclass SVM for Online Automatic Detection of Power Quality Disturbances. *IEEE Trans. Smart Grid* **2018**, *9*, 3018–3028. [CrossRef]

127. Dalai, S.; Dey, D.; Chatterjee, B.; Chakravorti, S.; Bhattacharya, K. Cross-Spectrum Analysis-Based Scheme for Multiple Power Quality Disturbance Sensing Device. *IEEE Sens. J.* **2014**, *15*, 3989–3997. [CrossRef]

128. Thirumala, K.; Pal, S.; Jain, T.; Umarikar, A.C. A Classification Method for Multiple Power Quality Disturbances Using EWT Based Adaptive Filtering and Multiclass SVM. *Neurocomputing* **2019**, *334*, 265–274. [CrossRef]

129. Mahela, O.P.; Shaik, A.G. Recognition of Power Quality Disturbances Using S-Transform Based Ruled Decision Tree and Fuzzy C-Means Clustering Classifiers. *Appl. Soft Comput.* **2017**, *59*, 243–257. [CrossRef]

130. Bravo-Rodríguez, J.C.; Torres, F.J.; Borrás, M.D. Hybrid Machine Learning Models for Classifying Power Quality Disturbances: A Comparative Study. *Energies* **2020**, *13*, 2761. [CrossRef]

131. Li, Y.; Saha, T.K.; Krause, O.; Cao, Y.; Rehtanz, C. An Inductively Active Filtering Method for Power-Quality Improvement of Distribution Networks with Nonlinear Loads. *IEEE Trans. Power Deliv.* **2013**, *28*, 2465–2473. [CrossRef]

132. Che, L.; Liu, X.; Shuai, Z. Optimal Transmission Overloads Mitigation Following Disturbances in Power Systems. *IEEE Trans. Ind. Inform.* **2019**, *15*, 2592–2604. [CrossRef]

133. Song, Z.; Zhang, Z.; Komurcugil, H.; Lee, C.H.T. Controller-Based Periodic Disturbance Mitigation Techniques for Three-Phase Two-Level Voltage-Source Converters. *IEEE Trans. Ind. Inform.* **2021**, *17*, 6553–6568. [CrossRef]

134. Alkahtani, A.A.; Alfalahi, S.T.; Athamneh, A.A.; Al-Shetwi, A.Q.; Mansor, M.B.; Hannan, M.A.; Agelidis, V.G. Power Quality in Microgrids Including Supraharmonics: Issues, Standards, and Mitigations. *IEEE Access* **2020**, *8*, 127104–127122. [CrossRef]

135. Tennakoon, S.; Perera, S.; Robinson, D. Flicker Attenuation—Part II: Transfer Coefficients for Regular Voltage Fluctuations in Radial Power Systems with Induction Motor Loads. *IEEE Trans. Power Deliv.* **2008**, *23*, 1215–1221. [CrossRef]

136. Bollen, M.H.J. The Influence of Motor Reacceleration on Voltage Sags. *IEEE Trans. Ind. Appl.* **1995**, *31*, 667–674. [CrossRef]

137. Gnacinski, P.; Peplinski, M.; Murawski, L.; Szelezinski, A. Vibration of Induction Machine Supplied with Voltage Containing Subharmonics and Interharmonics. *IEEE Trans. Energy Convers.* **2019**, *34*, 1928–1937. [CrossRef]

138. Hasan, S.; Nair, A.R.; Bhattarai, R.; Kamalasadan, S.; Muttaqi, K.M. A Coordinated Optimal Feedback Control of Distributed Generators for Mitigation of Motor Starting Voltage Sags in Distribution Networks. *IEEE Trans. Ind. Appl.* **2020**, *56*, 864–875. [CrossRef]

139. Wang, Y.J. Analysis of Effects of Three-Phase Voltage Unbalance on Induction Motors with Emphasis on the Angle of the Complex Voltage Unbalance Factor. *IEEE Trans. Energy Convers.* **2001**, *16*, 270–275. [CrossRef]

140. Gnacinski, P.; Mindykowski, J.; Peplinski, M.; Tarasiuk, T.; Costa, J.D.; Assuncao, M.; Silveira, L.; Zakharchenko, V.; Drankova, A.; Mukha, M.; et al. Coefficient of Voltage Energy Efficiency. *IEEE Access* **2020**, *8*, 75043–75059. [CrossRef]

141. Liu, J.; Pillay, P.; Douglas, H. Wavelet Modeling of Motor Drives Applied to the Calculation of Motor Terminal Overvoltages. *IEEE Trans. Ind. Electron.* **2004**, *51*, 61–66. [CrossRef]

142. Hasan, S.; Muttaqi, K.M.; Bhattarai, R.; Kamalasadan, S. A Coordinated Control Approach for Mitigation of Motor Starting Voltage Dip in Distribution Feeders. *2018 IEEE Ind. Appl. Soc. Annu. Meet. IAS 2018* **2018**, 1–6. [CrossRef]

143. Anwari, M.; Hiendro, A. New Unbalance Factor for Estimating Performance of a Three-Phase Induction Motor with Under- and Overvoltage Unbalance. *IEEE Trans. Energy Convers.* **2010**, *25*, 619–625. [CrossRef]

144. Hasan, S.; Gurung, N.; Muttaqi, K.M.; Kamalasadan, S. Electromagnetic Field-Based Control of Distributed Generator Units to Mitigate Motor Starting Voltage Dips in Power Grids. *IEEE Trans. Appl. Supercond.* **2019**, *29*, 1–4. [CrossRef]

145. Mohammadi-Ghazi, R.; Marzouk, Y.M.; Büyüköztürk, O. Conditional classifiers and boosted conditional Gaussian mixture model for novelty detection. *Pattern Recognit.* **2018**, *81*, 601–614. [CrossRef]

146. Gu, X.; Yang, S.; Sui, Y.; Papatheou, E.; Ball, A.D.; Gu, F. Real-time novelty detection of an industrial gas turbine using performance deviation model and extreme function theory. *Measurement* **2021**, *178*, 109339. [CrossRef]

147. Wang, H.; Jiang, K.; Shahidehpour, M.; He, B. Reduced-Order State Space Model for Dynamic Phasors in Active Distribution Networks. *IEEE Trans. Smart Grid* **2020**, *11*, 1928–1941. [CrossRef]

148. Fink, O.; Zio, E.; Weidmann, U. Novelty detection by multivariate kernel density estimation and growing neural gas algorithm. *Mech. Syst. Signal Process.* **2015**, *50–51*, 427–436. [CrossRef]

149. Surace, C.; Worden, K. Novelty detection in a changing environment: A negative selection approach. *Mech. Syst. Signal Process.* **2010**, *24*, 1114–1128. [CrossRef]

150. Naghavi-Nozad, S.A.; Amir-Haeri, M.; Folino, G. SDCOR: Scalable density-based clustering for local outlier detection in massive-scale datasets. *Knowl.-Based Syst.* **2021**, *228*, 107256. [CrossRef]

151. Shah, A.; Azam, N.; Ali, B.; Khan, M.T.; Yao, J. A three-way clustering approach for novelty detection. *Inf. Sci.* **2021**, *569*, 650–668. [CrossRef]

152. Salehi, M.; Arya, A.; Pajoum, B.; Otoofi, M.; Shaeiri, A.; Rohban, M.H.; Rabiee, H.R. ARAE: Adversarially robust training of autoencoders improves novelty detection. *Neural Netw.* **2021**, *144*, 726–736. [CrossRef] [PubMed]

153. La-Grassa, R.; Gallo, I.; Landro, N. OCmst: One-class novelty detection using convolutional neural network and minimum spanning trees. *Pattern Recognit. Lett.* 2021, *in press*. [CrossRef]

154. Hu, W.; Hu, T.; Wei, Y.; Lou, J.; Wang, S. Global Plus Local Jointly Regularized Support Vector Data Description for Novelty Detection, IEEE Trans. *Neural Netw. Learn. Syst.* 2021, 1–13, *in press*. [CrossRef]

155. Martinez-Viol, V.; Urbano, E.M.; Kampouropoulos, K.; Delgado-Prieto, M.; Romeral, L. Support vector machine based novelty detection and FDD framework applied to building AHU systems. In Proceedings of the 2020 25th IEEE International Conference on Emerging Technologies and Factory Automation (ETFA), Vienna, Austria, 8–11 September 2020; Volume 1, pp. 1749–1754. [CrossRef]

156. Mataifa, H.; Krishnamurthy, S.; Kriger, C. Volt/VAR Optimization: A Survey of Classical and Heuristic Optimization Methods. *IEEE Access* **2022**, *10*, 13379–13399. [CrossRef]

157. Guirguis, D.; Aulig, N.; Picelli, R.; Zhu, B.; Zhou, Y.; Vicente, W.; Iorio, F.; Olhofer, M.; Matusik, W.; Coello, C.A.; et al. Evolutionary Black-Box Topology Optimization: Challenges and Promises. *IEEE Trans. Evol. Comput.* **2020**, *24*, 613–633. [CrossRef]

158. Zhao, Q.; Li, C. Two-Stage Multi-Swarm Particle Swarm Optimizer for Unconstrained and Constrained Global Optimization. *IEEE Access 2* **0220**, *8*, 124905–124927. [CrossRef]

159. Xu, X.; Yan, X.; Sheng, C.; Yuan, C.; Xu, D.; Yang, J. A Belief Rule-Based Expert System for Fault Diagnosis of Marine Diesel Engines. *IEEE Trans. Syst. Man Cybern. Syst.* **2020**, *50*, 656–672. [CrossRef]

160. Choi, K.; Jang, D.; Kang, S.; Lee, J.; Chung, T.; Kim, H. Hybrid Algorithm Combing Genetic Algorithm with Evolution Strategy for Antenna Design. *IEEE Trans. Magn.* **2016**, *52*, 1–4. [CrossRef]

Review

Digital Twin in Electrical Machine Control and Predictive Maintenance: State-of-the-Art and Future Prospects

Georgios Falekas and Athanasios Karlis *

Department of Electrical and Computer Engineering, Democritus University of Thrace, 67100 Xanthi, Greece; gfalekas@ee.duth.gr
* Correspondence: akarlis@ee.duth.gr

Abstract: State-of-the-art Predictive Maintenance (PM) of Electrical Machines (EMs) focuses on employing Artificial Intelligence (AI) methods with well-established measurement and processing techniques while exploring new combinations, to further establish itself a profitable venture in industry. The latest trend in industrial manufacturing and monitoring is the Digital Twin (DT) which is just now being defined and explored, showing promising results in facilitating the realization of the Industry 4.0 concept. While PM efforts closely resemble suggested DT methodologies and would greatly benefit from improved data handling and availability, a lack of combination regarding the two concepts is detected in literature. In addition, the next-generation-Digital-Twin (nexDT) definition is yet ambiguous. Existing DT reviews discuss broader definitions and include citations often irrelevant to PM. This work aims to redefine the nexDT concept by reviewing latest descriptions in broader literature while establishing a specialized denotation for EM manufacturing, PM, and control, encapsulating most of the relevant work in the process, and providing a new definition specifically catered to PM, serving as a foundation for future endeavors. A brief review of both DT research and PM state-of-the-art spanning the last five years is presented, followed by the conjunction of core concepts into a definitive description. Finally, surmised benefits and future work prospects are reported, especially focused on enabling PM state-of-the-art in AI techniques.

Keywords: electrical machines; predictive maintenance; digital twin; artificial intelligence; Industry 4.0; data handling; life cycle

Citation: Falekas, G.; Karlis, A. Digital Twin in Electrical Machine Control and Predictive Maintenance: State-of-the-Art and Future Prospects. *Energies* **2021**, *14*, 5933. https://doi.org/10.3390/en14185933

Academic Editor: Anouar Belahcen

Received: 18 August 2021
Accepted: 15 September 2021
Published: 18 September 2021

Publisher's Note: MDPI stays neutral with regard to jurisdictional claims in published maps and institutional affiliations.

1. Introduction

Research activity in Electrical Machine (EM) Predictive Maintenance (PM) observes renewed interest in recent years as industrial and commercial applications diversify and expand into novel areas, while their role in them becomes more prominent. Conventionally, the sector's efforts have been focused on Squirrel Cage (SC) Induction Motors (IMs) and conventional rotor Synchronous Machines (SMs), owing to their domination of motor and generator applications, respectively. Nowadays, other types of EMs observe an increase in their relative usage in lieu of SCIMs and SMs, such as Wound-Rotor (WR) IMs (especially as generators in wind turbines), Permanent Magnet (PM) SMs (mainly due to electric vehicles), multiphase Alternating Current (AC) machines, and Switched or Synchronous Reluctance Machines (SRMs or SyncRMs). This shift is attributed to new materials, design and control architectures, the exponential expansion of generators in Renewable Energy Sources (RES), and specialized requirements of new applications such as spatial availability (power density), efficiency targets and fault-tolerant systems. Moreover, substantial increment in computing power, sensor advancements, artificial intelligence and Internet-of-Things (IoT) applications enable the discovery of new research avenues with novel approaches and powerful tools both in laboratory tests and industrial (on-site) applications, promoting a combination of different techniques.

EMs typically constitute the core part of their application and thus dictate its entire condition and performance. As electromechanical processes, they are prone to different faults of various severity. Breakdown during runtime leads to severe economical and safety repercussions (i.e., significant repair costs, unscheduled production halt, increased man-hours, missed deadlines) while faulty operation significantly reduces efficiency and is a safety hazard. Consequently, timely and correct fault diagnosis is of paramount importance to the industry by way of scheduling necessary repairs during downtime and before breakdown. In addition, online Condition Monitoring (CM), which is the main PM operation, ensures that processes run with optimal efficiency, further cutting down on operating costs and needed reserves. These efforts are further reinforced by novel cost-effective sensors, Data Acquisition (DAQ) and evaluation techniques, making EM diagnostics a significantly profitable venture in industry.

Great reviews, such as [1–6], concerning state-of-the-art PM methods and their application have been published in recent years, addressing techniques with their applications and comparison. Literature highlights a need for evaluation benchmarks and new combinations. The authors of this work hypothesize that this endeavor fits the paradigm of creating a Digital Twin of a part or the entirety of the studied system, as it was coined by Grieves [7] and expanded upon in recent years [8]. While the next-generation-Digital Twin (nexDT) concept is increasingly explored in recent works, the authors noted a lack thereof concerning specialized EM CM and control, thus inspiring this review. This work reviews DT literature concerning EM CM concepts (rotating machinery, electromechanical systems, PM applications) spanning the last five years, to evaluate its usability in EM CM and control going forward in Industry 4.0., as well as retrofitting and reverse engineering applications of older machines.

The main aim of this review is to combine literature descriptions of the nexDT, consider fellow researchers' concerns, emphasis, and preliminary work and then merge them with the well-established PM paradigms into a definitive description for the realization and usage of the nexDT in EM PM, in an effort to synchronize research efforts. Future prospects include the conjunction of a sensor setup and its connection to a Digital Twin via the appropriate interface and techniques, in order to set the stage for designing a state-of-the-art CM test bench with the goal of making it portable for industrial application. Surmised benefits are three-fold: establishing and validating the sector's cutting-edge approach, extracting and unifying manufacturing and operation data, and combining the latest proposed techniques of various sciences. Research has come to the consensus that a combination of methodologies is needed for achieving the optimal diagnostic procedure, namely obtaining a correct assessment with the minimum measurements in minimum time. According to the latest literature, DT is the prime candidate in this effort and the industrial and research state provides the perfect storm for its development at the current time. It is surmised that such a framework can facilitate future research efforts and aid benchmarking. To the best of the authors' knowledge and research, a focused review of EM DT has not been yet realized. The DT concept is picking up speed in the latest decade and is predicted to become the new paradigm in industry. Its synergy with EMs is even higher compared to various industrial applications, constituting its exploration mandatory. Ref. [9] provides a specialized review concerning EMs, explaining the concepts introduced.

All citations and studied work were found via the Google Scholar database by employing the following search keywords: digital twin; digital twin electrical machine; digital twin electrical machine fault diagnosis; digital twin electrical machine fault diagnosis review; digital twin machine fault; digital twin fault diagnosis; industry 4.0; digital twin software. This search yielded a low number of relevant works pertaining to usage of DT in EM CM and PM compared to broader DT concepts. While previous literature research into predictive maintenance seldom refers to abstract DT concepts, no established parallels are drawn. Choice of keywords was based on the DT review papers' (discussed below) methodology of selecting relevant work, in addition to a further limitation of including the keyword "digital twin" in search. This was done to establish that specifically searching PM DT will not guide the user to pertinent work. The authors noted a lack of relevant literature even when expanding the search terms to a broader spectrum. While general consensus regarding the concept can be derived from the results, we surmise that a specific review will enable and accelerate related work, serving as a starting point for novice and experienced researchers alike.

This work is divided into three major parts: review of DT literature as explained above, a brief report on EM PM state-of-the-art as provided by the related reviews mentioned above, and finally the conjoined report on most important aspects and suggestions regarding the two sectors, complete with groups and visuals. To the best of their ability, the authors made an effort not to repeat the reviewed work, but rather provide an outline and create a web of relevant citations.

2. Literature Review

In a broader context, the aim of a DT is to aid in optimization problems. Ref. [10] provides an excellent insight in robust design optimization and emerging technologies, specialized for EMs. As stated by the work, design optimization requires multi-disciplinary analysis and multi-pronged investigation of the system, areas where a DT excels. The main open challenge in state-of-the-art is accurate CM of an EM, as methodologies largely depend on accurate values to produce acceptable results. Simplified models due to software, hardware or knowledge limitations may hinder otherwise productive algorithms. The second open challenge of benchmarking and evaluation of novel AI techniques stems from the main, as it depends on the quality of data produced. Thus, the capability of a DT to generate a virtual copy of the system can prove invaluable to literature in both design and real-time monitoring stages.

According to [11], limitations in concurrent CM of industrial EMs have been exposed. Core problems concern modeling and DAQ fidelity, both of which are in the forefront of DT research. Secondary advances required are in EM degradation mechanisms, allowing for faster and more robust PM. Novel AI techniques are yet to be accepted in the field, requiring extensive testing, hindered by conventional modeling techniques. Literature concludes that combination of different CM methodologies is required in order to get a complete and reliable overview of the system. Common surveys (with their respective mediums) include: insulation testing (partial discharges), electrical testing (current spectrum), flux analysis (stray flux). The latest tendencies concern transient analysis. Steady-state analysis employs FEM to model the machine in greater detail. Focus has shifted into flux analysis, due to sensory advancements and its richer harmonic content [4]. Cutting edge approaches investigate AI and Fuzzy Cognitive Maps (FCM) decision making to distinguish fault indications. Latest trends in depth can be found in the cited work. We surmise that the DT can solve the combination of the aforementioned challenges and is a worthwhile venture in both literature and industry. Details are discussed in the framework presentation.

2.1. Digital Twin Reviews

Five major DT reviews [8,12–15] were chosen to be studied in the context of this work. The primary reason is their latest publishing date, while the secondary reason is their review of older DT reviews. Each review takes a slightly different approach. The goal is to provide an outline of the latest concept while discussing similarities and deviations. This work focuses on DT usage and applications in EM PM. Readers are encouraged to investigate the mentioned reviews for broader DT coverage. Ref. [16] provides a clear and categorical review of the three different digital replica variations researchers, including the above reviewers, have adopted: Digital Model (DM), Digital Shadow (DS), and Digital Twin. A clear data-flow categorization is used, as depicted in Table 1. Modeling methodology is not and should not be limited in the scope of this categorization.

This categorization is necessary in the context of EM PM, to clearly classify methodologies. One key example is the usage of DS for pure diagnosis, AI training and data logging, while the DT can also be integrated in control. Some proposed PM methodologies require the usage of bidirectional automatic data flow, meaning they should use the term

Digital Twin instead of Digital Shadow. Researchers are encouraged to abide by the above distinction.

Table 1. Digital construct categorization as established by [16] and adopted by broader literature.

Construct	Physical to Virtual Data Flow	Virtual to Physical Data Flow
Digital Model	manual	manual
Digital Shadow	automatic	manual
Digital Twin	automatic	automatic

What Is Digital Twin?

The concept's first appearance is contributed to M. Grieves, in his course on "product lifecycle management" in 2003. In his whitepaper in 2014, Grieves defines the DT [7]. In 2012, the DT concept is reevaluated by the National Aeronautics and Space Administration (NASA). After this point in time, the DT concept begins to rise exponentially in popularity. Since then, researchers have provided numerous definitions and concerns, making the DT an increasingly complete concept, which serves today's research needs. Some of the most concise and inclusive definitions are provided in chronological order, to be discussed in the scope of EM PM. The provided definitions include only historical ones in M. Grieves' and NASA for the sake of clarity, and some of those reviewed and provided by the studied papers. Only definitions that insert additional clarifications are included.

M. Grieves defines the DT in his whitepaper [7] as a "Virtual representation of what has been produced", after having discussed it in detail in previous years. This broader definition is meant to encapsulate the complete concept of the DT in the scope of production. NASA has defined the DT in greater detail in 2012 [17], catering to a more specific use, as interpreted by Tao et al. [8]: "The DT is a multiphysics, multiscale, probabilistic, ultrafidelity simulation that reflects, in a timely manner, the state of the corresponding twin based on the historical data, real-time sensor data, and physical model". New additions in definitions are mentioned by Gabor et al. [18]: "The DT is a special simulation, built based on the expert knowledge and real data collected from the existing system, to realize a more accurate simulation in different scales of time and space", Chen [19]: "A digital twin is a computerized model of a physical device or system that represents all functional features and links with the working elements", Zhuang et al. [20]: "Virtual, dynamic model in the virtual world that is fully consistent with its corresponding physical entity in the real world and can simulate its physical counterpart's characteristics, behavior, life, and performance, in a timely fashion", Liu et al. [21]: "The DT is a living model of the physical asset or system, which continually adapts to operational changes based on the collected online data and information and can forecast the future of the corresponding physical counterpart", Zheng et al. [22]: "A DT is a set of virtual information that fully describes a potential or actual physical production from the micro atomic to the macro geometrical level", Xu et al. [23]: "Simulates, records and improves the production process from design to retirement, including the content of virtual space, physical space and the interaction between them", Madni [24]: "A DT is a virtual instance of a physical system (twin) that is continually updated with the latter's performance, maintenance, and health status data throughout the physical system's life cycle", and Kannan and Arunachalam [25]: "Digital representation of the physical asset which can communicate, coordinate and cooperate the manufacturing process for an improved productivity and efficiency through knowledge sharing".

Core contributions of each definition are highlighted in bold. NASA's definition, originally given for a flying vehicle [17], is one of the most inclusive to start. The authors' vision, as described in the paper, is close to today's reality. In summary, the DT is a simulation with the following characteristics:

- Multiphysics, meaning cooperation of different system descriptions, such as aerodynamics, fluid dynamics, electromagnetics, tensions etc.;

- Multiscale. The DT simulation should adapt to the required depth in real time. Users can zoom into the component of a component, up to a complete view of the DT;
- Probabilistic, based on models derived from state-of-the-art analyses on each building block, to predict the future and follow the same description protocol as the real twin;
- Ultrafidelity, offering unlimited precision down to the lowest possible level. This is, of course, a compromise, often a tradeoff for computational power and time.

Originally proposed as a three-dimensional model [7], the five-dimensional extension model proposed by [8] has attracted a lot of attention in literature, according to the studied reviews ([12,15]) and reviewed work. The dimensions (initial three and extended two) are:

1. Physical;
2. Virtual;
3. Service;
4. Data;
5. Connections.

The two extra dimensions (4 and 5) are important enough in differentiating the DT from previous work to warrant a categorization akin to i.e., the physical space. It is important to note that the DT is not only a virtual representation of an object, but can encapsulate an entire process i.e., a complete diagnostic procedure, depicting the necessary equipment, data acquisition, flow and handling, connections, and algorithms. This iteration, however, can often produce more confusion than results. Researchers are encouraged to proceed as they see fit, following the established DT paradigm. It is, however, important to emphasize digitization of services and processes along with objects [19]. Correlation between these layers is depicted in Figure 1.

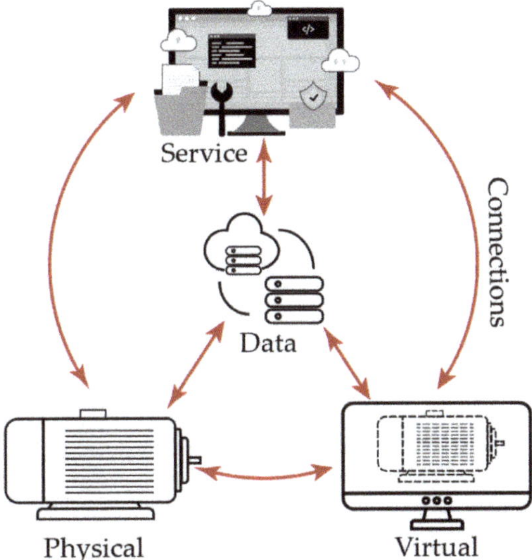

Figure 1. Five-dimensional DT model as proposed in literature [8].

The concept of the DT began in manufacturing. Operational DTs are the state-of-the-art in literature. Stemming from the previous statements, one can make a case for which portion of a system's lifetime is included in the simulation. The obvious answer is "all of them". To elaborate a bit further, we can split a system's lifetime in the following phases, as expertly studied by [15] and expanded upon in [26,27], shown in Figure 2:

1. Design phase;
2. Manufacturing phase;

3. Service phase;
4. Retire phase.

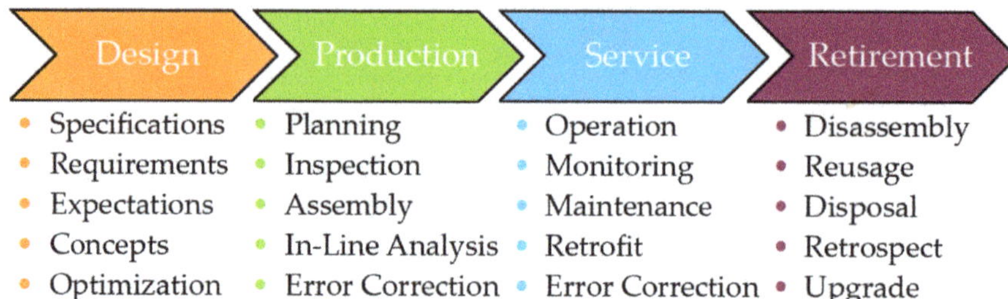

Figure 2. Summary of a product lifecycle and relevant operations.

PM is of course interested in the service phase. However, a recurring problem in relevant literature is the existence of little to no manufacturing data, which can prove invaluable for state-of-the-art diagnostics. Algorithms often must expend time and resources in evaluating systems and gathering preliminary data. Access to a DT generated in the design and manufacturing phase of the system can enable instantaneous PM methodologies, as well as strengthen existing ones with a slew of data [25]. Ref. [15] concurs that researchers often neglect the retire phase of the system. Integrating it in the DT study can provide valuable information for next generation manufacturing, completing the circle. Retrofitting older machines will be discussed in the conclusion. It is imperative that DT manifestations span the entirety of the system's lifecycle [24].

Data are considered the core of the DT, since it is this bidirectionality and processing that defines it. DT data comes from three major sources:

1. Historical data;
2. Real-time sensors;
3. Models.

DT data can be in a multitude of forms, such as physical sensor signals, virtual signals, manuals, tables, data banks. Localization can simultaneously be in the system itself, in adjacent (ancillary) systems which may or may not be part of the DT itself (although their data is), and the cloud. Furthermore, these data can be raw (i.e., voltage, current, flux, counts, dimensions) or processed (i.e., health indexes, state values, clustered or labeled). Therefore, adequate handling is of paramount importance. Data scale corresponds to the Big Data definition, which in itself is the sweeping trend in industry according to the reviews.

The most important aspect of the DT regarding PM is its definition as a "living" model [21]. The main concept of this "life" is updating the system itself to reflect the physical twin, meaning health state and history, dynamic behavior, efficiency and performance, and links with outside elements. One very challenging aspect of DT is keeping these calculations and representations "in a timely manner" as many works have stated [20]. The simplest definition of this is doing the necessary processes (handled by a CPU) before the next necessary update, according to required fidelity.

2.2. Surveyed Literature

In this section, the chosen relevant papers are reviewed shortly in ascending chronological order regarding their publication date. A brief overview of each work is given, followed by a general discussion regarding research state of the art and primary concerns.

Modeling an EM is a complex, multi-level task. The dual electro-mechanical behavior of the system, in addition to its high symmetry, requires heterogenous skills in definitively achieving a robust and reliable model. Ref. [28] provides an insight into how DT technology can model and solve EMs and their challenges in an industrial environment. The EM modeling problem stems from their application since the earliest days of modern industry, where inadequate provision of data sheets can skew modeling efforts. Furthermore, high dependence on load and environmental conditions results in a highly nonlinear and stochastic asset. Logging an EM's features requires a considerable testing effort without ensuring minimum uncertainty. Most present drives do not provide an accurate parametrization of the controlled machines. Goals are reduced sensor costs, minimized invasion and estimation accuracy. The paper proposes a reduced order FEM model of an IM. Design parameters include real-time application while preserving the adequate accuracy, managing this accuracy-computational complexity tradeoff, and realizing the DT as a virtual sensor. Proof of concept is provided via current density and thermal modeling through measurements, while highlighting the benefits of an optimized cooling system, which is frequently overused due to non-reliable temperature readings. Finally, real-time implementation is evaluated.

Ref. [29] presents a DT application of an automotive braking system PM. Although not directly related to EMs, the principle of using DT in PM remains unchanged. It is important to note that all such works combine different modeling formalisms (FEM, mathematical) and dimensionalities (0D, 2D, 3D) integrated into one master model. In this regard, universal and combinable software should be used. This work uses Modelica models and ANSYS Simplorer FEM simulations. The authors observe an increasing interest from industry giants (Siemens, GE) in employing DTs for predictive maintenance. The resulting physics-based part of the model can be subjected to various failure modes to simulate its response, while combination with machine learning algorithms can trigger pre-emptive maintenance and optimize operational downtime, while also being trained by the simulated data.

Ref. [30] provides a practical approach in industrial DT employment. The proposed application is focused on machine reconditioning projects, which involve a reverse engineering phase and short commission times due to lack of data and production timelines, respectively. The guiding principle is conforming to Industry 4.0 practices while completing the work of retrofitting to validating the machine. In order to be productive, the DT realization should pay off the extra time required to build it. It is important to note that the old machine retrofitting company typically is not the one that built the machine. The concept of Virtual Commissioning is proposed, which provides the evaluation party with a DT of the project under scrutiny, reducing travel time and costs while providing the commissioner with preliminary confidence before the real-world application. It is stated that the DT was started to be used more extensively in the early 2010s and brought upon a new wave in modelling and simulation. Since the DT realization depends on the complexity of the system itself, it is viable (scaled) for both simple and complex projects.

The studied application is the reconditioning of a core making machine used for foundry sand cores for the automotive industry, with retrofitting of the old Siemens PLC and HMI devices. The allocated time was dictated by the maximum possible stop in production, four weeks. The DT was successfully built, validated and used in validation during this time period. The used software is Simumatik3D. The simulation engines are 3D graphics, physics, and logic. Importantly, the complete model can be controlled by the same industrial controllers exactly like the real twin. The four application phases are:

1. Part analysis and reverse engineering;
2. Development;
3. Virtual Commissioning and Validation;
4. Physical Reconditioning.

Ref. [31] utilizes the DT approach to model a physics-based Remaining Useful Life (RUL) prediction model for an offshore wind turbine power converter. While the model itself is based on literature proven methodologies, the DT aspect benefits can be summarized as such: Combination of the SCADA DAQ system with physics-based models can enable medium- and short-term predictions to accompany SCADA long-term data availability, increasing the prediction accuracy by a significant margin. Each system is modeled in the appropriate setting (computational, numerical, model, FEM) and then combined via the DT framework. This enables the integration of the large number of components present in the system, while ensuring optimal modeling of each part and varying degree of precision. This situation enables operators and owners to make the important decisions of every aspect of the turbine (end-of-warranty review, inspections, life extension, re-powering, retrofitting, decommissioning) remotely and with better visualization. Furthermore, the DT platform "converts big data into manageable small data and presents it as high-level performance indicators that influence the decisions of O&M planning and execution". Once again, the DT method is employed to merge the real and virtual twins to achieve the best possible control and maintenance methods, while enabling the operators with the optimal Human Machine Interface. In this scenario, uses extend to evaluating the weather conditions prediction of the SCADA system to include operator transit planning and other miscellaneous uses indirectly related to the system maintenance. Future work of this and every proposed DT framework is two-fold: optimize the twins' correlation and discover new uses and merges for neighboring equipment and services.

Ref. [32] from Siemens AG provide the definition of the DT simulation in three major principles: linked collection (of all available data during manufacturing and service life), evolution mechanism (tracking any and all changes to the system and keeping a history), and behavior description coupled with solution provision (evaluation and decision-making aspect of the DT). The authors highlight the need to support any and all stakeholders' (manufacturers to end users) interests in forming the next generation DT, meaning a balance between information obscuring and usability. The course is to follow the current paradigm of supporting design tasks and validating system properties while completing the merge of the physical and virtual world. The key takeaway of this work is the separation of manufacturing and usage data. While corporate secrets and espionage are an important matter, combination of a product's entire life cycle data will provide researchers with a great source of data. "Digital artefacts" terminology is stated to be any and all data structs (simulations, measurements, descriptions, values). These are linked via a "Knowledge Graph", to be accessible by any stakeholder at any time. Physics-based models are used to predict the future via differential equations, while builds give an insight into the inner workings of the studied system.

The second part of the work focuses on planning the next generation DT. The first important idea is that the DT can become a product in and of itself, providing additional functionalities to consumers, and following the design principle of normal product features. The second insight is the choice of location depending on application, ranging from embedded logic to a cloud-based service on demand. The authors conclude with a brief maintenance analysis, stating that the applied logic is the deviation of the two models, and the fact that today's maintenance data are disconnected from the models and the sensor data. To summarize, the DT is the gateway to realizing new services with low effort. The stated paradigm is far from complete, and the need for additional research, benchmarks and applications is highlighted. The two main challenges are connecting the various aspects and structuring the model parts.

Refs. [33–36] make a case for the importance of the DT synergistically with Computerized Numerical Control (CNC). CNC revolutionized the industry and provided us with smarter manufacturing. The tackled problem is the dependence of simulations on user provided data and manual records, which block instantaneous response and reduce accuracy. Provision of a consistent, accurate model (via real-time and reliable data-mapping) between design and operation will take advantage of the full capabilities of CNC, namely self- sensing, prediction, and maintenance. Related work evaluation concludes with the following three approaches:

1. Interdisciplinary, mature software collaboration. Literature protagonists include MATLAB/Simulink, ANSYS and ADAMS.
2. High Level Architecture (HLA), as established by IEEE and recommended by NATO through STANAG 4603.

3.   Unified Modeling Languages (UML), such as Modelica.

Finally, a complete Modelica DT model is proposed, comprised of the physical, digital, and mapping (connection) layers. The work is finished with the common statement that DT research still requires extensive application and validation and proposing future work. The development of the suggested model is in three parts, with the second part advancing the theory. In the third work, a hybrid application for a milling head's cutting tool maintenance is performed, providing extensive results. Continuation of the above work in 2021 concerns a model consistency retention, which has been steadily gaining more traction in the nexDT research. The need for the DT to follow the performance attenuation of its real counterpart is highlighted. The authors propose a method for achieving this purpose based on their previous work, and a brief review of state-of-the-art attenuation parameters (to be chosen). A rolling guiderail is taken as a case study. Results are promising and fit the next generation DT paradigm, but as the authors themselves state, further validation is required.

Ref. [37] presents a clear paradigm for the DT. Industry 4.0 is in the process of transforming its environment into a "networked system of systems", making automation more "production friendly by being more reconfigurable and adaptable" via modular architectures. This approach is what the DT capitalizes and expands on, reshaping the pre-existing methodologies to fit the new plant environment while providing human technicians with sophisticated tools and a clear UI.

The focus of this work lies in enabling the human engineer by considering him/her as a part of the framework to increase his/her efficiency on the industrial floor. The DT can serve as a back end for future visual aiding technologies. The paper highlights modularity and describes the approaches of modeling the DT parts (component-, skill-, function-based approach). A proof of concept is provided by modeling a part using the capabilities and built-in library of Modelica while using the AutomationML data exchange protocol. Once again, the DT serves in realizing the multi-disciplinary aspect of any industrial system. Human aid can be provided in two ways:

1.   Training personnel in a simulated environment.
2.   Providing a visual aid to serve as a manual and a library during downtime.

In this work, a good overview of approaches and the HMI aspect of the DT is provided.

Ref. [38] provides an excellent proof of concept for the employment of DT in mechanical maintenance. The focus of the study is an aero-space engine main shaft bearing, which is both the key component of the engine and its weakest link, corresponding to the typical EM fault. The work begins with a quick state-of-the-art review in bearing maintenance and then proceeds to explain the integration of the DT approach. The proposed maintenance technique follows the tested modality separation (as excellently depicted in [39]), while the DT approach emphasizes data handling and processing.

The proposed model contains three kinds of elements: physical entity, virtual mode and service system. The dimensionality of the model is five, including the three entities plus DT data and connections. Furthermore, the model classifies the data via clustering techniques to assign data labels. Then, the DT can be used both in fault prediction via data feeding and fault diagnosis via real-time visualization. The process of creating the DT of the bearing is the following, illustrated in Figure 3:

1.   Geometric model, constructing its spatial representation.
2.   Physical model, inserting the physical attributes to the representation.
3.   Behavior model, integrating the reaction with the environment.
4.   Constraint model, realizing the boundaries of all operations.

This guarantees that the virtual twin operates within the scope of the real twin and thus the information between the twins can be interactively integrated. The virtual aspect of the DT is further explored via training a Neural Network to classify the state of the bearing as depicted in its virtual twin.

Finally, the implementation of creating and validating the repair process of the virtual world inquires additional research. The authors propose validating the repair process via integrating it in the DT and performing the actual physical repair only when the simulation results in an acceptable outcome. This aspect of the DT can be integral in assuring the maximum cost reduction of the maintenance process and has up until now only been a speculation.

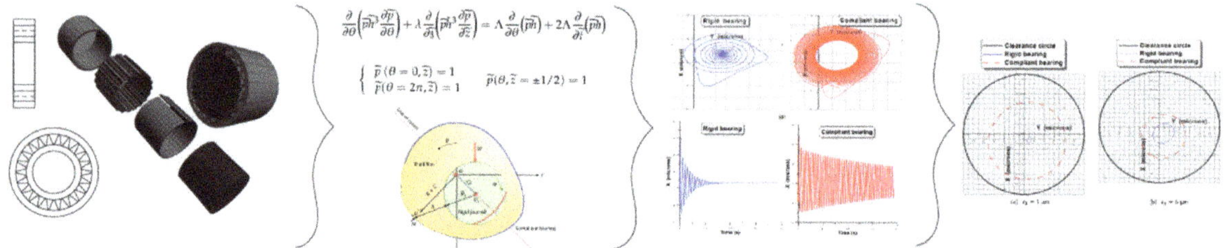

1. 2D/3D Modeling

- Dimensions
- Angles
- Arrangement

2. Differential Equations

- Mass - Density
- Laws - Fields
- Part Behavior

3. Fault Mechanisms

- Interactions
- Environment
- Total Behavior

4. Physical Constraints

- Tensions
- Materials
- Boundaries

Figure 3. Basic illustration of the creation of one complete part of the DT, as described in [38]. An important aspect of the DT is the 3D representation of the object, including physical flaws. Incorporating geometries with differential equations systems allows for greater freedom in simulation. Typical usage of FEM software. Images for illustration purposes only, courtesy of [40].

Ref. [41] proposes a theoretical DT approach in monitoring a PMSM in an EV, based on its casing temperature. The authors developed and trained an ANN and a Fuzzy Logic DT with the same inputs and outputs. An important aspect that is presented in this work is the simulation of the EV driving cycle using MATLAB Simulink and its repercussions on the motor maintenance, including RUL, time to refill bearing lubricant, and motor temperatures. An EV emphasizes the interdisciplinary nature of EM health monitoring and maintenance, being on a narrow, highly synergistic and "hostile" environment for the machine. According to the DT guideline of being modular and user-friendly, this theoretical model can be expanded and applied in a wide range of uses and users, from EV manufacturers to service companies and individual users.

The authors of [23] provide a clear and concise usage of the DT as literature suggests it should be used, combining manufacturing and operation. This work focuses on Deep-Neural Network (DNN) diagnostics assisted by the DT approach using Deep Transfer Learning (DTL) while also presenting a case study in a car body-side production line to compare the traditional and DT-based methodology. The aforementioned combination, a guideline of DT evolution, is achieved in two parts:

- The manufacturer of the system constructs a DT in their software of choice as they deem appropriate. The DT is validated and then the physical system is constructed.
- Data are amassed and learning algorithms are trained during this process. The resulting physical system is evaluated and DTL is used as the connecting utility between the real and virtual twin. The DT is used as a complete evaluation, real-time, and simulation system.

This work's contribution can be summarized as: applying the (sometimes theoretical or untested) intelligent diagnostic methods in the real environment and extending the diagnosis period from operation only to the full life cycle. This last part is a novel approach to using the DT to also optimize the manufacturing process.

The authors provide a quick, targeted explanation of the DT concept in industry, followed by a thorough review of DNN applications in predictive maintenance. The proposed work excellently combines the advantages of both DTL and DT technology, as explored and suggested by the relevant literature and the scientific consensus.

Ref. [42] presents a DT reference model for rotor unbalance diagnosis. Once again, the authors highlight the need to further investigate how to properly construct a DT to accurately represent the physical system. The trust of industry giants such as ANSYS, Oracle, SAP, Siemens and GE is confirmed. One important aspect that is proposed in this work is a model-updating algorithm, so that the virtual twin can better simulate its real counterpart. Physical systems include wear and tear. A good DT procedure should reflect this issue, which can be used in both RUL estimation during manufacturing and real-time conditioning.

The work includes a brief physical model and smart sensing description followed by the corresponding DT construction using the three blocks of digital modeling, data analytics and knowledge base. The novel model updating strategy is proposed. Finally, the proposed strategy is employed in a rotor simulation combined with a physical measurement using additional masses as rotor imbalances. The results are satisfactory and prove the usefulness of the model update.

Ref. [43] illustrates a DT proof of concept in prognostics with Low Availability Run-to-Failure (RTF) data. The retrofit solution is readily available and low-cost in the form of a Raspberry Pi accelerometer mounted to the studied machine (a drill) and feeding data into MATLAB Simulink, where a model is realized. The work provides a brief but concise explanation of Condition-Based Maintenance (CBM) and RUL prediction models. DAQ is performed by the PLC of the drilling machine and the retrofitted Pi accelerometer. The employed prognostic technique is the Exponential Degradation Model, a stochastic approach which is suited to low data availability. In addition, it is a parametrized model and can be applied to a population of machines, enhancing its diagnostic capability via their comparison. In this work, the DT is used only as a "watchdog agent", meaning as a real-time evaluator. The objective is to bring together the concepts of DT and CBM applied to rotating machinery. In the broader aspect of DT in industry, it includes adaptation to retrofitting, simple software-hardware synergy and another proof of concept.

The authors of [44] provide a thorough analysis of the RUL estimation and how this approach can be integrated in DTs. The three pillars of PM are:

1. Data extraction and processing;
2. Maintenance knowledge modeling; and
3. Advisory capabilities

All of them conform to the application of the DT in the industry. In addition, as is with expansion of the DT capabilities under the exponential growth of computational powers and sensor precision in the scope of Industry 4.0, the authors suggest the "simultaneous consideration of economic and stochastic dependence aiming at determining the optimal trade-off between reducing the RUL of components and decreasing maintenance set-up costs". The caveat of conventional RUL methodologies is mentioned again, referring to their dependence on historical data. The DT is the state-of-the-art approach in combatting this gap. The authors conclude that, due to increasing complexity, RUL estimations should be made in the component level. An optimal DT should be able to conform to the required fidelity, with the known trade-off between it and computational time. Furthermore, this work also classifies the components into three useful (in making the approach modular) categories, illustrated in Figure 4:

1. Black Boxes. These models only include inputs and outputs without knowledge of the inner workings of the component. Can be used as placeholders during the making process.
2. Grey Boxes, using theoretical data to complete the component's effects and results. Can be used in cases of missing data or choice of not including a sensor, while not having a previous iteration of it.
3. White Boxes. The exact functionality and inner workings of the model are known. These are the current final iteration of the component model and include update protocols, have undergone validation and closely resemble the real component.

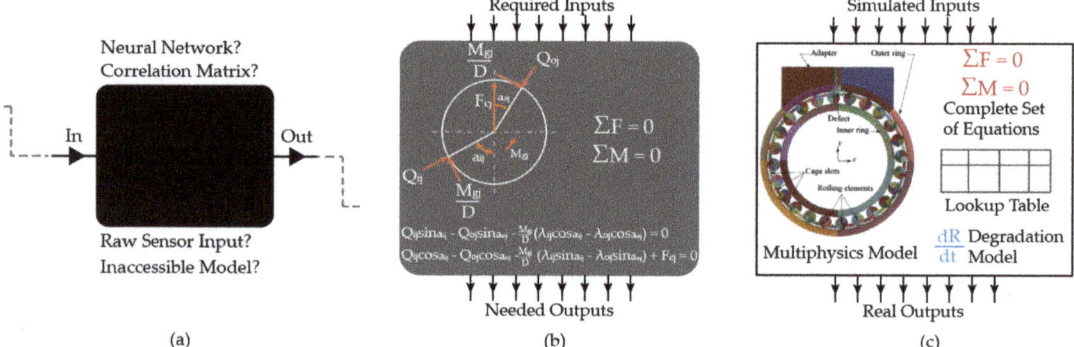

Figure 4. Example of a model bearing component in the three different "box" iterations. (**a**) Black Box; (**b**) Grey Box; (**c**) White Box. Black boxes are unknown to the model and can draw their solution from several sources. Grey boxes are models describing the mechanics and nature of the component, using theoretical data to complete it. White boxes are fully known to the user, customized for the application and simulate the component with near perfect accuracy. Analysis derived from [45], FEM model in (**c**) courtesy of [46].

The proposed approach is separated in four, common in the DT literature stages:

1. Advanced physical modeling of the machine. Kinematic and dynamic characteristics of the model, including positions of virtual sensors, only where and if they are required.
2. Synchronous simulation tuning in order to as closely as possible resemble their real counterparts.
3. Simulation using real input data and comparison with real output data.
4. Combination of the Real and Virtual Twin in a comparison. This is the final stage of the DT usage.

The software used in this endeavor is the OpenModelica, due to its correlation with the DT undertaking. A case study is presented, calculating the RUL of a six-axis robotic structure used for welding tasks. The authors conclude that the results are satisfactory while mentioning the methodology's caveats, and this process can be part of a more generic PM framework.

An important takeaway from this work is the suggestion that DT in industry and literature will become exponentially better as more and more actors propose, design, build and evaluate DTs. Industry giants depict an interest in this undertaking, while the current trend of Industry 4.0 is the perfect ground and timing for such an approach, facilitating and encouraging this effort.

Ref. [47] presents the challenges of developing a DT for RES generation. The author confirms the essentiality of DTs in optimizing the design and reliability of energy systems. In addition, the literature consensus that DTs have no serious strategy and comprehensive strategy yet is further elaborated. DFIGs with Power Electronic converters may have solved the controllability issue, but come with additional impacts on the functionality, lifetime and reliability of the system, since they provide additional interactions. The author suggests that an optimization of manufacturing and maintenance in these systems is of paramount importance since any improvement will reflect on the energy production and reliability, the greatest issue of this century, according to many experts. The DT, along with its many benefits, is a prime candidate as a solution.

The author presents an exhaustive study and realization of a large DFIG digital twin, accompanied by a sophisticated test bench. The process handles conventional EM CM challenges such as converter-fed operation and the accompanying harmonics. This issue is of paramount importance since the magnetic flux can provide a plethora of information about the machine and accurate modeling and calculation facilitate the process. The undertaking combines the important aspects of DT, namely multi-physical system modeling and data acquisition and handling. Conventional simulation models have been developed, but the combined model is in a reduced state-space, deeming it impossible to calculate all parameters with a high enough precision. The approach combines NNs, FEM and simulation modeling.

The proposed DT includes the novel approach of including stochastic modification of internal machine properties, which traditionally are very challenging to compute or model. Modeling is separated into four levels:

- Level 1 is the physical model of the generator and its electromagnetic parameters, following the existing literature.
- Level 2 enhances the previous levels with additional analytical models, resulting in a better estimation of the machine. This point has also been explored.
- Levels 3 and 4 realize the flux calculation, which is an extensively discussed and complicated procedure. Levels 2–4 utilize the novel DT methodology benefits of enhancing the conventional model with interdisciplinary approaches.
- Level 5 is the final level and employs the connection of this output to the "back" of the machine to create the closed loop model.

The author has taken the DT approach to a further step via tackling the modification problem with intelligent algorithms, meaning optimizing the DT with its own capabilities. This is an exemplar usage of the next generation DT approach. Finally, the comprehensive multi-physical model is converted into a "true" DT by combining internal calculations with real test-bench data to estimate the real behavior of the machine. It is important to note that this process requires training, since EM deterioration is not a spontaneous process but a gradual one, and a proper DT should reflect it. According to this work and literature consensus, this is the biggest challenge in realizing the novel DT approach, as suggested. In conclusion, the DT should include an ultimate understanding of system characteristics. This approach is one of the most extensive yet.

Ref. [48] proposes the interesting idea of creating a DT for one of the fundamental machines in industry, the Intelligent Machine Tool (IMT). This intelligence originates from the notion that the new IMT is "no longer limited to the operation of machining" but includes "features of multifunctionality, integration, intellectualization and environmental friendliness". These machines should follow the Industry 4.0 protocols to realize the faster creation of better, cheaper products while facilitating the product DT manifestation via appropriate measurements stemming from the IMT itself. This work follows the next generation DT paradigm in creating the IMT DT, meaning the task separation in physical, virtual and connection layers. The authors provide an on-point review of the DT in the industry of interest and explain the tackled challenges.

Previous work on the IMT DT has been focused on theoretical design or just the data analysis. This work provides a clear design paradigm. The main takeaways are the focus on the HMI and the mapping, since the IMT is an actuator for the creation of another product. IMT DT data falls into the category of Big Data, which is expected but also confirmed by the work. Finally, the authors present two experiments which confirm the great degree of optimization received from this endeavor. The IMT DT is a key component of connecting the manufacturing and the operation process through the data it handles, and thus should be a focus of the nexDT paradigm.

Ref. [49] is one of the first EM-related works encountered in this search. It proposes a precomputed FEM model originating from the machine geometry, fed with online measurements, which is the natural approach in considering an EM DT according to the state-of-the-art. The main benefit over non-DT methods is the consideration of difficult-to-compute and/or speculated EM quantities such as local flux, bar current and torque, in addition to asymmetries. This work also comments on the ambiguity of DT in literature.

A summary of this work follows. Industry uses dq models of machines due to their real-time simulation capability which stems from quantity speculation. Real-time monitoring requires a model which can be computed with contemporary CPUs faster than the real machine. FEM models typically offer the highest fidelity in a trade-off with high computational times. The authors considered two hybrid (with FEM) approaches to combine the benefits of high accuracy and computational cost. The two models are Magnetic Equivalent Circuits (MEC) and lumped circuit models. The resulting model combines MEC with FEM and is named Combination of Finite Element with Coupled Circuits (CFE-CC). Preliminary results show a close resemblance to the real twin, making this model a prime candidate for being the virtual twin. The authors present three strong points:

1. It considers space harmonics and magnetic imbalances, due to FEM modeling.
2. It can support any number of electrical circuits, following the modularity requirement of the DT.
3. It can compute a distribution of power losses inside the machine, which is paramount to achieving a true DT and a major challenge of research.

State-space reduction is achieved by using a 2D approach in the center of the machine, while the rotor skewing, and coil ends are tackled by altering the computed induction matrices. DAQ is guided by the necessary time step (thus choice of CPU) as discussed previously, while physical quantities are measured by common techniques, namely encoder and PCB. The entirety of the virtual layer is realized using the SimPowerSystems library of MATLAB Simulink. Finally, the work compares the proposed DT with real machine measurements, confirming its feasibility and more accurately its accuracy and computational efficiency.

Ref. [50] offers an in-depth analysis of a DT component model, namely an improvement on the well-established Γ-equivalent circuit of an Induction machine. The main focus lies in improving the loss distribution in rotor and stator to more closely resemble the results of a FEM model while using an equivalent circuit. Results show that the developed model provides a good alternative to FEM (and thus computationally expensive) modeling and can be used as a component of an IM DT without dramatically increasing its cost. The work continues with the model proposal, which is then fitted to the FEM model in 33 different steady states. While dynamic comparison is quite tough due to the solver integration with the control software requirement, the proposed model is meant to be used in a dynamic simulation. The comparison between the proposal and the FEM model is done in a different state so as to compare the result when it had not been used in fitting. Results are promising and this model should be thoroughly considered in coupling iron losses to the thermal analysis inside a complete DT of an IM.

The authors of [51] follow state-of-the-art discussed DT paradigms to present case studies of energy conversion systems. The five-dimensional model is accepted, while the importance of the "living" model is highlighted. A concise explanation of the concept is presented, followed by relevant tools and process flow. The first main part of the work discusses potent applications for the DT, namely: industrial robotics and virtual testbeds in manufacturing, EV design, CM and control, wind turbine PM, and finally telescopes. These applications are prime candidates for DTs and, apart from telescopes, are in the spotlight. The authors provide a brief but concise explanation to be tied with the next section, modeling methods. Total product lifecycle is discussed, with pertinent work concerning each phase, namely modeling and optimization, energy conversion, maintenance and service, diagnostics, and finally control. Finally, implementation examples are given, concluding with an excellent Strengths-Weaknesses-Opportunities-Threats (SWOT) analysis of the DT application.

3. Discussion

Reviewed work is discussed in terms of adhering to the proposed definitions, starting from the basic Model—Shadow—Twin ambiguity. Results are summarized in Table 2. Many authors acknowledge these differing capabilities of digital representation but refer to, i.e., virtual sensors (namely DS) as DTs. Proper naming of each work is surmised to guide readers to the exact scope of the proposed technology, thus speeding up research.

Table 2. Characterization of reviewed work regarding data flow automation definitions.

Ref.	Char.	Comment
[28]	DS	Thorough usage of proposed DS capabilities. Role in control is acknowledged in definition, but proposed model serves only as virtual sensor. Usage of data in automatic control is implied, but not enforced. Modifying virtual control values in any way should be reflected in real twin to be a DT.
[29]	DM	Excellent demonstration of multiphysics, multiscale simulation with software cooperation. Real twin control methods are simulated but not accessible from virtual twin. Measurements are done once before constructing the model and no real-time connections are mentioned, rendering work as DM. Model is used to predict RUL and wear-tear, in addition to AI training. PM target is achieved.
[30]	DT	Inclusive realization of DT capabilities. Digital verification. Physical controllers fully cooperative with DT software, displaying the DT's role in control. Operators controlled DT via physical console. Results of application case include upgrades to the machine through retrofitting project, enabled by its emulation.
[31]	DS	Authors duly describe methodology as in DT "Framework". Focus in one part of system, emphasizing fidelity choice. Operational data from SCADA system into physical-based models with no control through software—DS. Virtual sensors used to optimize real locations. DT paradigms cleanly recalled.
[32]	DT	Original coin of "nexDT". Proposition of DT paradigm, no application. Agreement on all aspects of proposed DT definition. Broader designation but encapsulates all aspects.
[33–36]	DT	Proposed DT model of CNC Machine Tool. CNCMT is excellent candidate for first era DTs as it can facilitate creation of projects' DTs. Cited work explores all aspects of proposed DT methodology and provides real-time interconnection between cyber and physical space. Extensive application case study and report.
[37]	DM	Authors provide proof of concept using Modelica based model along with inputs and outputs, acknowledging minimality. Description of work fits DM as fault and sensor models exist in simulation. No data receiving from real sensors or back-transfer of control is mentioned. Goal is UI/Testbench for operator.
[38]	DT	Work follows five-dimensional DT model and all life-cycle phases. Focus is on data handling for PM. DT is ancillary. Excellent proposition of control usage in PM. DT chooses and validates maintenance work before guiding repair robot. Exemplary for PM DT.
[41]	DS	PMSM, IGBT, battery models followed by NN is encased in DT. Receives real sensory input but no role in control as user decides following steps post observation. Directly useful without bloat. "i-DT" term is mentioned for NN-enhanced model.
[23]	DT	Work follows paradigm of including all life-cycle phases and relevant rules and subsystems. Control not directly realized but physical entity operates according to simulation without intervention, thus being deemed an automatic data transfer process.
[42]	DS	Work suggests automatic optimization of physical system based on analysis of virtual twin, but application only employed in diagnostics. Sensory input from physical system. Model is continuously updated.
[43]	DS	Drilling machine DT proof of concept. Sensory input and comparison to real system, status updated but used only in RUL prediction. No back adjustments.

Table 2. *Cont.*

Ref.	Char.	Comment
[44]	DS	Advisory nature proposed DT is highlighted. Sensory input from real twin. Great paradigm of DT usage in PM as transfer of data back to real twin is of limited usefulness. All other aspects excellently covered.
[47]	DS	Author presents the most complete combination of physical test bench and digital representation out of all reviewed works. Digital twin is self-optimizing and receives extensive input but does not automatically change real twin, receiving the designation of DS. All other paradigms of DT included in work.
[48]	DT	Another approach to Intelligent CNCMT. Sensors update digital model, complete with intelligent algorithms. After validation, simulated data are returned to real twin to optimize and complete work.
[49]	DS	Full life-cycle model of EM. Sensory output with model updating. PM operations achieved but no back transfer of information. Core of work is comparison and establishment of better base model.
[50]	DM	Acknowledged proposition of model to be used in DT. Standalone work is a DM of EM as it has no sensory input yet.
[51]	DT	Authors acknowledge bidirectional, seamless data transfer and role in control, designating their approach as a DT. While multiple applications are discussed, the consensus fits the proposed DT paradigm.

Authors of [28] provide numerous DT definitions in their introduction. While DT capabilities are reflected, many are ambiguous in terms of the proposed differing characterization. In this work's introduction, we mention that "Modeling methodology is not and should not be limited in the scope of this categorization". A DS is not lesser to a DT; it only provides a different service. Ref. [29] acknowledges the fact that their proposed DT is model based, reinforcing the notion that researchers discern the different applications of each proposed characterization. It is only a naming matter.

Incorporating DTFs in PM encourages the usage of DS, namely a living model reflecting the changes in the system in real time without the necessity of back-transfer of information. Models are not efficient in PM barring their usage as a subsystem and not representing a complete diagnostic technique. DTs are of course encouraged but not strictly needed. Great examples of automatic correction provided by the virtual twin are the ones characterized as DT. A classification summary regarding the number of each framework is given in Figure 5.

DTF CLASSIFICATIONS

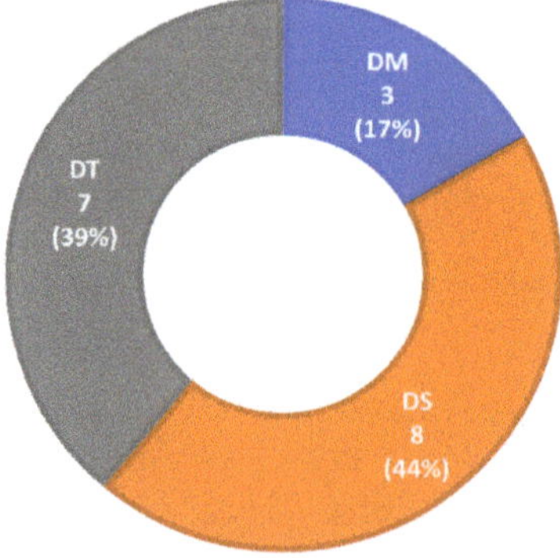

Figure 5. Population of each classification of DTF reviewed in this work.

The expectation that most of the proposed work is categorized as DS is matched, since it is the most useful framework of depicting in-use state-of-the-art diagnostic techniques. Researchers are now essentially matching the concepts of PM and DT into one complete merger. This, however, is no strict requirement or correlation. The different aspects of each subcategory are discussed in Table 3.

Table 3. Primary applications according to the different aspects of each classification.

Classification	Works	Applications	Overview
Digital Model	[29,37,50]	Manufacturing, Design, Training	Non-living model, snapshot from DTF. Modeling of subsystems, steady-state operations, low computational cost applications. Conventional modeling currently employed.
Digital Shadow	[28,31,41–44,47,49]	Condition Monitoring, Predictive Maintenance, DAQ, Optimization, UI	Living footprint of physical system. Complete visualization, tracking, DAQ, UI. Optimized and complete methodology combination in state-of-the-art CM and PM.
Digital Twin	[23,30,32–36,38,48,51]	All of the above plus Control, Industrial IoT, Retrofit, Reconfiguration, Operation	Living simulation of physical system. Control integration and changes reflection. Boundaries and constraints. Complete lifecycle depiction and cloud storage. Industry 4.0 paradigm.

3.1. Is This Classification Useful for Literature?

One presented problem in this approach is if this categorization enhances or hinders research efforts. As of now, the primary obstacle in broader DT adoption or evolution is its ambiguity in literature. The proposed classification resolves one aspect and guides researchers to pertinent work. However, we presume larger confusion, should most researchers choose not to adopt it, or introduce further ambiguity in classifying relevant work. Furthermore, one could argue that all relevant work is essentially a digital twin concept, and this classification introduces unnecessary complexity. Finally, one could acquire the notion that since each classification encapsulates the previous one, DM is lesser than DS and DS is lesser than DT, thus shadowing their work in a DS in favor of another DT. We argue that this fact perfectly encapsulates the "multiscale" pillar; if the DS is perfectly appropriate for the objective, a DT would only offer an unnecessary complexity.

Proposed Solution

We propose the adoption of the broader term: "Digital Twin Framework" to encapsulate all relevant work. We surmise it sufficiently conveys the concept to solve the aforementioned argument. Furthermore, we encourage the further adoption of the DM–DS–DT classification, as it has already been explored by researchers and found to be helpful, with the added requirement that the DS reflects the evolution of the real twin in time ("living" model), in addition to the data flow requirement. This means that a DTF, which automatically receives real-time sensory input but does not include i.e., aging or fault progression mechanisms, is classified as a DM instead of a DS. The essence of a model is its depiction of a particular effect or mechanism, while a DS reflects the complete state of the visualized system. The minimum requirements of each classification are given in Table 4.

Table 4. Improved categorization of DTF by necessary requirements.

Requirement of	Digital Model	Digital Shadow	Digital Twin
Automatic Data Flow	no	Physical to Virtual	Bidirectional
Time-Varying	no	yes	yes

3.2. Proposed Complete Definition of Digital Twin Framework

Stemming from state-of-the-art literature, we propose a complete definition of the Digital Twin Framework from the scope of PM. This definition combines the core qualities of the DT from its inception to present, as mentioned in the reviewed work.

3.2.1. Life Cycle

Broad adoption of a well-designed DTF aims to solve one of the main problems encountered in CM and PM: lack or adequate handling of data. Managing the complete lifecycle of a product is of paramount importance in industry. There exists an information gap in a product's lifecycle, mainly between the production and the service phase, hindering provision of adequate after-sales services while driving up their cost. Furthermore, the retire phase of a product is all but ignored in relevant literature, allowing older generation issues to pass onto the next when they could have been easily avoided [27]. These problems are depicted clearer in Figure 6.

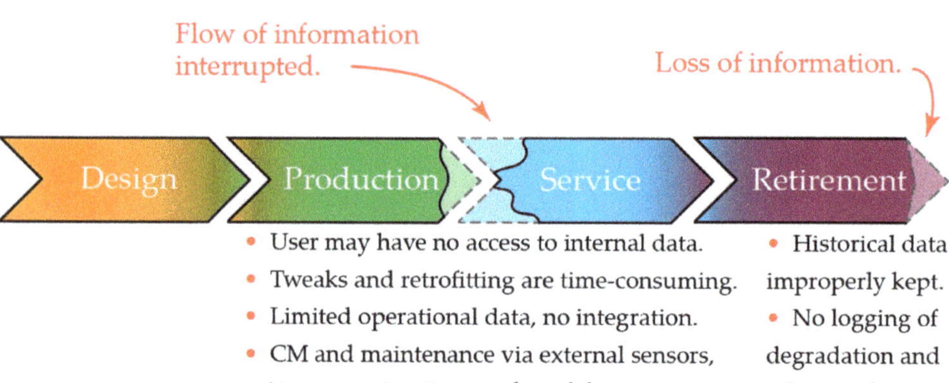

Figure 6. Illustration of data flow problems in the conventional product lifecycle.

The proposed DTF includes all four lifecycle domains. Design and production are handled by the provider and stored in the DTF in the appropriate form (which will be discussed below). Data are integrated into the DTF itself, using the "Box" paradigm. It is imperative that providers can withhold corporate and technological secrets. The obscurity associated with this fact can be overcome via providing a Black or Gray Box in the DTF, safely stored in the provider's cloud server (and accessed directly by the DTF). Retire phase data is stored in the DTF, which is perpetual and not destroyed along with the physical twin, as long as storage is provided. These data can be used to improve next models in the design phase. Most importantly, historical and usage data, mainly degradation and aging, is of paramount importance to the PM industry and will accelerate research efforts immensely.

The proposed lifecycle paradigm manifests into a circular lifecycle, as depicted in Figure 7 and is the final guideline of this work. The most important contributions are mentioned.

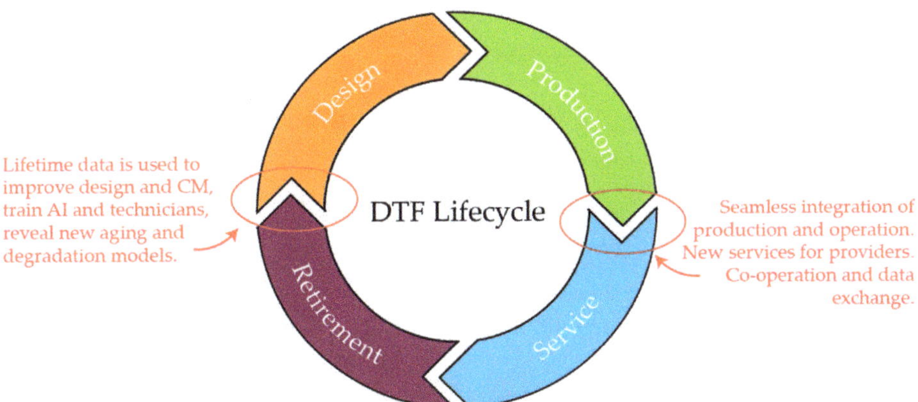

Figure 7. Proposed DTF lifecycle. The data layer allows for storage and improvement, making the lifecycle essentially circular.

3.2.2. Five-Dimensional Digital Twin Framework

The proposed DTF follows the five-dimensional paradigm discussed in the introduction. We surmise that these dimensions are exactly needed for a proper understanding, build, and integration of the DTF in current and future work.

Physical: encompasses the system hardware. In typical CM applications such as EMs, the machine is the core of the physical dimension and is built upon with sensors, controllers, and other needed hardware. One important aspect to consider is the duality of these components; EMs are purely physical as they are electro-mechanical conversion systems fulfilling one purpose. Sensors and controllers, on the other hand, are in both the physical and the virtual dimension. Their hardware, physical indicators and logical operations are in the physical dimension. However, they offer a clear virtual footprint to be tapped into by the simulation. We advise that DTFs should not create virtual twins of these "digitally enabled" components and rather treat them as existing in both dimensions, saving on both computational power and complexity.

Virtual: its core is a mirror representation of the core physical system to the best of our ability. Creating this mirror is an iterative process, discussed below. Models follow the "Box" paradigm and are multilevel regarding their fidelity, built upwards. Toward "digitally enabled" components, the virtual dimension encompasses and displays their software part directly. The virtual dimension is differentiated from the closely related Data and Service dimensions via its pure usage as a representation medium, mirroring the real world's mechanisms and laws, written themselves as virtual twins.

Data: this dimension bears no physical representation (barring the actual storage hardware which is not a concern of the DTF) or virtual workings. The DTF data are the collective information of values and their physical meaning. Its purpose is feeding the physical and virtual dimensions with information to be worked by their mechanisms. It can be found on-premises and on-cloud and comes from the three different source iterations as discussed above: historical, real-time, and predicted. Finally, data can be found in three different forms: structured, semi-structured, and unstructured [52]. Data used to be part of the virtual dimension but became its own due to its complexity and different workings in the DTF. The DTF aims to provide Product Embedded Information (PEI), meaning all necessary data are included in the framework.

Services: include any and all ancillary utilities of the DTF. "Twinning" means to create something identical, while services are additional inclusions. New approaches such as the DTF enable novel services and completely revamp older ones, warranting their own dimension. The service dimension can be regarded as a toolbox for the DTF; examples include the UI, CM, AI platforms, training utilities etc. Services are not twins per the discussed definition; they are products of the DTF. Conventional services were not included in the limitations of what was defined as "the system".

Connections: similar to the data concept, connections are the realization of connecting the different parts of the DTF. These include connections between models and mechanisms, blocks and boxes, different DTFs, services, hosts, providers and users. Akin to the data dimension, connections play a pivotal role in the DTF and concern state-of-the-art research and techniques. Their primary role is the translation of different data forms into the operation appropriate. Connections can be physical (cables, tubes, shafts) and virtual (links, decryptions, translators).

Authors of [53] have proposed an eight-dimensional version of the DTF, which follows the same paradigm but analyzes the data-focused dimensions further. We deem the five-dimensional model as the maximum necessary complexity for adoption and broader appeal.

3.2.3. Creating the DTF Iteratively

We mentioned that creation of a DTF is an iterative process. There currently is no optimal way to approach it, but literature agrees on this iterative principle, expertly demonstrated in [38,44,47,49]. The proposed process combines and expands upon the paradigm followed in these works.

Core Model: Ref. [38] offers a concise methodology for the core model. The first step is to create the geometric representation of the real twin, following dimensions, orientation, and all geometric qualities. Then, these qualities are given physical attributes such as weight, density, and other material qualities. The third proposed step is creating the behavior model, the interactions between components and the environment, following the laws of physics and the virtual laws in the simulation. Finally, a constraint model is realized, giving the model the boundary of the physical world. At this stage, our model is what we referred to as a Digital Model. It depicts an object and mechanism in steady state. This is also the first step in the other mentioned works.

Enhancement: the core model is enhanced with DTF exclusive techniques such as prediction models, fault progressions and probability mechanisms. The latter exceptionally describe the physical world and are materialized via state-of-the-art technology such as ANNs, Fuzzy Logic, and general AI techniques. The exclusivity refers to the edge this approach gives to the DTF compared to conventional, autonomous to each other approaches.

Model Optimization: the model is now upgraded with better behavioral descriptions, higher fidelity models and more extensive analyses. Focus is dependent on the objective of each DTF and is not strictly enforced. For example, a DS aiming to calculate the thermal strain of a pump should have analytical thermal modeling and can manage with simpler electrical or hydraulic models. Following in this concept, our model is optimized according to its purpose; the core model is the same in every similar system, but the optimization can differ dramatically. An example is [47] Levels 2–4 or [44] Step 2.

Data Validation: following the clear-cut [44] steps, the next iteration is validating the virtual representation model with real output data. Holding on the comparison with the real twin for the next step, data validation provides larger freedom of changes and has no time constraints. Large system DTFs are often accompanied by integration constraints and time limits. The proposed DTF should resemble as closely as possible the real twin according to theory.

Real Twin Validation: the proposed model is connected to the real twin and validated with comparison methods. Changes done to the model in this stage should be limited to data tables and weight constants, due to the aforementioned issues.

The complete vision of the proposed DTF is illustrated in Figure 8.

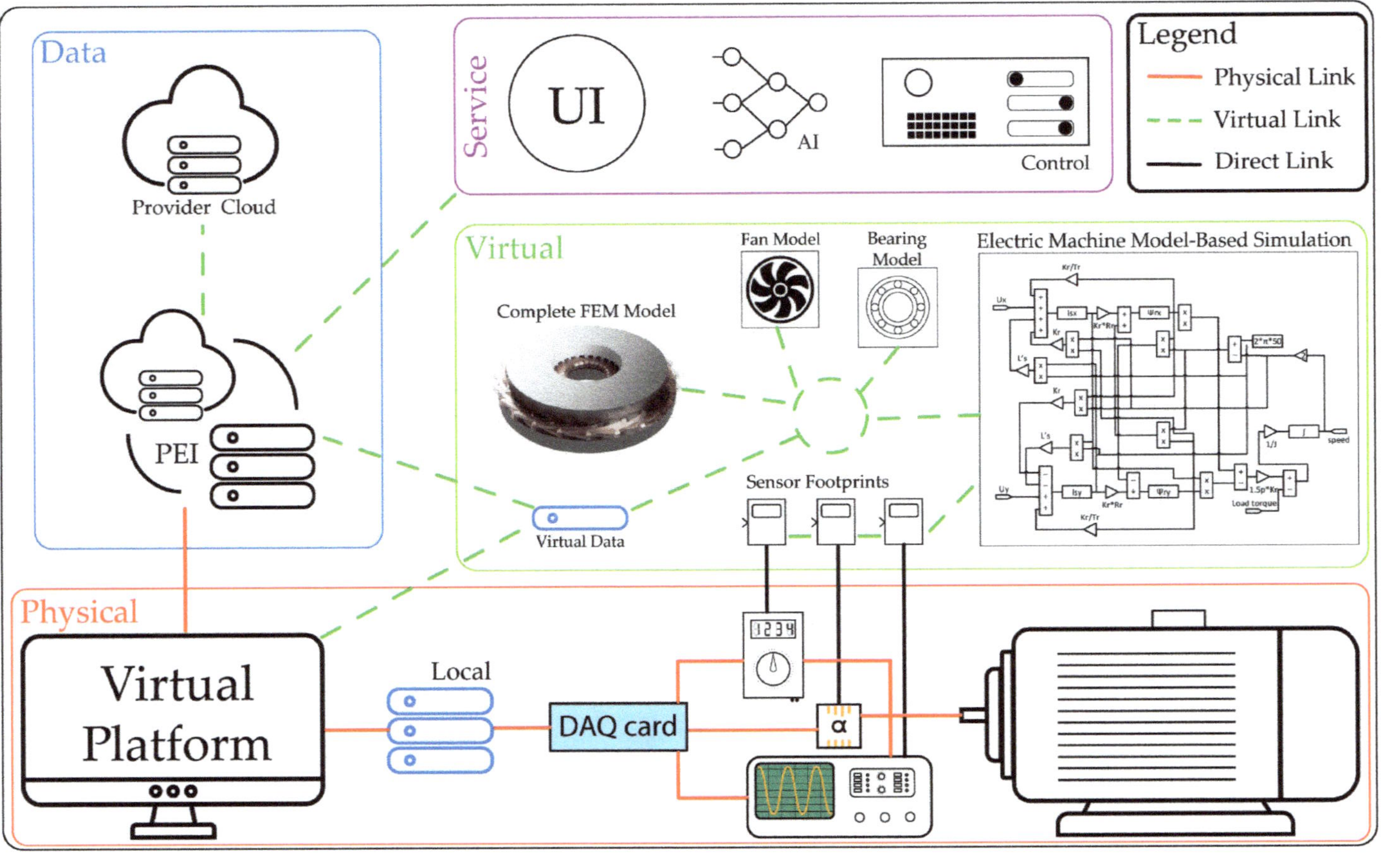

Figure 8. Complete visualization of proposed five-dimensional DTF model, following all paradigms. Sensors and CPU exist in both physical and virtual space. Virtual links established by software links. Physical connections are either wired or wireless. The Simulink model illustration from [54], FEM icon from [55].

3.2.4. Software

DTFs can be realized in numerous software packages provided by large competitors. We incentivize usage of the researcher's preferences, as each package specializes in different aspects of the DT. Instead, focus of research ought to be in integrating the final frameworks into a seamless, plug-and-play package able to cooperate with various mediums. To that end, we surmise that the general guideline for DTF integration is akin to the one proposed by [30].

3.3. Contribution of the DTF in Industry

Finally, this work aims to discuss the purpose of the DTF in industrial state-of-the-art, focusing on CM and more especially PM. In short, the DTF ventures to tap into the data availability experienced in today's IoT, namely:

- Serve as a data integrator for a CPS in one PEI package;
- Translate and optimize the data into usable form for IoT technologies;
- Preserve post-processing information for future endeavors.

Modeling and simulating the physical system have already reached an adequate level and have been in the works for the last 20 years. CNC has been a staple in manufacturing since the 2000s. The cornerstone of the DTF and the reason for its rapid advancement today is data handling and integration. State-of-the-art capabilities in CPUs and AI techniques both enable and benefit from the DTF. Thus, focus should be split into two major offensives:

1. Combining the pre-existing models and analyses;
2. Integrating Big Data and AI technologies in the DTF.

Models and theories can of course be updated, especially in the second generation of DTFs (assuming today's work launches the first), after having the historical data to improve progression and prediction mechanisms significantly, which is in our opinion the only lacking (compared to modeling and theories) aspect of PM in CM. A review on Big Data and comparison with DTs can be found in [52]. We once again propose the "build-upon" DTF paradigm. Enable IoT integration via assuring that PEI is structured data, handled internally. Focus appears to be in feeding raw measurement data in hybrid neural networks such as ANFIS, since literature suggests that they provide the best results [56]. Concerning the DTF, even raw data input becomes structured when exiting a proper DTF.

Advanced market and user needs warrant facilitation of more customized products in concordance with smarter manufacturing. Smart sensors and IoT integration are pre-existing in most modern shop floors and machines. The DTF's purpose is to interconnect this pre-existing foundation and, in unison with big data technologies, build the virtual representation of the system. The challenge is combining these heterogenous devices and organizational structures [57] in a uniform framework.

A real example concerns employing the discussed hierarchical structure. Computer-assisted Design (CAD) models are the basis of creating the DT for each part of the system. FEM analysis is optional but can greatly enhance the effects of the second part, the behavioral model, typically realized with differential equations. After the creation and validation of each subsystem, the complete representation of the system is connected to its physical counterpart via the sensor footprints. The data layer is then constructed via calculations and experiments, followed by the services after sufficient data handling and accumulation. Connections are handled by frameworks such as AutomationML. Interconnection schemes are left to the discretion of the user/client and bear no importance bar adhering to literature and industry consensus, such as the work discussed in this paper. Different generic architectures have been proposed and research remains to determine the feasibility and contribution of each. Core challenges in every studied work include identifying the basic structures and relationships and encapsulating the critical details of each component. An in-depth methodology adhering to the proposed paradigms can be found in [57].

Real applications can be found in studied work such as the CNC machine tool [33–36] and industrial machine retrofitting [30]. Our future work pertains to the creation of a ship generator and propulsion system DT and PM of an industrial machine in a factory shop floor adhering to industrial IoT and the Industry 4.0 paradigm.

3.4. Proposed Definition

The presented DTF paradigm can be summarized as a hierarchical approach, both microscopically (in-DT) and macroscopically (DT cooperation), in addition to its iterative aspect. In the context of Power and Energy Systems, we expect the DTF to fit in adjacent sectors, such as EVs and Micro-Grids, which are built in the same philosophy and face the same challenges [58]. Our complete definition of the term "Digital Twin" follows:

"The Digital Twin is an organic multiphysics, multiscale, probabilistic simulation that can represent the physical counterpart of a system in real-time, based on the bi-directional flow and complete volume of product-embedded information, encapsulating the full lifecycle data to facilitate knowledge sharing and integration."

We deem the above definition as complete regarding the discussed requirements in classifying a DT, given in the shortest context possible. We highlight two aspects of the above definition as potentially ambiguous and in need of further clarification, namely:

- Organic *(adjective, formal):* consisting of different, interconnected parts; happening in a natural way [59]. This single world encapsulates two of the most important aspects of the proposed DTF but can be unintuitive to some; we deem the definition better for it.
- **Can** represent ... in real time: the DTF *can* be used with ultrafidelity to delve deeper into a single mechanism, forgoing the requirement of real-time computational capability. However, to be classified as a DT, the framework should be able to tap into the "multiscale" quality: sacrifice fidelity for real-time simulation. Otherwise, it is a DM.

The DTF is a broader term, encouraged to convey the potential or intended usage of the proposed system in a way equivalent to the new IoT/Industry4.0 paradigm, missing one or more qualities of the true DT (that may or may not be added in the future). Furthermore, the prefix "i-DT", as in "intelligent-DT", which conveys the usage of AI in the framework (not a classification requirement), is encouraged as it provides an additional layer of information. Finally, the "nexDT" term, while excellent in this establishment era of the concept, is of limited use up until this new paradigm is established, and thus has a finite lifespan so as to not be included in the definition.

4. Conclusions

In the final section of this work, we address the open challenges pertaining to DTF realization in EM CM and PM:

An important but indirect usage of an established DTF in industry is the safety and risk assessment of the system. Worker safety is one of the pillars of CM, as heavy machinery faults can endanger the lives of personnel and material damages. There is no extensive research yet, but the excellent work of [60] follows the discussed paradigms, provides extensive coverage and proposes a DT catering to the issue.

Following on the ambiguity of literature, there exists no benchmark or general guideline concerning the creation of an industry standard DT. While this paper aims to introduce this full concept, future work is needed to establish a working prototype and framework which can be validated and criticized by fellow researchers. Afterwards, the DTF itself can aid in methodology benchmarking, which is another CM research topic in dire need of evolution.

Information technologies are the main enabling factor of the DTF. As we mentioned before, DT elements fall into the category of Big Data and require new approaches and validations thereof. Analysis and review of these approaches is deemed out of scope of this work but remains an open challenge. Relevant, extensive work in DT data handling can be found in [61–63].

DT offers a unique opportunity for AI integration into CM in the form of appropriate data structures and connections. AI integration is the focus of state-of-the-art PM research as decision-making offers the biggest upgrade value, since conventional methodologies already are at peak performance [63].

DT technology further enables a better UI, especially through AR [64]. The data integration and sensor technology allow for seamless UI improvements through ancillary systems. While secondary work (post realization and validation), this approach will combine the DT and the human element, encapsulating, training, maintenance, control, and validation, main endeavors of the DTF.

Our final statement for this work is that the DTF is observed to be the state-of-the-art approach in the scientific community. Enabling technologies and guiding principles are intertwined with modern Electrical Engineering and Power Systems concepts such as RES, EVs, Distribution Networks, EM PM, and manufacturing. The Internet of Everything era and Industry 4.0 call for co-operation of research and technologies. We surmise that the DTF is the connecting catalyst for this next generation of industry. The first step is to establish the guiding principles for DTF realization catering to the necessities of each sector. This work aims to explain DTF integration with the requirements of the EM PM community, a brief analysis of which can be found in [65]. Future prospects include working proof of concept and its usage in state-of-the-art research, namely: benchmarking, validation, AI training, combination of methodologies, and commissioning time reduction.

Author Contributions: Conceptualization, G.F. and A.K.; methodology, G.F.; software, G.F.; validation, A.K.; formal analysis, A.K.; resources, G.F.; writing—original draft preparation, G.F.; writing—review and editing, G.F. and A.K.; visualization, G.F.; supervision, A.K.; project administration, A.K. Both authors have read and agreed to the published version of the manuscript.

Funding: This research received no external funding.

Conflicts of Interest: The authors declare no conflict of interest.

References

1. Gritli, Y.; Bellini, A.; Rossi, C.; Casadei, D.; Filippetti, F.; Capolino, G.A. Condition monitoring of mechanical faults in induction machines from electrical signatures: Review of different techniques. In Proceedings of the 2017 IEEE 11th International Symposium on Diagnostics for Electrical Machines, Power Electronics and Drives (SDEMPED), Tinos, Greece, 29 August–1 September 2017; pp. 77–84. [CrossRef]
2. Capolino, G.A.; Romary, R.; Hénao, H.; Pusca, R. State of the art on stray flux analysis in faulted electrical machines. In Proceedings of the 2019 IEEE Workshop on Electrical Machines Design, Control and Diagnosis WEMDCD, Athens, Greece, 22–23 April 2019; pp. 181–187. [CrossRef]
3. Riera-Guasp, M.; Antonino-Daviu, J.A.; Capolino, G.A. Advances in electrical machine, power electronic, and drive condition monitoring and fault detection: State of the art. *IEEE Trans. Ind. Electron.* **2015**, *62*, 1746–1759. [CrossRef]
4. Antonino-Daviu, J. Electrical monitoring under transient conditions: A new paradigm in electric motors predictive maintenance. *Appl. Sci.* **2020**, *10*, 6137. [CrossRef]
5. Henao, H.; Capolino, G.A.; Fernandez-Cabanas, M.; Filippetti, F.; Bruzzese, C.; Strangas, E.; Pusca, R.; Estima, J.; Riera-Guasp, M.; Hedayati-Kia, S. Trends in fault diagnosis for electrical machines: A review of diagnostic techniques. *IEEE Ind. Electron. Mag.* **2014**, *8*, 31–42. [CrossRef]
6. Lopez-Perez, D.; Antonino-Daviu, J. Application of Infrared Thermography to Failure Detection in Industrial Induction Motors: Case Stories. *IEEE Trans. Ind. Appl.* **2017**, *53*, 1901–1908. [CrossRef]
7. Grieves, M. Digital twin: Manufacturing excellence through virtual factory replication. *White Pap.* **2014**, *1*, 1–7.
8. Tao, F.; Zhang, H.; Liu, A.; Nee, A.Y.C. Digital Twin in Industry: State-of-the-Art. *IEEE Trans. Ind. Inform.* **2019**, *15*, 2405–2415. [CrossRef]
9. Kande, M.; Isaksson, A.J.; Thottappillil, R.; Taylor, N. Rotating electrical machine condition monitoring automation-A review. *Machines* **2017**, *5*, 24. [CrossRef]
10. Orosz, T.; Rassõlkin, A.; Kallaste, A.; Arsénio, P.; Pánek, D.; Kaska, J.; Karban, P. Robust design optimization and emerging technologies for electrical machines: Challenges and open problems. *Appl. Sci.* **2020**, *10*, 11–13. [CrossRef]
11. Lee, S.B.; Stone, G.C.; Antonino-Daviu, J.; Gyftakis, K.N.; Strangas, E.G.; Maussion, P.; Platero, C.A. Condition Monitoring of Industrial Electric Machines: State of the Art and Future Challenges. *IEEE Ind. Electron. Mag.* **2020**, *14*, 158–167. [CrossRef]
12. Fuller, A.; Fan, Z.; Day, C.; Barlow, C. Digital Twin: Enabling Technologies, Challenges and Open Research. *IEEE Access* **2020**, *8*, 108952–108971. [CrossRef]
13. Lim, K.Y.H.; Zheng, P.; Chen, C.H. A state-of-the-art survey of Digital Twin: Techniques, engineering product lifecycle management and business innovation perspectives. *J. Intell. Manuf.* **2020**, *31*, 1313–1337. [CrossRef]

14. Qi, Q.; Tao, F.; Hu, T.; Anwer, N.; Liu, A.; Wei, Y.; Wang, L.; Nee, A.Y.C. Enabling technologies and tools for digital twin. *J. Manuf. Syst.* **2021**, *58*, 3–21. [CrossRef]

15. Liu, M.; Fang, S.; Dong, H.; Xu, C. Review of digital twin about concepts, technologies, and industrial applications. *J. Manuf. Syst.* **2021**, *58*, 346–361. [CrossRef]

16. Kritzinger, W.; Karner, M.; Traar, G.; Henjes, J.; Sihn, W. Digital Twin in manufacturing: A categorical literature review and classification. *IFAC-PapersOnLine* **2018**, *51*, 1016–1022. [CrossRef]

17. Glaessgen, E.H.; Stargel, D.S. The digital twin paradigm for future NASA and U.S. Air force vehicles. In Proceedings of the 53rd AIAA/ASME/ASCE/AHS/ASC Structures, Structural Dynamics and Materials Conference: Special Session on the Digital Twin, Honolulu, HI, USA, 23–26 April 2012. [CrossRef]

18. Gabor, T.; Belzner, L.; Kiermeier, M.; Beck, M.T.; Neitz, A. A simulation-based architecture for smart cyber-physical systems. In *Proceedings of the 2016 IEEE International Conference on Autonomic Computing (ICAC)*; IEEE: Piscataway, NY, USA, 2016; pp. 374–379.

19. Chen, Y. Integrated and intelligent manufacturing: Perspectives and enablers. *Engineering* **2017**, *3*, 588–595. [CrossRef]

20. Zhuang, C.; Liu, J.; Xiong, H. Digital twin-based smart production management and control framework for the complex product assembly shop-floor. *Int. J. Adv. Manuf. Technol.* **2018**, *96*, 1149–1163. [CrossRef]

21. Liu, Z.; Meyendorf, N.; Mrad, N. The role of data fusion in predictive maintenance using digital twin. In *Proceedings of the AIP Conference Proceedings*; AIP Publishing LLC: Melville, NY, USA, 2018; Volume 1949, p. 20023.

22. Zheng, Y.; Yang, S.; Cheng, H. An application framework of digital twin and its case study. *J. Ambient Intell. Humaniz. Comput.* **2019**, *10*, 1141–1153. [CrossRef]

23. Xu, Y.; Sun, Y.; Liu, X.; Zheng, Y. A Digital-Twin-Assisted Fault Diagnosis Using Deep Transfer Learning. *IEEE Access* **2019**, *7*, 19990–19999. [CrossRef]

24. Madni, A.M.; Madni, C.C.; Lucero, S.D. Leveraging digital twin technology in model-based systems engineering. *Systems* **2019**, *7*, 7. [CrossRef]

25. Kannan, K.; Arunachalam, N. A digital twin for grinding wheel: An information sharing platform for sustainable grinding process. *J. Manuf. Sci. Eng.* **2019**, *141*, 021015. [CrossRef]

26. Kiritsis, D.; Bufardi, A.; Xirouchakis, P. Research issues on product lifecycle management and information tracking using smart embedded systems. *Adv. Eng. Inform.* **2003**, *17*, 189–202. [CrossRef]

27. Söderberg, R.; Wärmefjord, K.; Carlson, J.S.; Lindkvist, L. Toward a Digital Twin for real-time geometry assurance in individualized production. *CIRP Ann. Manuf. Technol.* **2017**, *66*, 137–140. [CrossRef]

28. Toso, F.; Favato, A.; Torchio, R.; Carbonieri, M.; De Soricellis, M.; Alotto, P.; Bolognani, S. Digital Twin Software for Electrical Machines. Master's Thesis, Universita' Degli Studi di Padova, Padova, Italy, 2020.

29. Magargle, R.; Johnson, L.; Mandloi, P.; Davoudabadi, P.; Kesarkar, O.; Krishnaswamy, S.; Batteh, J.; Pitchaikani, A. A Simulation-Based Digital Twin for Model-Driven Health Monitoring and Predictive Maintenance of an Automotive Braking System. In Proceedings of the 12th International Modeling Conference, Prague, Czech Republic, 15–17 May 2017; Volume 132, pp. 35–46. [CrossRef]

30. Ayani, M.; Ganebäck, M.; Ng, A.H.C. Digital Twin: Applying emulation for machine reconditioning. *Procedia CIRP* **2018**, *72*, 243–248. [CrossRef]

31. Sivalingam, K.; Sepulveda, M.; Spring, M.; Davies, P. A Review and Methodology Development for Remaining Useful Life Prediction of Offshore Fixed and Floating Wind turbine Power Converter with Digital Twin Technology Perspective. In Proceedings of the 018 2nd International Conference on Green Energy and Applications (ICGEA), Singapore, 24–26 March 2018; pp. 197–204. [CrossRef]

32. Rosen, R.; Boschert, S.; Sohr, A. Next Generation Digital Twin. *ATP Mag.* **2018**, *60*, 86–96. [CrossRef]

33. Luo, W.; Hu, T.; Zhu, W.; Tao, F. Digital twin modeling method for CNC machine tool. In Proceedings of the 2018 IEEE 15th International Conference on Networking, Sensing and Control (ICNSC), Zhuhai, China, 27–29 March 2018; pp. 1–4. [CrossRef]

34. Luo, W.; Hu, T.; Zhang, C.; Wei, Y. Digital twin for CNC machine tool: Modeling and using strategy. *J. Ambient. Intell. Humaniz. Comput.* **2019**, *10*, 1129–1140. [CrossRef]

35. Luo, W.; Hu, T.; Ye, Y.; Zhang, C.; Wei, Y. A hybrid predictive maintenance approach for CNC machine tool driven by Digital Twin. *Robot. Comput. Integr. Manuf.* **2020**, *65*, 101974. [CrossRef]

36. Wei, Y.; Hu, T.; Zhou, T.; Ye, Y.; Luo, W. Consistency retention method for CNC machine tool digital twin model. *J. Manuf. Syst.* **2021**, *58*, 313–322. [CrossRef]

37. Vathoopan, M.; Johny, M.; Zoitl, A.; Knoll, A. Modular Fault Ascription and Corrective Maintenance Using a Digital Twin. *IFAC-PapersOnLine* **2018**, *51*, 1041–1046. [CrossRef]

38. Liu, Z.; Chen, W.; Zhang, C.; Yang, C.; Chu, H. Data Super-Network Fault Prediction Model and Maintenance Strategy for Mechanical Product Based on Digital Twin. *IEEE Access* **2019**, *7*, 177284–177296. [CrossRef]

39. Swana, E.F.; Doorsamy, W. Investigation of Combined Electrical Modalities for Fault Diagnosis on a Wound-Rotor Induction Generator. *IEEE Access* **2019**, *7*, 32333–32342. [CrossRef]

40. Bou-Saïd, B.; Lahmar, M.; Mouassa, A.; Bouchehit, B. Dynamic performances of foil bearing supporting a jeffcot flexible rotor system using FEM. *Lubricants* **2020**, *8*, 14. [CrossRef]

41. Venkatesan, S.; Manickavasagam, K.; Tengenkai, N.; Vijayalakshmi, N. Health monitoring and prognosis of electric vehicle motor using intelligent-digital twin. *IET Electr. Power Appl.* **2019**, *13*, 1328–1335. [CrossRef]

42. Wang, J.; Ye, L.; Gao, R.X.; Li, C.; Zhang, L. Digital Twin for rotating machinery fault diagnosis in smart manufacturing. *Int. J. Prod. Res.* **2019**, *57*, 3920–3934. [CrossRef]

43. Cattaneo, L.; MacChi, M. A Digital Twin Proof of Concept to Support Machine Prognostics with Low Availability of Run-To-Failure Data. *IFAC-PapersOnLine* **2019**, *52*, 37–42. [CrossRef]

44. Aivaliotis, P.; Georgoulias, K.; Chryssolouris, G. The use of Digital Twin for predictive maintenance in manufacturing. *Int. J. Comput. Integr. Manuf.* **2019**, *32*, 1067–1080. [CrossRef]

45. Cao, H.; Niu, L.; Xi, S.; Chen, X. Mechanical model development of rolling bearing-rotor systems: A review. *Mech. Syst. Signal Process.* **2018**, *102*, 37–58. [CrossRef]

46. Singh, S.; Köpke, U.G.; Howard, C.Q.; Petersen, D. Analyses of contact forces and vibration response for a defective rolling element bearing using an explicit dynamics finite element model. *J. Sound Vib.* **2014**, *333*, 5356–5377. [CrossRef]

47. Ebrahimi, A. Challenges of developing a digital twin model of renewable energy generators. In Proceedings of the 2019 IEEE 28th International Symposium on Industrial Electronics (ISIE), Vancouver, BC, Canada, 12–14 June 2019; pp. 1059–1066. [CrossRef]

48. Tong, X.; Liu, Q.; Pi, S.; Xiao, Y. Real-time machining data application and service based on IMT digital twin. *J. Intell. Manuf.* **2020**, *31*, 1113–1132. [CrossRef]

49. Bouzid, S.; Viarouge, P.; Cros, J. Real-time digital twin of a wound rotor induction machine based on finite element method. *Energies* **2020**, *13*, 5413. [CrossRef]

50. Mukherjee, V.; Martinovski, T.; Szucs, A.; Westerlund, J.; Belahcen, A. Improved analytical model of induction machine for digital twin application. In Proceedings of the 2020 International Conference on Electrical Machines (ICEM), Gothenburg, Sweden, 23–26 August 2020; pp. 183–189. [CrossRef]

51. Rassõlkin, A.; Orosz, T.; Demidova, G.L.; Kuts, V.; Rjabtšikov, V.; Vaimann, T.; Kallaste, A. Implementation of digital twins for electrical energy conversion systems in selected case studies. *Proc. Est. Acad. Sci.* **2021**, *70*, 19–39. [CrossRef]

52. Qi, Q.; Tao, F. Digital Twin and Big Data towards Smart Manufacturing and Industry 4.0: 360 Degree Comparison. *IEEE Access* **2018**, *6*, 3585–3593. [CrossRef]

53. Stark, R.; Fresemann, C.; Lindow, K. Development and operation of Digital Twins for technical systems and services. *CIRP Ann.* **2019**, *68*, 129–132. [CrossRef]

54. Saifulin, R.; Pajchrowski, T.; Breido, I. A Buffer Power Source Based on a Supercapacitor for Starting an Induction Motor under Load. *Energies* **2021**, *14*, 4769. [CrossRef]

55. Baranov, G.; Zolotarev, A.; Ostrovskii, V.; Karimov, T.; Voznesensky, A. Analytical model for the design of axial flux induction motors with maximum torque density. *World Electr. Veh. J.* **2021**, *12*, 24. [CrossRef]

56. Stefenon, S.F.; Freire, R.Z.; dos Santos Coelho, L.; Meyer, L.H.; Grebogi, R.B.; Buratto, W.G.; Nied, A. Electrical insulator fault forecasting based on a wavelet neuro-fuzzy system. *Energies* **2020**, *13*, 484. [CrossRef]

57. Jiang, H.; Qin, S.; Fu, J.; Zhang, J.; Ding, G. How to model and implement connections between physical and virtual models for digital twin application. *J. Manuf. Syst.* **2021**, *58*, 36–51. [CrossRef]

58. Elmouatamid, A.; Ouladsine, R.; Bakhouya, M.; El Kamoun, N.; Khaidar, M.; Zine-Dine, K. Review of control and energy management approaches in micro-grid systems. *Energies* **2021**, *14*, 168. [CrossRef]

59. Hornby, A.S.; Cowie, A.P. *Oxford Advanced Learner's Dictionary*; Oxford University Press: Oxford, UK, 1995; p. 1428.

60. Bevilacqua, M.; Bottani, E.; Ciarapica, F.E.; Costantino, F.; Di Donato, L.; Ferraro, A.; Mazzuto, G.; Monteriù, A.; Nardini, G.; Ortenzi, M.; et al. Digital twin reference model development to prevent operators' risk in process plants. *Sustainability* **2020**, *12*, 1088. [CrossRef]

61. 61. Riku Ala-Laurinaho Sensor Data Transmission from a Physical Twin to a Digital Twin. Master's Thesis, Aalto University, Espoo, Finland, 2019; p. 105.

62. Angrish, A.; Starly, B.; Lee, Y.S.; Cohen, P.H. A flexible data schema and system architecture for the virtualization of manufacturing machines (VMM). *J. Manuf. Syst.* **2017**, *45*, 236–247. [CrossRef]

63. He, Y.; Guo, J.; Zheng, X. From Surveillance to Digital Twin: Challenges and Recent Advances of Signal Processing for Industrial Internet of Things. *IEEE Signal Process. Mag.* **2018**, *35*, 120–129. [CrossRef]

64. Zhu, Z.; Liu, C.; Xu, X. Visualisation of the digital twin data in manufacturing by using augmented reality. *Procedia CIRP* **2019**, *81*, 898–903. [CrossRef]

65. Aivaliotis, P.; Georgoulias, K.; Alexopoulos, K. Using digital twin for maintenance applications in manufacturing: State of the Art and Gap analysis. In Proceedings of the 2019 IEEE International Conference on Engineering, Technology and Innovation (ICE/ITMC), Valbonne Sophia-Antipolis, France, 17–19 June 2019. [CrossRef]

Review

Additive Manufacturing and Topology Optimization of Magnetic Materials for Electrical Machines—A Review

Thang Pham [1,*,†], **Patrick Kwon** [2,†] **and Shanelle Foster** [1,*,†]

[1] Department of Electrical and Computer Engineering, Michigan State University, East Lansing, MI 48824, USA
[2] Department of Mechanical Engineering, Michigan State University, East Lansing, MI 48824, USA; pkwon@egr.msu.edu
* Correspondence: phamtha1@egr.msu.edu (T.P.); hogansha@egr.msu.edu (S.F.); Tel.: +1-517-355-5234 (T.P.)
† Current address: College of Engineering, Michigan State University, 428 S. Shaw Lane, East Lansing, MI 48824, USA.

Abstract: Additive manufacturing has many advantages over traditional manufacturing methods and has been increasingly used in medical, aerospace, and automotive applications. The flexibility of additive manufacturing technologies to fabricate complex geometries from copper, polymer, and ferrous materials presents unique opportunities for new design concepts and improved machine power density without significantly increasing production and prototyping cost. Topology optimization investigates the optimal distribution of single or multiple materials within a defined design space, and can lead to unique geometries not realizable with conventional optimization techniques. As an enabling technology, additive manufacturing provides an opportunity for machine designers to overcome the current manufacturing limitation that inhibit adoption of topology optimization. Successful integration of additive manufacturing and topology optimization for fabricating magnetic components for electrical machines can enable new tools for electrical machine designers. This article presents a comprehensive review of the latest achievements in the application of additive manufacturing, topology optimization, and their integration for electrical machines and their magnetic components.

Keywords: additive manufacturing; three-dimensional printing; topology optimization; magnetic materials; soft magnetic materials; permanent magnets; electrical machines

Citation: Pham, T.; Kwon, P.; Foster, S. Additive Manufacturing and Topology Optimization of Magnetic Materials for Electrical Machines—A Review. *Energies* **2021**, *14*, 283. https://doi.org/10.3390/en14020283

Received: 20 December 2020
Accepted: 4 January 2021
Published: 6 January 2021

Publisher's Note: MDPI stays neutral with regard to jurisdictional claims in published maps and institutional affiliations.

1. Introduction

The electrical machine is considered a key part in electric drives, which account for approximately 50% to 70% of electricity usage in the EU and the United States [1]. Its applications include, but are not limited to, compressors, HVAC systems, power tools, generators, electric and hybrid vehicles, elevators, and MAGLEV trains. In the last decade, there have been consistent efforts from both the US Department of Energy and the EU to advance the design and development of future generations of electrical machines that positively impact the environment and reduce greenhouse gas emissions [1,2]. The next generation electrical machines include designs with high efficiency and power density; however, another important aspect is their environmentally friendly construction, including aspects such as minimal material waste and recyclability.

Additive manufacturing (AM), also known as 3D printing, is an emerging manufacturing technology that can potentially enable and facilitate development toward the next generation electrical machines. AM provides key advantages over traditional manufacturing methods. AM can reduce material waste and scrap parts associated with many traditional manufacturing processes. The 3D printing process, in general, recycles unused raw materials such as powder and wire filament [3], potentially achieving full use of the raw material. Recycling and reusing the raw material are critical to reduce cost, especially for high-cost raw materials such as permanent magnets. Also, recent technological advancements in AM allow the use of a wide range of materials, including copper [4,5], ceramics, and magnetic materials [6]. These materials are key for manufacturing electrical machine components.

One of the most important advantages is that AM requires minimal tooling and additional processing techniques to fabricate complex topologies. For very complex shapes, AM technologies may provide the most economical and expedited means for fabrication of small quantities. Topology optimization (TO) has been used in many application areas to identify novel designs that reduce weight without compromising mechanical integrity. TO determines the optimal way of distributing a single or multiple materials in a defined design space. Complex designs often result from TO. Application of TO is seen in the design of the rotor core of a switched reluctance machine [7]. TO is also used to find the optimal distribution of the permanent magnet and the iron rotor core in permanent magnet machines [8,9].

One drawback of TO designs is the manufacturability of the optimized solutions. This has hindered the adoption of TO toward electrical machine design. However, recent applications of AM toward electrical machine components, especially in ferromagnetic materials and permanent magnet, have revitalized the adoption of topology optimization. As AM can manufacture almost any complex topology, it has become clear that TO and AM have high levels of synergy and can be used in parallel to facilitate the development of next generation electrical machines.

There is limited literature on the integration between AM and TO for magnetic components in electrical machines. In [10], a combined magnetic-structural TO is applied for the design of a rotor core of a surface mount permanent magnet machine. The optimized rotor core is then 3D printed using high silicon steel. In [11], permanent magnets with multiple magnet grades are proposed for a surface mounted machine to reduce manufacturing cost without penalizing machine performance. Though AM was proposed for fabrication, TO was not applied; however, it could be used to identify optimal distribution of magnet grades.

There are works discussing the current state of additive manufacturing for electrical machines and their components, including magnetic materials and windings [12]. In [13], applications of AM technologies are broadly discussed for components of electrical machines, including iron cores, windings and insulation systems, magnets, and heat management/exchanger systems. For each component, ref. [13] provides a broad view around the performance of the 3D printed components and where they are compared to traditionally manufactured components. In [14], the advantages of AM technologies are discussed toward the construction and assembly side of the electrical machines.

In this paper, applications of additive manufacturing and topology optimization toward magnetic components for electrical machines are reviewed. Fundamental concepts regarding AM, especially for magnetic materials, are mentioned first to set the stage for discussing the integration of topology optimization in the later part of the paper. Also featured in more detail is the current state of the art in integration of additive manufacturing and topology optimization, especially toward iron cores and permanent magnets in electrical machines. These case studies highlight the novel integration between these emerging technologies and show their potential in future design of electrical machines.

2. Additive Manufacturing of Soft Magnetic Materials

Soft magnetic materials are characterized with low intrinsic coercivity, typically below 1000 A/m, and can be easily magnetized or demagnetized [15]. As the iron cores are responsible for the guidance and improvement of the main flux created by the continuously moving magnetic field, there are some criteria in the selection of soft magnetic materials during the design phase. The following characteristics are considered to be key for the iron cores: magnetic saturation J_s, intrinsic coercivity H_c, relative permeability μ_r, hysteresis loss density p_h, dynamic loss p_e, and yield strength [16]. For electrical machines, the iron cores are traditionally made of either steel laminations or soft magnetic composites (SMCs).

Steel laminations are typically formed from iron alloyed with silicon, nickel, cobalt, and other additives. To form the desired stator and rotor geometry, steel laminations are usually punched, either with mechanical or laser cutting technique. They are then stacked, welded, or bolted together to form the iron cores. It is well recognized that degradation of the magnetic properties can occur with mechanically handling techniques [17]. Thus, the magnetic properties of the stacked iron core may be very different from the properties of the mother coil. Another concern with stacked iron cores is waste of materials associated with the traditional manufacturing process. For segmented or complete laminated cores, the amount of steel waste due to cutting and punching of laminations can range between 50% to 80% [18]. This low use ratio signals a waste in producing iron cores. Additionally, forming the iron cores for machine topologies that require a 3D flux path can become a challenge with laminations. Thus, laminations are typically seen in iron cores where the preferred flux paths are parallel with the in-plane lamination directions.

In contrast to stacked iron cores, SMCs are selected for machine topologies where easy flux flow in three directions is preferred [19]. These machine topologies can include axial flux machines, tubular linear machines, or claw pole machines. Since iron cores made from SMCs are produced by compacting and molding iron particles into desired shapes, it may require less mechanical handling and post-processing steps. Thus, mechanical processing steps such as stamping and welding may not be required. Another advantage of SMCs in comparison with laminated steel is that it has lower eddy current loss at high excitation frequencies. At excitation frequency of 1000 Hz and above, the eddy current loss of SMC core is much lower compared to laminated steel core. This benefits SMCs for electrical machine designs where high speed operation is a requirement. There are, however, notable challenges regarding application of SMCs. They are subjected to high hysteresis loss, high intrinsic coercivity, low relative permeability, and low yield strength. Typical yield strength value of SMCs is below 20 MPa, while for lamination steel the typical value is around 350 MPa. For high speed electrical machines, rotor cores made from SMCs can be subjected to high von Mises stress, which is undesirable.

The recent proliferation in studies regarding AM for soft magnetic materials aim at providing alternative materials for fabricating the iron cores. In general, these studies try to exploit key features of AM, at the same time improving the performance of the printed soft magnetic materials that can be potentially used in the cores. Some of the early work in application of AM for fabricating iron cores focus on demonstration of AM as an easy mean of manufacturing complex geometries, Figure 1. Multiple demonstrations of additive manufacturing for a rotor core of a synchronous reluctance motor is shown in [20,21]. Here, the fabrication of the prototype rotor cores is achieved with two 3D printing techniques, fused deposition modeling (FDM) and selective laser melting (SLM) without any use of molding or tooling. In [22], the surface of the rotor core of a switched reluctance machine that resembles the structure of a honeycomb is analyzed. The use of the honeycomb structure was shown numerically to improve both the torque ripple and the leakage flux associated with the machine.

In electrical machines, the three commonly seen soft magnetic materials are iron-cobalt (FeCo), iron-nickel (FeNi), and iron-silicon (FeSi) alloys. As previous studies highlighted AM capability in fast prototyping of complex iron core geometries, other research in AM of soft magnetic materials focus on laying foundations for printing these iron alloys. These foundations include magnetic, mechanical, and microstructural characterization of the printed iron alloys, as well as their relationships to the printing parameters.

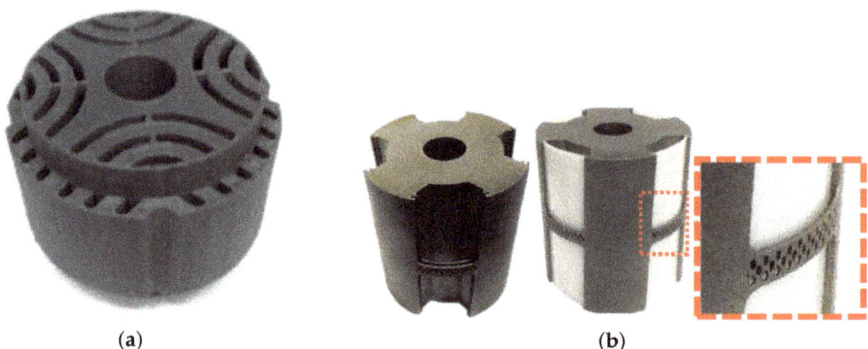

Figure 1. Additive manufactured rotor cores. These structures were fabricated without molding and tooling. (**a**) Printed rotor core for a synchronous reluctance machine [21]; (**b**) Printed rotor core for a switched reluctance machine [22].

2.1. Iron-Cobalt (FeCo)

One of the most attractive properties of FeCo is that it has the highest magnetic saturation compared to other soft magnetic iron alloys, with J_s value settling around 2.4 T. However, the conventional production of FeCo iron cores is subjected to the high material cost of cobalt, the low workability of the iron-cobalt alloy, and the additional requirement of heat treatment of the stacked iron core. This limits the use of FeCo iron cores, especially for cost-sensitive applications. As a result, reducing the challenges associated with the production of FeCo iron cores via additive manufacturing is of high interest among AM research groups. Efforts in printing FeCo iron cores have been shown via the applications of 3D screen printing and laser engineered net shaping technologies (LENS). It is reported in [6] that FeCo fabricated with 3D screen printing achieves magnetic induction comparable with commercial FeCo alloy with 15–20% cobalt content. Higher magnetic induction and saturation can be achieved if the porosity level in 3D screen-printed FeCo cores is reduced. Further comparison between screen-printed FeCo and commercial laminated FeCo shows that printed cores have higher iron loss, making its magnetic performance less appealing.

FeCo parts printed with LENS technology, ref. [23] show good potential, with achieved magnetic saturation settling around 2.2 to 2.3 T, within 10% in comparison to commercial FeCo alloyed with vanadium. When as-built FeCo core is heat-treated, its maximum relative permeability increases approximately three-fold, while its intrinsic coercivity drops by two thirds, as shown in Figure 2. Here, the annealing process leads to the development of a bimodal grain size characteristic, where coarse grains with average grain size around 200 to 600 μm are surrounded by finer grains (around 2 μm in size). By tuning the printing parameters or mixing additives into the iron alloy starting powder, it is possible to 3D print FeCo cores with even further attractive properties [24,25]. These explorations in AM of FeCo alloys show promise in overcoming the workability issues associated with the conventional mechanical processing of FeCo cores, while achieving DC magnetic characteristics close to commercial products.

2.2. Iron-Nickel (FeNi)

Iron-nickel alloys, in comparison to iron-cobalt alloys, have a much higher maximum relative permeability, more than 100,000, while saturates at a much lower J_s value, usually between 0.7 and 1.6 T, depending on the nickel wt.% content. Two laser-based AM processes, SLM and LENS, are typically seen in 3D printing of iron-nickel alloys. Reported results on fabricated Fe−30%Ni and Fe−80%Ni showed great potential in achieving magnetic saturation M_s comparable to commercial FeNi at the same nickel wt.% [26]. Analysis on the relationship between magnetic characteristics and 3D printing parameters found that for printed FeNi alloys, magnetic saturation is significantly influenced by the laser power and the laser scan speed [27,28]. These printing parameters directly impact the grain size and the density of the fabricated parts, which in turn impacts the magnetic saturation. Optimization of the laser parameters, however, is required for FeNi with different percentages of Ni content to improve the magnetic saturation J_s value. As shown in [29], for Fe−30%Ni, SLM printed iron alloys show an increase of more than 20% in magnetic saturation when laser speed is increased, while for Fe−80%Ni, the magnetic saturation is slightly decreased as the laser scan speed increases [28]. Other printing parameters such as laser scan width or the number of scan passes have been found to have a low impact on magnetic saturation J_s, thus optimization of these parameters may not be necessary [30].

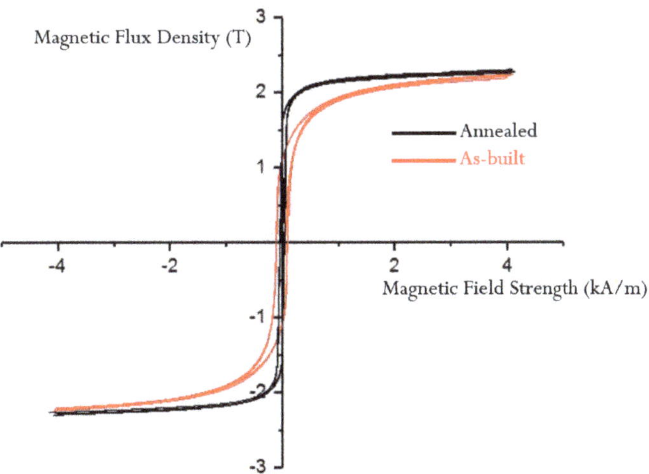

Figure 2. Comparison between quasi-static hysteresis loops of as-built FeCo and annealed FeCo. Figure is adapted from [23].

One of the major issues with the FeNi processed with either SLM or LENS is the high intrinsic coercivity H_c. The measured coercivities of the fabricated FeNi alloys range between 80 A/m to 3000 A/m [31,32], which are much higher than typical values of intrinsic coercivity found in commercial FeNi alloys. High intrinsic coercivity indicates a high hysteresis loss associated with printed FeNi alloys. Additionally, high intrinsic coercivity can have a negative impact toward maximum relative permeability, which is one of the main features of iron-nickel electrical steel. Reduction of intrinsic coercivity in printed iron-nickel is thus important. Analysis has shown that reduction in intrinsic coercivity can be achieved by reducing the porosity level as well as microstructural defects in the printed parts. This can be done by optimization of the laser power and laser scan speed as these parameters have direct influence on the cooling rate and exposure time of the molten pool. These, in turn, impact the defects, porosity, and density levels of printed parts [33]. Alternatively, the coercivity may also be reduced by blending FeNi alloys with additives such as vanadium or molybdenum as shown in [27].

2.3. Iron-Silicon (FeSi)

Considering performance per cost, iron-silicon electrical steel variants have high magnetic saturation, high maximum relative permeability, low intrinsic coercivity, low hysteresis loss, and low eddy current loss up to hundreds of Hz in excitation frequency. Variants of iron-silicon electrical steel are thus found in most iron cores used in electrical machines [16]. In pursuit of additively manufactured iron cores, most research and development activities for 3D printed ferromagnetic materials also focus on iron-silicon.

Similar to the 3D printing of iron-cobalt and iron-nickel, SLM is the most employed AM process for iron-silicon. In [34], SLM is proposed as an alternative method to produce iron-silicon with silicon content at 6.9%wt., which is brittle and challenging to produce with via conventional manufacturing method. Here, the investigation of the SLM printing parameters on the magnetic properties shows that there is a non-linear relationship between laser energy input and the relative permeability, intrinsic coercivity, and the total loss density of the printed iron-silicon. It is thus important to optimize the printing process to obtain optimal magnetic performance of printed iron-silicon. The nature of the SLM method, however, introduces defects and residual stresses on the microstructures of the printed parts, which hinders the magnetic properties of SLM iron-silicon. Compared to commercial iron-silicon lamination steel, the maximum relative permeability of as-built iron-silicon from the SLM process is lower [34,35]. Applying heat treatment to the as-built parts can help remove residual stresses and significantly improve the relative permeability as well as other magnetic properties of SLM iron-silicon [36]. In [37], the annealing process is shown to improve the maximum relative permeability of as-built parts, from an approximate value of 2000 to more than 24,000, which is on par with high performance iron-silicon steel laminations. Other magnetic properties, including total iron loss density, intrinsic coercivity, and saturation are also positively impacted via the annealing process.

Another interesting characteristic of the SLM process is that it introduces grain elongation in the build direction of the printed parts. As a result, iron-silicon fabricated using SLM can have high levels of magnetic anisotropy [38]. Additionally, higher laser energy input can even change the crystallographic texture of the printed iron-silicon, leading to the formation of Goss texture also known as cube-on-edge texture, which is seen in grain-oriented electrical steel [39]. This suggests that SLM can be potentially used as an alternative approach in producing grain-oriented iron-silicon, which in turns can be used for applications such as transformers or large electrical machines.

To avoid the effect of residual stress caused by the local melting due to the laser energy source as in SLM, other AM techniques have also been explored. In [6], FeSi sample is prepared using 3D screen printing and then compared with commercially available FeSi lamination steel. In this AM process where the powder is held together via binder, the printed part is heat-treated uniformly upon printing completion. The magnetic induction and relative permeability at low magnetic field strength are comparable to commercial FeSi steel. However, the magnetic saturation of screen-printed iron-silicon is lower than commercial lamination equivalent, owing to the low density and high porosity level of printed parts.

Binder jet printing (BJP) is another AM technique that does not use laser as an energy source. In [40], the BJP process is used to prepare iron-silicon samples with superior relative permeability in comparison to commercial soft magnetic composites (SMCs). Maximum relative permeability of BJP iron-silicon can be improved to more than 10,000 under post-processing heat treatment [41]. As a laser is not used to melt the powder particles together, there is no grain elongation associated with the BJP process. Thus, an advantage of the BJP process is that it can produce iron-silicon with low level of magnetic anisotropy. This is shown in [42], where the three-dimensional magnetic characterization confirms that BJP iron-silicon can achieve a low level of magnetic anisotropy, similar to SMCs. As a result, iron cores prepared with BJP process can be suitable for applications where easy flux flow in three dimensions is preferred. Hysteresis loss of BJP iron-silicon can be further reduced with the addition of boron into the starting powder [43,44]. This can be explained by the increase of 40% in average grain size of BJP iron-silicon with the addition of boron.

2.4. Performance of Additively Manufactured Soft Magnetic Materials

There are promising results at this early stage of AM for soft magnetic materials for electrical machines. AM allows freedom in design of magnetic cores, which can increase the performance of electrical machines. Additionally, magnetic properties of 3D printed materials, especially iron-silicon, are improving and reaching the levels of many commercial electrical steel laminations as well as SMCs, see Figure 3. Maximum relative permeability of printed iron-silicon is high and comparable to iron-silicon steel lamination, especially when the printed sample is heat-treated, Table 1. As the AM iron-silicon samples undergo post-processing heat treatment steps, their grain size can significantly improve, which in turn leads to higher magnetic induction and permeability.

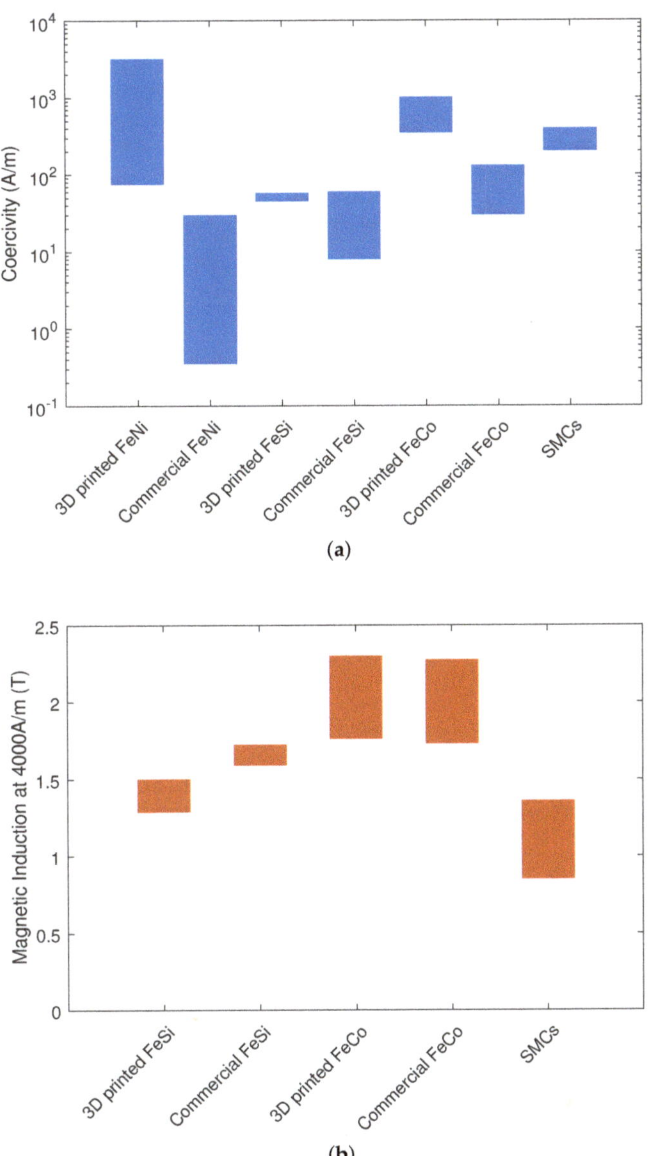

Figure 3. Comparison between additively manufactured and commercial electrical steel. (**a**) Comparison between reported coercivity of 3D printed electrical steels, commercial electrical steels, and SMCs; (**b**) Comparison between reported magnetic induction at 4 kA/m, B_{40}, of 3D printed electrical steels, commercial electrical steels, and SMCs.

Table 1. Comparison of grain size and maximum relative permeability between as-built and heat-treated printed soft magnetic materials [34,37,40,43,45].

	Grain Size (µm)	Max Relative Permeability μ_{max}
As-built	10–200	1200–5500
Heat-treated	400–1000	24,000–31,000

Application of AM for soft magnetic materials can also provide the capability to manipulate the crystallography of fabricated parts on demand. As the crystal structure of the soft magnetic material changes to the desired requirements, the magnetic properties of the soft magnetic materials would also change. In [46], gas atomized silicon powder coated with nickel layer is used for printing soft magnetic cuboid sample. The SLM fabricated sample shows changes in the crystallographic structure, switching from bcc structure, which is prevalent for iron-silicon, to fcc structure, which is typically seen for iron-nickel alloy, Figure 4. Tuning the printing parameters along with customized feedstock powder can thus offer unique opportunities in achieving customized magnetic properties.

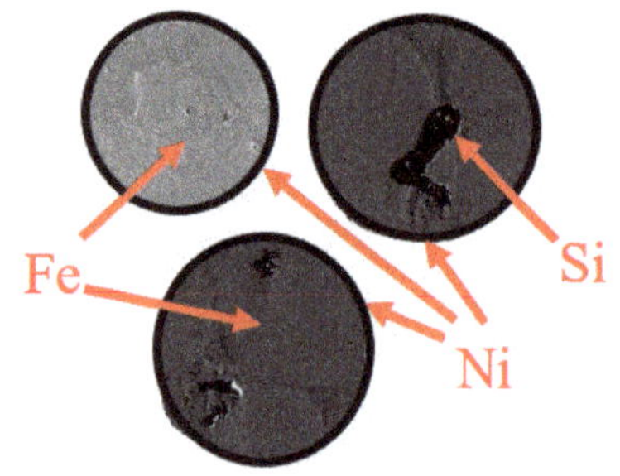

(**a**) Illustration of nickel layer coating on iron-silicon, gas atomized powder.

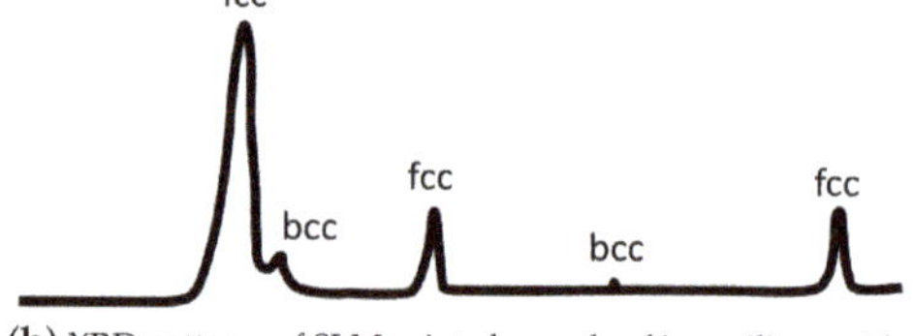

(**b**) XRD patterns of SLM printed sample of iron-silicon with nickel coatings.

Figure 4. Illustration of impacts of using nickel iron silicon coated powder on the crystallographic structure of printed samples. Figures are adapted from [46].

2.5. Challenges

Iron loss characteristics of printed FeSi is still a major concern. At DC or quasi-static excitation, additive manufactured iron-silicon has been shown to achieve hysteresis loss comparable to iron-silicon laminations [43]. The intrinsic coercivity of SLM-processed and BJP-processed FeSi is nearly 4 times lower than SMCs. At low excitation frequency of 50 Hz, the total iron loss density of additively manufactured iron-silicon is between 2 and 6 W/kg at 1 T. This puts printed iron-silicon in line with SMCs in terms of total iron loss, and a bit higher in comparison to iron-silicon lamination steels. These iron loss values at low frequency range, however, are promising providing that this is still at an early stage of applying AM for soft magnetic materials. As the excitation frequency increases, however, the total iron loss of 3D printed iron-silicon increases exponentially [43]. The resistivity and eddy current loss of 3D printed iron-silicon must be reduced. In [47], a rotor core of a switched reluctance machine is printed and compared directly with a laminated rotor core, Figure 5. Experimental results show that at base speed, 3D printed machine is about 20% lower in efficiency in comparison to the benchmark laminated machine, Figure 5b. Higher operating speed leads to even higher eddy current loss and higher reduction in efficiency. Eddy current loss remains a concern with 3D printed iron core, especially when the core is printed as bulk.

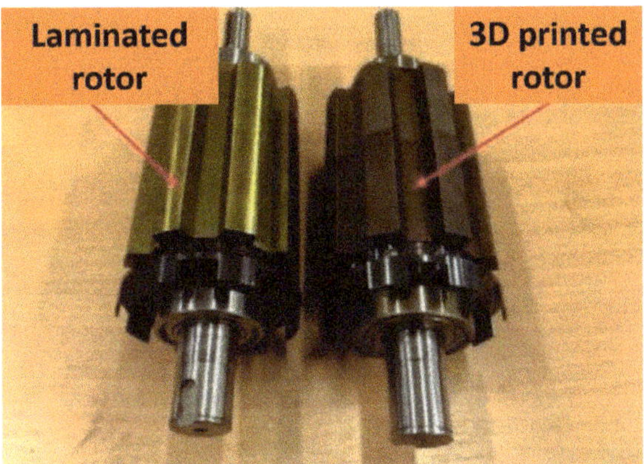

(**a**) Laminated rotor and 3D printed rotor of a switched reluctance machine.

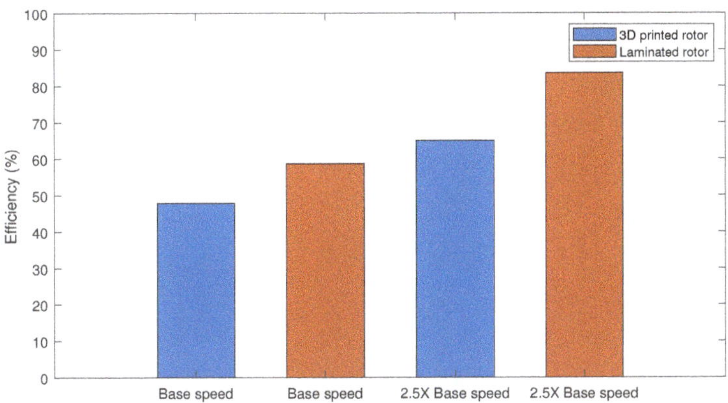

(**b**) Differences in efficiency of conventional laminated machine and electrical machine with 3D printed rotor core.

Figure 5. Illustration of eddy current loss impacts on efficiency of 3D printed electrical machine. Figures are adapted from [47].

Some strategies have been suggested for improving eddy current loss within printed ferromagnetic materials. These strategies take advantage of unique capabilities of additive manufacturing. In [45,48], AM is proposed to fabricate iron-silicon cores with complex cross-section geometries. Figure 6 shows an example of an iron-silicon rod made of multiple parallel thin plates of 400 µm thickness, where struts are added to connect the plates together. This complex cross-section geometry disrupts the paths of the eddy current, which leads to the reduction of eddy current loss when compared to the solid iron-silicon rod. Printed iron-silicon with Hilbert cross-section geometry can also result in a lower eddy current loss. Another example of complex cross-section geometries is shown in [45], where the layering technique using different materials is used to 3D print iron alloy. Here, layers of iron and iron-aluminum alloy are stacked back to back in an alternative fashion, as shown in Figure 7. Due to the differences in the resistivity between the iron alloy and the iron-aluminum alloy, the eddy current is restricted in each layer, which leads to a lower eddy current loss and total iron loss. At the excitation frequency of 100 Hz, the measured eddy current loss shows a reduction of 15 times compared to the bulk printed iron alloy. Tuning layering materials and thickness can help with further reduction in eddy current loss, which is significant for when applying such materials for high-speed electrical machine applications.

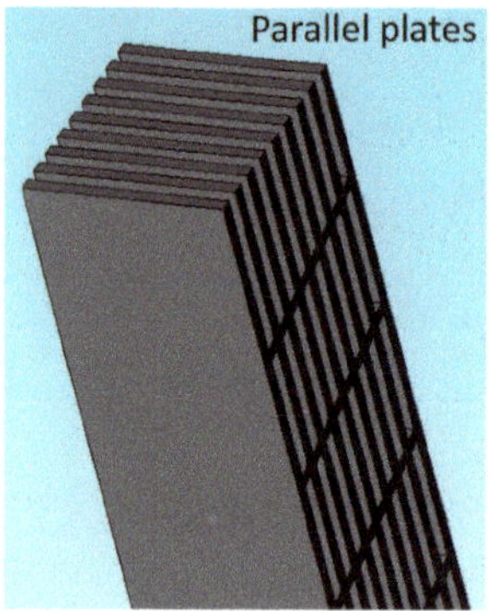

Figure 6. Examples of geometry structure in additive manufactured iron-silicon that can help with reducing eddy current loss. Printed iron-silicon with parallel plates or Hilbert cross-section can have a significant reduction in eddy current loss. Illustration adapted from [48].

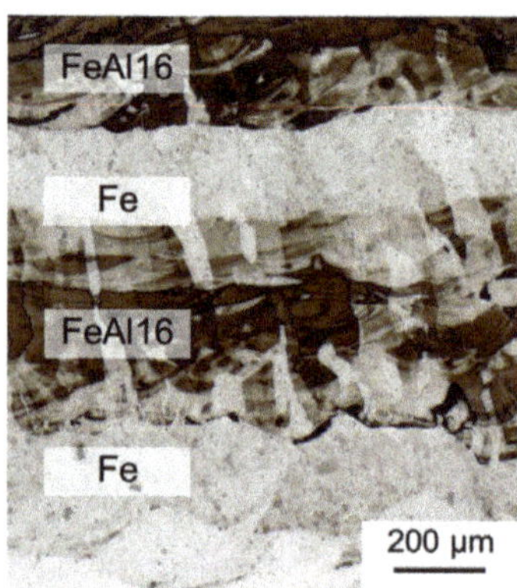

Figure 7. Optical microscopy cross-section image of printed soft magnetic sample. Here iron layer and iron-aluminum alloy layer are stacked back to back [45]. This printing strategy is implemented to reduce eddy current loss.

There are still many challenges in fabricating high performance ferromagnetic materials for electric motors and generators. The electromagnetic and mechanical properties of 3D printed materials, although advancing rapidly, still need to match or exceed the current properties of many existing commercial equivalents. Judging at the performance development of SMCs a couple of decades ago [19], and the current performance of SMCs today [49], it is encouraging knowing that the performance of additively manufactured ferromagnetic materials will continue to improve. Material characterization is thus important, especially in the development of the correlation between the 3D printing parameters, the microstructure, the crystallographic texture, and the desired properties of printed ferromagnetic parts. Other efforts in optimization of the starting powder/filament, recovery of unused material, and improvement in printing bed size, and build rate can contribute to cost reduction of the printed parts [50], furthering the application of AM beyond just prototyping. Another concern typically seen in additively manufactured parts is the surface roughness, which can potentially lead to the 3D printed iron cores with rough surface. A poor contact between the stator core and the housing of the electrical machine can change the contact thermal resistance between the core and the housing, which in turns can lead to negative impacts in the heat transfer between them [51].

3. Additive Manufacturing of Permanent Magnets

There are three main requirements for permanent magnets used in electrical machines:

1. high functioning magnetic properties,
2. stability under high temperature and
3. corrosion resistance.

Regarding the magnetic properties, the important characteristics include remnant flux density, B_r, intrinsic coercive force, H_{ci}, and the maximum energy product $(BH)_{max}$. Permanent magnets typically seen in electrical machines are NdFeB, AlNiCo, SmCo, and ferrite magnets. For NdFeB-based magnets, rare earth element such as Dy is used to help enhancing the intrinsic coercive force of the magnets [52], especially at temperature of 180 °C and higher [53,54]. This is beneficial to permanent magnet machines operated in high temperature settings as magnet flux remnant flux density is less susceptible to thermal degradation. However, the use of critical rare earth elements such as Dy and Nd can increase the cost of NdFeB-based magnets, and constrain the supply. By reducing the use of rare earth elements, it is possible to lower the magnet cost and diversify the magnet supply [54].

The production process of sintered permanent magnets includes crushing molten or strip cast alloys into fine powder, and aligning the powder particles under strong external magnetic field. The aligned powder particles are then compacted via cold isostatic pressing or other pressing techniques. The green part is then sintered, heat-treated, and machined to desired geometries. For bonded magnets, magnetic alloy powders are mixed with polymers and then formed to desired shapes by compression or injection molding. Additive manufacturing can potentially allow both sustainability and cost reduction in manufacturing magnets as well as increasing the magnet supply chain by reducing mechanical steps, and allowing for complex geometries without use of molds. At this current stage, the magnetic properties of 3D printed NdFeB bonded magnets are highly comparable to those found in commercial bonded NdFeB magnets [55]. These commercial bonded magnets are typically manufactured with injection molded (IM) method. The magnetic properties of printed magnets, however, are dependent on the printing technology, the density of the printed parts, and the chemical composition of the filament or powder mixture used in 3D printing.

3.1. Status on Additively Manufactured NdFeB Magnets

Powder bed fusion, binder jetting, and material extrusion methods are three main technologies that have been investigated at fabricating NdFeB magnets. Big Area Additive Manufacturing (BAAM) is a special technology within the category of material extrusion, developed at Cincinnati Inc. together with ORNL's Manufacturing Demonstration Facility, has also been used to fabricate NdFeB magnets [56]. One major advantage of BAAM is that it is capable of printing large scale parts of volume up to 26 cubic meters [57].

The powder bed technology with SLM shows promise in printing magnets at above 90% density [58]. In this laser-based AM process, the laser can introduce significant cracks and residual stresses in the printed magnets which may negatively impact the magnetic and mechanical properties of the magnets. Thus, optimization of the laser parameters, such as proper selection on the scan speed and laser power, can lead to optimal performance of magnetic properties of printed magnets. Additionally, optimization of the laser scanning pattern can also help with the practical distribution of heat within the printed magnet, which can help improve its overall the performance. Here, the reported magnetic polarization of the laser-based printed magnet is around 0.55 T. Other performance characteristics of the printed magnet, however, are not included.

The material extrusion and binder jetting technologies, which do not use laser as an energy source, are widely investigated for printing NdFeB magnets. Magnetic properties of NdFeB produced via BJP have similar properties compared to NdFeB fabricated via fused deposition modeling (FDM), a technology within material extrusion, as shown in [59]. They share similar performance in density, remnant flux density, and intrinsic coercivity [60]. Magnets fabricated via BJP have a density value around $3.5\,\text{g/cm}^3$, which is approximately above 40% of the NdFeB theoretical density. One of the current challenges in fabricating magnets via BJP is to increase the volume content of NdFeB powder in the printed parts, which can help improving the magnet density and the remnant flux density. Infiltration of NdCuCo and PrCuCo alloys toward as-built BJP magnets can help improve density and mechanical integrity of the magnets and the intrinsic coercive force, as shown in [61]. However, there is a slight tradeoff of remnant flux density with the inclusion of these additives.

NdFeB magnets fabricated via BAAM, however, outperform magnets prepared with the other printing technologies [62]. Density of BAAM magnets is approximately 20% higher in comparison with BJP magnets, while remnant flux density and the maximum energy product are 20% and 40% better, respectively. The magnetic properties of BAAM magnets can further be improved with a higher volume percentage of the magnet powder used in the printing mixture [63]. A direct comparison with commercial injection molded NdFeB magnets in [56] shows competitive performance of BAAM magnets, in terms of intrinsic coercive force, remnant flux density, and magnet energy product. Thermal performance of BAAM magnets also shows similar behavior when compared to injection molded magnets, as shown in Figure 8. Here, degradation of performance of BAAM magnet as a function of temperature shows similar negative trend as ambient temperature increases. A brief comparison of the current status of magnetic properties for different additive manufacturing technologies as well as injection molded magnets is shown in Table 2.

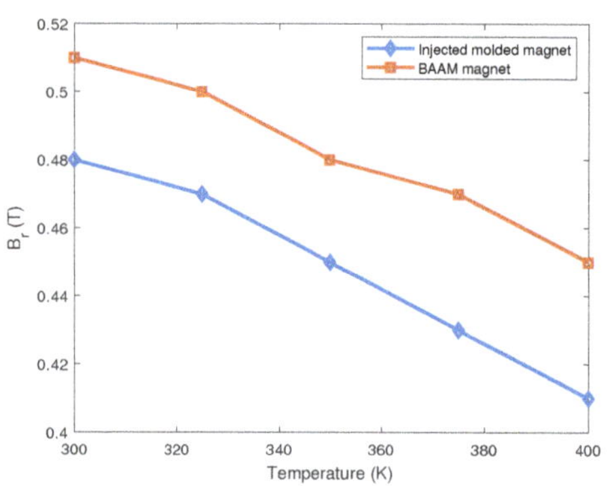

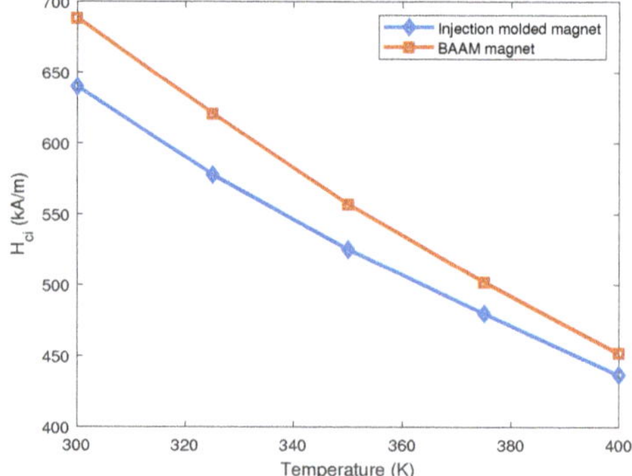

(**a**) Remnant flux density as function of ambient temperature   (**b**) Intrinsic coercive force as function of ambient temperature

Figure 8. A comparison on thermal degradation of performance between commercial injection molded magnet and BAAM fabricated magnet. Figures are plotted using data in [56].

Table 2. Comparison of characteristics between commercial bonded magnets and additive manufactured NdFeB magnets [55,56,59,62,63].

Manufacturing Method	ρ (g/cm^3)	B_r (T)	H_{ci} (kA/m)	$(BH)_{max}$ (MGOe)
IM	3.85–5.7	0.22–0.68	135–463	3.3–9.4
BJP	3.3–3.86	0.3–0.42	700–1100	2.4–3.8
BAAM	4.9–5.2	0.51–0.58	688–708	5.3–5.47
FDM	3.53	0.3	990	NA

3.2. Status on Other Additively Manufactured Magnets

Research on additively manufactured SmCo magnets is still at the early stage, so far focusing on printing magnets made of recycled magnet powder. The authors in [64] shows the process on producing magnet filament, made of recycled SmCo magnet powder and polylactic acid (PLA) plastic, for 3D printing bonded magnet. The produced magnet filament, however, has low remanent flux density, below 0.1 T. This is due to the low volume percentage loading of the magnet powder (less than 20%), in the production of the magnet filament. Increasing the loading of the magnet powder can potentially lead to magnet filament with higher remnant flux density and maximum energy product.

On the other hand, investigation on additively manufactured AlNiCo magnets shows promising results toward application in electrical machines. The 3D printed AlNiCo magnets, via LENS technology, can achieve maximum energy product $(BH)_{max}$ with values between 48% and 66.7% in comparison to commercial sintered or cast AlNiCo magnets [65]. Regarding the intrinsic coercive force of printed AlNiCo magnets, it is equivalent to commercial AlNiCo magnets, varying between 140 kA/m to 160 kA/m. The remanent flux density, B_r, of printed AlNiCo magnets varies between 0.75 T and 0.92 T, which is approximately 10 to 15% lower than typically values found in commercial AlNiCo. These early findings show the competitiveness of 3D printing as an alternative AlNiCo permanent magnet manufacturing process. Here, optimization of the printing process, as well as further exploration in material composition can potentially produce additively manufactured AlNiCo with magnetic properties exceeding commercial equivalents. One of the key advantages of AlNiCo magnets is that the $(BH)_{max}$ value of AlNiCo can stay relatively constant at temperatures up to 300 °C. Thus, the ability to have an alternative supply chain of AlNiCo magnets is beneficial to electrical machine applications where high operating temperature is a concern.

4. Topology Optimization for Magnetic Structures

The shaping of the magnetic structures for electrical machines can be generally categorized into two groups: (1) conventional shaping and optimization techniques, and (2) topology optimization. For the conventional shaping techniques, mathematical models and sensitivity analysis are typically used on a pre-selected machine geometry template [66,67]. The computational and analytical efforts are often intensive to improve the accuracy of the calculation of the airgap flux density, torque components, and the magnetic flux density distribution [68]. Conventional optimization, which is typically based on evolutionary multi-objective optimization algorithms, further refines the shape of the magnetic structures to improve the machine performance. As a result, the derivation of uniquely shaped magnetic structures for electrical machines can be slow.

Emerging from structural optimization, TO is increasingly applied in magnetic devices [69], and subsequently in design of electrical machines, especially at the component level such as magnetic cores [70] and permanent magnets [71]. In contrast to conventional optimization, TO can generate an initial geometry template from scratch with less analytical modeling [72]. In general, TO investigates optimal distribution of single or multiple materials within a defined design space [73]. Compared to conventional optimization shaping techniques, it offers additional flexibility in optimizing the geometry of the magnetic components for attaining the desired performance. Thus, TO can yield unique shapes that are generally not realizable with conventional optimization approach.

4.1. Topology Optimization Shaping Techniques

In seeking unique and optimal geometry either made of single or multiple materials, TO can employ approaches such as the on/off method and the density-based method. For the on/off method, the design space is typically divided into cells or put into a grid-like structure. Each cell is then represented by a variable, such as normalized density ρ_n, which can be assigned a value of zero or one, as illustrated in Figure 9a. Zero and one indicate the absence and presence of material, respectively. The pattern of the material distribution can be determined via selection of objective functions and use of evolutionary multi-objective or gradient-based algorithm. Thus, in a finalized topology optimized design via on/off method, it typically has an unconventional geometry.

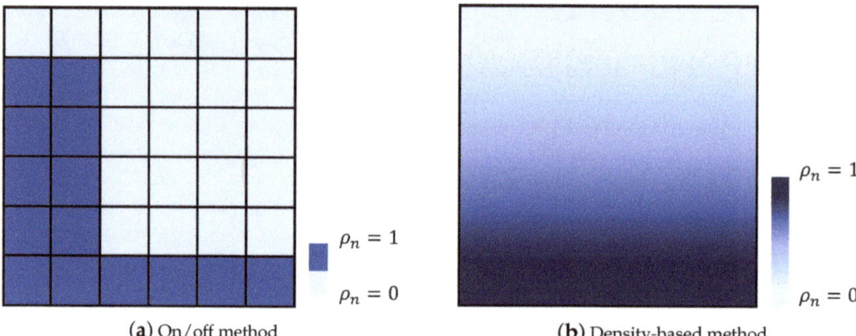

(**a**) On/off method

(**b**) Density-based method

Figure 9. Illustrations of two topology optimization methods for magnetic cores. Here, ρ_n is the normalized density variable under optimization.

An on/off TO is modified to optimally distribute the soft magnetic material for a rotor core of an interior permanent magnet machine and then to smooth the shape of the design is presented as shown in Figure 10. Here, the TO algorithm in [8] first uses a genetic-based method to find an optimal solution in the global search space. The solution is then smoothed out via the use of a gradient-based method in the local search space. The illustration of the modified algorithm is shown in Figure 11.

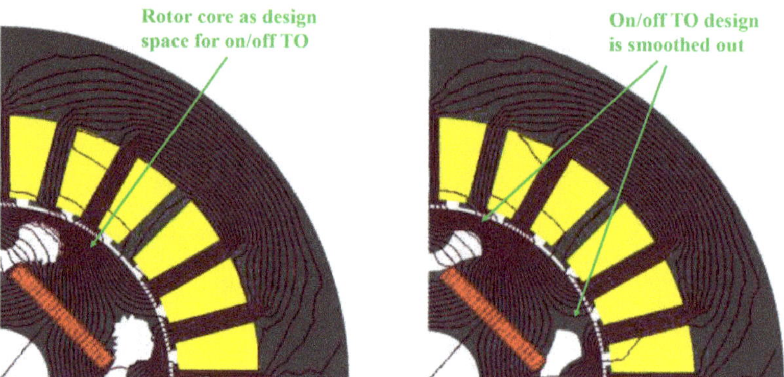

Figure 10. Example of a rotor core of a permanent magnet machine achieved via on/off TO algorithm. The TO design is smoothed out. Figure is adapted from [8].

The on/off TO method can also be applied to find the optimal distribution of multiple materials such as iron, copper, and permanent magnet. In [74], a multi-material on/off TO algorithm is used to maximize the force acted on the plunger of a permanent magnet linear actuator. The TO optimal design achieves a unique structure compared to the original design and an additional increase of 40% in average force, as shown in Figure 12.

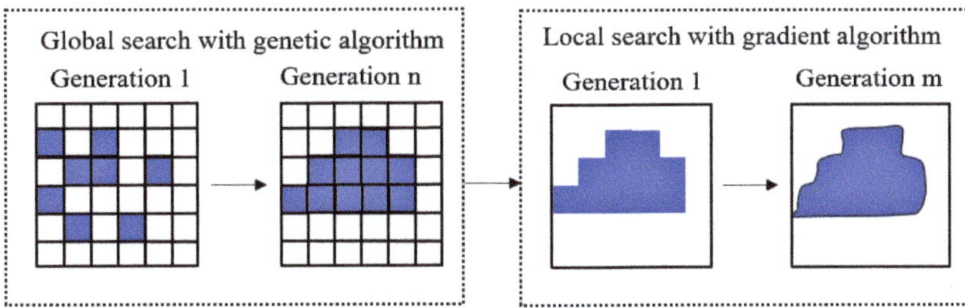

Figure 11. Example of an on/off TO algorithm where a genetic-based algorithm is combined with a gradient-based algorithm. The shape achieved with this algorithm is smoother. Figure is adapted from [8].

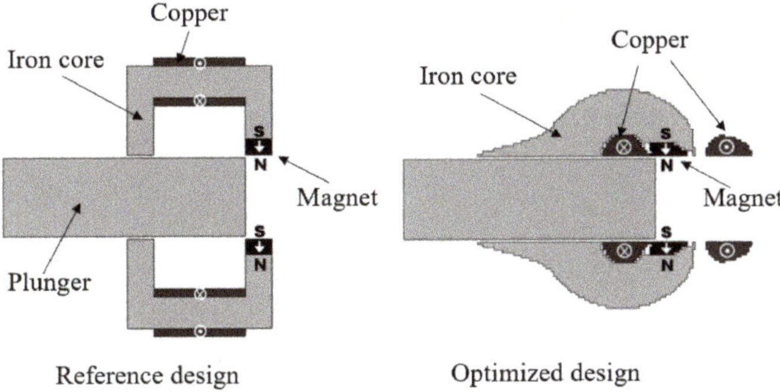

Figure 12. Illustration on the result of the on/off TO algorithm for multiple materials on a permanent magnet linear actuator. Figure is adapted from [74].

As the on/off method assigns the optimizing density variable to be binary, the density-based method assigns the density variable ρ_n a continuous value between zero and one, as illustrated in Figure 9b. A version of the density-based method is applied on a rotor core of a permanent magnet synchronous machine in [9]. Here, the exclusive magnetic-based topology optimization works toward maximizing the average torque while constraining the torque ripple and cogging, and finally leads to the unique design of the rotor core as shown in Figure 13. The white areas in the rotor core represent the permanent magnet, while the dark red areas and the dark blue areas represent iron and air, respectively. In between them are regions with intermediate density values, which are represented with different shades of colors.

Magnetic-based topology optimization may result in unique designs that achieve desired electromagnetic performance. These unique designs also present challenges that must be considered. Exclusively using magnetic-based TO algorithm results in designs that are only optimized for electromagnetic performance. Since mechanical stress of the structure is not included as an objective, the mechanical integrity of the magnetic-based TO design may become a concern. A magnetic-based TO design with a unique distribution of air pockets is shown in Figure 14. Here the TO algorithm, however, does not include mechanical stress analysis in the iron regions adjacent to the air pockets. This can lead to durability performance of the design under real operating condition where the adjacent iron regions can be subjected to stress values between 240 MPa to above 450 MPa [75]. Coupled structural and electromagnetic analysis can provide tradeoff solutions to address this concern. In [76], the density-based method is applied on rotors of wound field synchronous machines. The optimized rotors are achieved via the integration of both magnetic and structural topology optimization analysis.

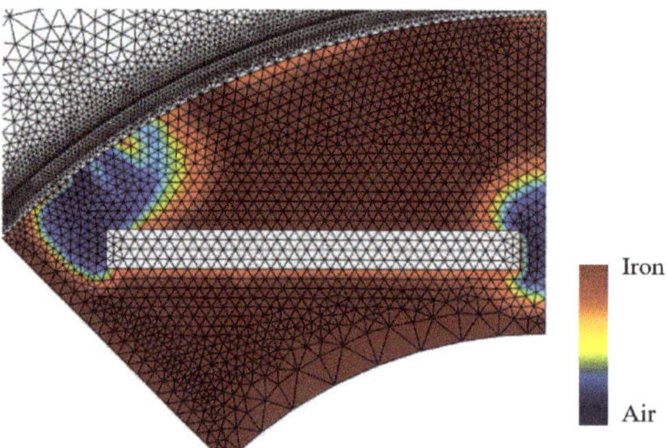

Figure 13. Illustration of density-based topology optimization for the rotor core of an interior permanent magnet synchronous machine, adapted from [9].

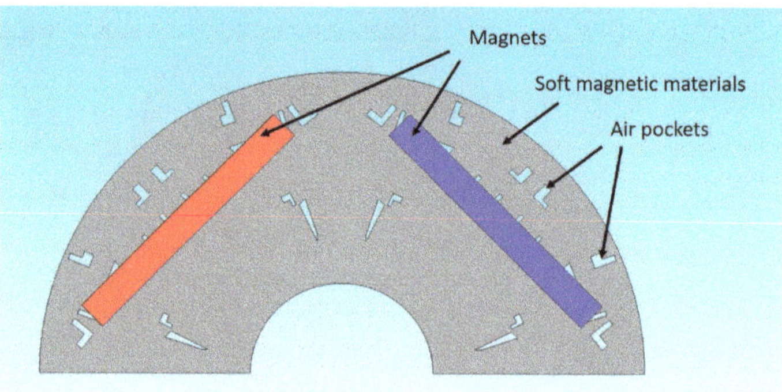

Figure 14. Example of magnetic-based topology optimization for the rotor core of an interior permanent magnet synchronous machine, adapted from [77]. Iron regions adjacent to air pockets and magnets can be subjected to high mechanical stress.

4.2. A Design Tool for Additive Manufacturing

As TO is increasingly adopted in developing unique geometries for electrical machines, manufacturability of the unique magnetic core designs is equally important. Design complexities may increase manufacturing cost for electrical steel laminations. Additionally, manufacturing methods including subtractive techniques can compromise the magnetic properties of punched laminations [17], leading to magnetic cores with inferior magnetic performance compared the mother coil. Thus, topology optimized designs in Figure 10 and Figure 14 may not be able to achieve the desired magnetic performance. Powder metallurgy can potentially be used as an alternative approach to fabricate such complex designs, as shown in Figure 15. However, the added cost of molding and tooling may become a concern.

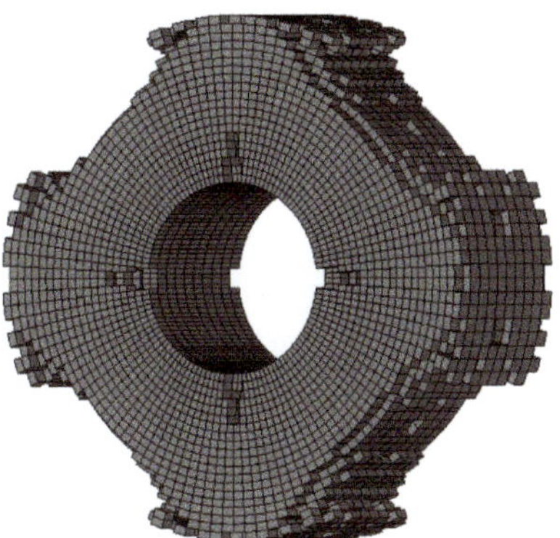

Figure 15. An example on topology optimization of a rotor core for a switched reluctance motor [78].

In application where non-homogeneous magnetic core is desired as in [79,80], powder metallurgy manufacturing approach is not a viable solution for production as it may reach the limits in fabricating such composite, non-homogenous structures. Similarly, designs of magnetic cores for electrical machines generated with the TO density-based method as in [9,76], may request material whose properties may not correspond to an available material [81]. Additive manufacturing can potentially overcome difficulties typically observed in conventional manufacturing methods, and in some cases is the only viable manufacturing solution [82].

Recent advancements in AM as well as the proliferation of its application in fabricating magnetic components for electrical machine have revitalized TO as an advanced design tool. The synergy between TO and AM can potentially lead to the development of magnetic components, whose properties and geometries are complex. Investigations in integration of TO toward AM in producing magnetic components for electrical machines have shown very promising results. In [10], a topology optimized design of a rotor core of a surface mount permanent magnet machine is additively manufactured via the SLM process, as shown in Figure 16. Here, the TO algorithm combines both the electromagnetic and structural optimization stages to achieve a rotor core geometry with 50% reduction in weight, at a tradeoff of less than 2% in average torque, while achieving maximum von Mises stress in the optimized rotor core well within the yield strength of the material. The result of this work highlights the exploitation of multi-physics TO as an advanced design tool for AM in developing new, unconventional electrical machines.

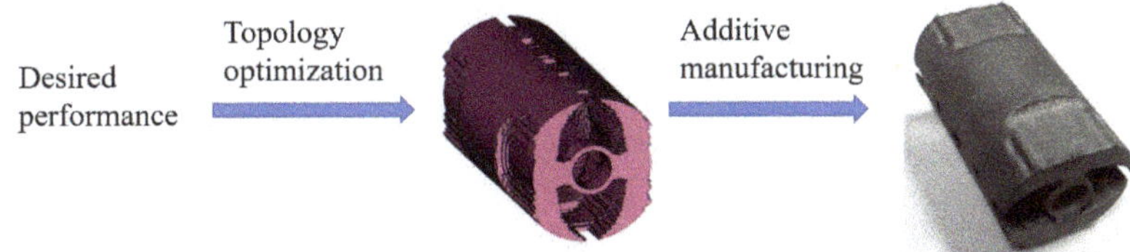

Figure 16. Optimized rotor core via coupled magnetic-structural TO. The design is then additively manufactured via SLM. Figure is modified from [10].

The integration of TO into AM is also seen in fabricating permanent magnet with unique shapes and structures. In [71], TO is implemented to generate the design of a permanent magnet such that it can provide magnetic flux density waveform close to the predefined external field. The design is then additively manufactured via the FDM process from a magnetic compound of NdFeB powder, ferrites, and polymers, Figure 17. In [11], permanent magnet made of multiple magnet grades is proposed for a surface mounted machine to reduce manufacturing cost without penalizing machine performance. The multi-grade magnet is optimized, and investigated via finite element analysis. Although TO is not implemented, the analysis suggests the combined use of TO and AM technologies in potentially producing lower cost magnets.

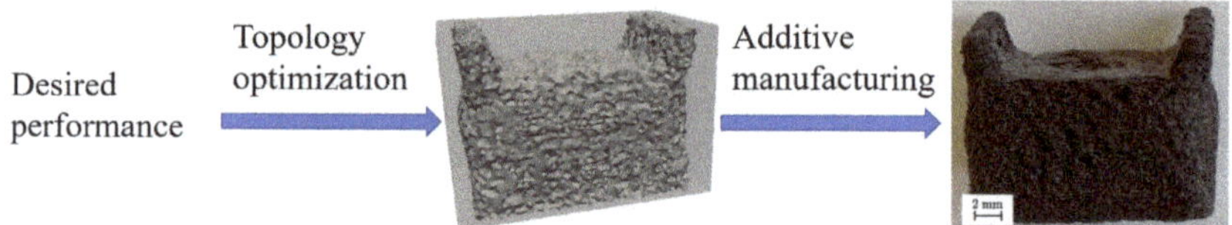

Figure 17. Topology optimized magnet. The design is then additively manufactured via fused deposition modeling. Figure is modified from [71].

5. Discussion

The implementation of AM technologies for magnetic components in electrical machines is still in its early stages. Common research themes for application of AM technologies for electrical machines focus either on understanding and improving the performance of individual components, or leveraging the freedom in design, an inherent characteristic of AM. An understanding of the characteristics of additively manufactured parts can help further the adoption of AM for electrical machines, as well as its integration with topology optimization as an enabling design tool.

5.1. Impacts of Defects Due to AM Process

Presently, there is opportunity to develop full density printed magnetic materials for electrical machines. Porosity can be observed in microstructural analysis of printed iron-silicon and printed magnets, as shown in Figure 18. Porosity, cracks, and residual stresses in the printing process typically hinder both the magnetic and mechanical properties in 3D printed parts. These defects reduce relative permeability, magnetic saturation, iron loss performance in printed soft magnetic materials, and decrease the remanent flux density and maximum energy product in printed permanent magnets. Printed iron cores and permanent magnets, if used in the rotor of electrical machines, can be susceptible to stress at high operating speed [83]. This can potentially lead to mechanical integrity issues for electrical machines. Enhancements of these magnetic and mechanical properties, however, are possible via the use of non-thermal method such as material infiltration or post-processing thermal methods [3,61].

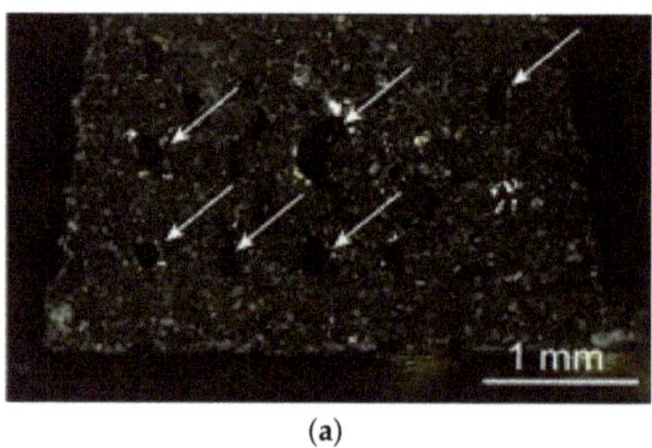

(a)

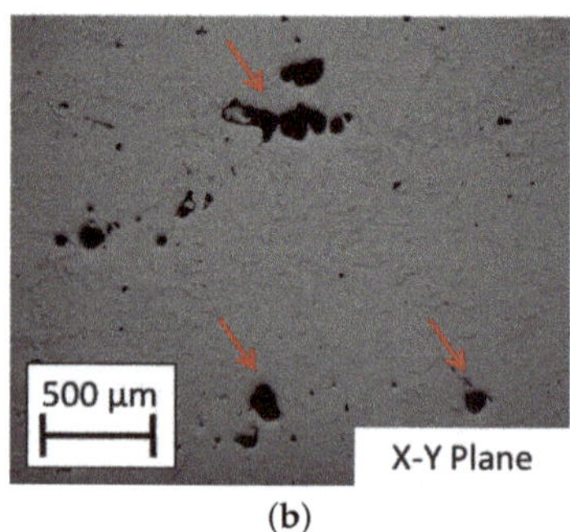

(b)

Figure 18. Images of additively manufactured NdFeB bonded magnets and iron-silicon. Here, black features indicated by the arrows in the images represent pores. (**a**) Cross-section of bonded NdFeB magnets fabricated with material extrusion [59]; (**b**) Optical micrograph of selective laser melted iron-silicon [84].

5.2. Multi-Material AM for Electrical Machines

At the current stage, the majority of AM technologies are employed for single components, using a single material for printing. The printed components are then assembled together with conventionally manufactured components to form the electrical machines [85,86]. Challenges in printing multiple materials are due to the dissimilarities in characteristics of the materials. Successful application of multi-material AM can open up opportunities for printing a complete electrical machine in a single step. Early exploration of multi-material AM for electrical machines show encouraging results, focuses on printing iron alloys, conductors, and insulation at the same time Figure 19. This proof of concept demonstrates the potential to reduce the production and assembly effort for electrical machines. The build time for the single material rotor core in Figure 16 is about 48 hours. Further research and development, however, is necessary for improving the printing process toward mass production [50].

(**a**) Transformer printed with multi-material process.
(**b**) Printed stator with multi-material process.

Figure 19. Examples of printing multiple key materials for electrical machines: iron core, conductors, and ceramic insulation [87].

5.3. Integration with Topology Optimization

As AM becomes a realization manufacturing method for topology optimization, there are still challenges remaining with combining these two enabling technologies. Since AM is a layer-based manufacturing technique, printed parts can suffer a degree of anisotropy. Electromagnetic properties can be slightly different in the build direction compared to the other directions. The formulation of the topology optimization algorithm has to account for the anisotropy effect when applying on a 3D optimization problem. Another concern associated with TO is the computation efforts. Optimization of just the permanent magnet rotor core in [88], via genetic algorithm, can take up to hundreds of generations to result in the Pareto front. Thus, significant computation effort would be required for of a multi-objective, multi-physics TO of multiple materials. The addition of complexities from the TO design can potentially prolong the printing time of the prototype.

6. Summary

This paper presents the current research and trends on applications of additive manufacturing, topology optimization, and their integration toward magnetic components in electrical machines. As an enabling technology, additive manufacturing provides topology optimization a method to fabricate unique 3D geometries that are challenged to be manufactured via conventional methods. Successful combinations of additive manufacturing and topology optimization on fabricating both the iron core and permanent magnet with desired performance show a great potential as a design tool for electrical machine designers. Capability of using multiple materials within additive manufacturing is another advantage that electrical machine designers can further exploit during the early machine design process. Recent results also show that additive manufacturing technologies have great potential as an alternative approach for realizing magnetic components for electrical machines. With the advancement in printing technology and the understanding between printing process and the magnetic properties of printed parts, there have been success toward printing of magnetic components. For soft magnetic materials, the current performance of the 3D printed parts is getting close to the parts manufactured using conventional methods. As for permanent magnets, performance of additively manufactured bonded NdFeB magnets is similar to the value seen in commercial counterparts. Additionally, results on printing magnets with good thermal performance as well as using producing magnet filament from recycled materials are encouraging. This provides the possibility to drive down the future cost of manufacturing high performance permanent magnets for electrical machine applications.

Author Contributions: Literature review, analysis, data collection, and the writing, T.P.; review and editing, P.K.; supervision, writing, and editing, S.F. All authors have read and agreed to the published version of the manuscript.

Funding: This research was funded by the Department of Navy, Office of Naval Research for sponsoring the project, under ONR Award Number N00014-18-1-2514.

Institutional Review Board Statement: Not applicable

Informed Consent Statement: Not applicable

Data Availability Statement: No new data were created or analyzed in this study. Data sharing is not applicable to this article.

Conflicts of Interest: The authors declare no conflict of interest. The funders had no role in the design of the study; in the collection, analyses, or interpretation of data; in the writing of the manuscript, or in the decision to publish the results.

Abbreviations

The following abbreviations are used in this manuscript:

AM	Additive Manufacturing
BAAM	Big Area Additive Manufacturing
BJP	Binder Jet Printing
FDM	Fused Deposition Modeling
FFF	Fused Filament Fabrication
IM	Injection Molded
LENS	Laser Engineered Net Shaping
SLM	Selective Laser Melting
SMC	Soft Magnetic Composites
TO	Topology Optimization

References

1. Commission Regulation (EU) 2019/1781. Available online: https://op.europa.eu/en/publication-detail/-/publication/218c6599-f734-11e9-8c1f-01aa75ed71a1/language-en (accessed on 1 October 2019)
2. Energy Department Awards \$22 Million to Support Next Generation Electric Machines for Manufacturing. Available online: https://www.energy.gov/articles/energy-department-awards-22-million-support-next-generation-electric-machines-manufacturing (accessed on 16 September 2015)
3. Gibson, I.; Rosen, D.; Stucker, B. *Additive Manufacturing Technologies*; Springer: New York, NY, USA, 2015.
4. Simpson, N.; Mellor, P.H. Additive Manufacturing of Shaped Profile Windings for Minimal AC Loss in Electrical Machines. In Proceedings of the 2018 IEEE Energy Conversion Congress and Exposition (ECCE), Portland, OR, USA, 23–27 September 2018; pp. 5765–5772. [CrossRef]
5. Lorenz, F.; Rudolph, J.; Wemer, R. Design of 3D Printed High Performance Windings for Switched Reluctance Machines. In Proceedings of the 2018 XIII International Conference on Electrical Machines (ICEM), Alexandroupoli, Greece, 3–6 September 2018; pp. 2451–2457. [CrossRef]

6. Lindner, M.; Werner, R. Hysteresis-model oriented test procedure for soft-magnetic properties of printed or laminated toroids. In Proceedings of the 2014 4th International Electric Drives Production Conference (EDPC), Nuremberg, Germany, 30 September–1 October 2014; pp. 1–8. [CrossRef]
7. Lee, J.; Seo, J.H.; Kikuchi, N. Topology Optimization of Switched Reluctance Motors for the Desired Torque Profile. *Struct. Multidiscip. Optim.* **2010**, *42*, 783–796. [CrossRef]
8. Hidaka, Y.; Sato, T.; Igarashi, H. Topology Optimization Method Based on On–Off Method and Level Set Approach. *IEEE Trans. Magn.* **2014**, *50*, 617–620. [CrossRef]
9. Hermann, A.N.A.; Mijatovic, N.; Henriksen, M.L. Topology optimisation of PMSM rotor for pump application. In Proceedings of the 2016 XXII International Conference on Electrical Machines (ICEM), Lausanne, Switzerland, 4–7 September 2016; pp. 2119–2125.
10. Garibaldi, M.; Gerada, C.; Ashcroft, I.; Hague, R. Free-Form Design of Electrical Machine Rotor Cores for Production Using Additive Manufacturing. *J. Mech. Des.* **2019**, *141*, 071401. [CrossRef]
11. McGarry, C.; McDonald, A.; Alotaibi, N. Optimisation of Additively Manufactured Permanent Magnets for Wind Turbine Generators. In Proceedings of the 2019 IEEE International Electric Machines Drives Conference (IEMDC), San Deigo, CA, USA, 12–15 May 2019; pp. 656–663.
12. Silbernagel, C. Investigation of the Design, Manufacture and Testing of Additively Manufactured Coils for Electric Motor Applications. Ph.D. Thesis, The University of Nottingham, Nottingham, UK, 2019.
13. Wu, F.; EL-Refaie, A.M. Toward Additively Manufactured Electrical Machines: Opportunities and Challenges. *IEEE Trans. Ind. Appl.* **2020**, *56*, 1306–1320. [CrossRef]
14. Wrobel, R.; Mecrow, B.C. A Comprehensive Review of Additive Manufacturing in Construction of Electrical Machines. *IEEE Trans. Energy Convers.* **2020**, *35*, 1054–1064 [CrossRef]
15. Krings, A.; Cossale, M.; Tenconi, A.; Soulard, J.; Cavagnino, A.; Boglietti, A. Characteristics comparison and selection guide for magnetic materials used in electrical machines. In Proceedings of the 2015 IEEE International Electric Machines Drives Conference (IEMDC), Coeur d'Alene, ID, USA 10–13 May 2015; pp. 1152–1157. [CrossRef]
16. Krings, A.; Boglietti, A.; Cavagnino, A.; Sprague, S. Soft Magnetic Material Status and Trends in Electric Machines. *IEEE Trans. Ind. Electron.* **2017**, *64*, 2405–2414. [CrossRef]
17. Bali, M. Magnetic Material Degradation Due to Different Cutting Techniques and its Modeling for Electric Machine Design. Ph.D. Thesis, Technischen Universitat Graz, Graz, Austria, 2016.
18. Libert, F.; Soulard, J. Manufacturing Methods of Stator Cores with Concentrated Windings. In Proceedings of the 2006 3rd IET International Conference on Power Electronics, Machines and Drives—PEMD 2006, Dublin, Ireland, 4–6 April 2006; pp. 676–680.
19. Jack, A.G.; Mecrow, B.C.; Dickinson, P.G.; Stephenson, D.; Burdess, J.S.; Fawcett, N.; Evans, J.T. Permanent-magnet machines with powdered iron cores and prepressed windings. *IEEE Trans. Ind. Appl.* **2000**, *36*, 1077–1084. [CrossRef]
20. Zhang, Z.; Jhong, K.J.; Cheng, C.; Huang, P.; Tsai, M.; Lee, W. Metal 3D printing of synchronous reluctance motor. In Proceedings of the 2016 IEEE International Conference on Industrial Technology (ICIT), Taipei, Taiwan, 14–17 March 2016; pp. 1125–1128. [CrossRef]
21. Zhang, Z.; Tsai, M.; Huang, P.; Cheng, C.; Huang, J. Characteristic comparison of transversally laminated anisotropic synchronous reluctance motor fabrication based on 2D lamination and 3D printing. In Proceedings of the 2015 18th International Conference on Electrical Machines and Systems (ICEMS), Pattaya, Thailand, 25–28 October 2015; pp. 894–897. [CrossRef]
22. Tseng, G.; Jhong, K.; Tsai, M.; Huang, P.; Lee, W. Application of additive manufacturing for low torque ripple of 6/4 switched reluctance motor. In Proceedings of the 2016 19th International Conference on Electrical Machines and Systems (ICEMS), Chiba, Japan, 13–16 November 2016; pp. 1–4.
23. Kustas, A.B.; Susan, D.F.; Johnson, K.L.; Whetten, S.R.; Rodriguez, M.A.; Dagel, D.J.; Michael, J.R.; Keicher, D.M.; Argibay, N. Characterization of the Fe-Co-1.5V soft ferromagnetic alloy processed by Laser Engineered Net Shaping (LENS). *Addit. Manuf.* **2018**, *21*, 41–52. [CrossRef]
24. Khajepour, M.; Sharafi, S. Characterization of nanostructured Fe–Co–Si powder alloy. *Powder Technol.* **2012**, *232*, 124–133. [CrossRef]
25. Geng, J.; Nlebedim, I.; Besser, M.; Simsek, E.; Ott, R. Bulk combinatorial synthesis and high throughput characterization for rapid assessment of magnetic materials: Application of laser engineered net shaping (lensTM). *JOM* **2016**, *68*, 1972–1977. [CrossRef]
26. Vacuumschmelze. *Soft Magnetic Materials and Semi-Finished Products*; Vacuumschmelze GMBH & Co. KG: Hanau, Germany. 2002.
27. Mikler, C.; Chaudhary, V.; Borkar, T.; Soni, V.; Choudhuri, D.; Ramanujan, R.; Banerjee, R. Laser additive processing of Ni-Fe-V and Ni-Fe-Mo Permalloys: Microstructure and magnetic properties. *Mater. Lett.* **2017**, *192*, 9–11. [CrossRef]
28. Zhang, B.; Fenineche, N.E.; Liao, H.; Coddet, C. Magnetic properties of in-situ synthesized FeNi3 by selective laser melting Fe-80%Ni powders. *J. Magn. Magn. Mater.* **2013**, *336*, 49–54. [CrossRef]
29. Mikler, C.; Chaudhary, V.; Soni, V.; Gwalani, B.; Ramanujan, R.; Banerjee, R. Tuning the phase stability and magnetic properties of laser additively processed Fe-30at%Ni soft magnetic alloys. *Mater. Lett.* **2017**, *199*, 88–92. [CrossRef]
30. Wang, X.; Wraith, M.; Burke, S.; Rathbun, H.; DeVlugt, K. Densification of W–Ni–Fe powders using laser sintering. *Int. J. Refract. Met. Hard Mater.* **2016**, *56*, 145–150. [CrossRef]
31. Zhang, B.; Fenineche, N.E.; Liao, H.; Coddet, C. Microstructure and Magnetic Properties of Fe–Ni Alloy Fabricated by Selective Laser Melting Fe/Ni Mixed Powders. *J. Mater. Sci. Technol.* **2013**, *29*, 757–760. [CrossRef]

32. Zhang, B.; Fenineche, N.E.; Zhu, L.; Liao, H.; Coddet, C. Studies of magnetic properties of permalloy (Fe–30%Ni) prepared by SLM technology. *J. Magn. Magn. Mater.* **2012**, *324*, 495–500. [CrossRef]

33. Shishkovsky, I.; Saphronov, V. Peculiarities of selective laser melting process for permalloy powder. *Mater. Lett.* **2016**, *171*, 208–211. [CrossRef]

34. Garibaldi, M.; Ashcroft, I.; Hillier, N.; Harmon, S.; Hague, R. Relationship between laser energy input, microstructures and magnetic properties of selective laser melted Fe − 6.9%wt Si soft magnets. *Mater. Charact.* **2018**, *143*, 144–151. [CrossRef]

35. Jhong, K.J.; Huang, W.C.; Lee, W.H. Microstructure and Magnetic Properties of Magnetic Material Fabricated by Selective Laser Melting. *Phys. Procedia* **2016**, *83*, 818–824. Photonic Technologies Proceedings of the LANE 2016, Fürth, Germany, 19–22 September 2016. [CrossRef]

36. Lammers, S.; Adam, G.; Schmid, H.J.; Mrozek, R.; Oberacker, R.; Hoffmann, M.J.; Quattrone, F.; Ponick, B. Additive Manufacturing of a lightweight rotor for a permanent magnet synchronous machine. In Proceedings of the 2016 6th International Electric Drives Production Conference (EDPC), Nuremberg, Germany, 30 November–1 December 2016; pp. 41–45. [CrossRef]

37. Garibaldi, M.; Ashcroft, I.; Lemke, J.; Simonelli, M.; Hague, R. Effect of annealing on the microstructure and magnetic properties of soft magnetic Fe-Si produced via laser additive manufacturing. *Scr. Mater.* **2018**, *142*, 121–125. [CrossRef]

38. Niendorf, T.; Leuders, S.; Riemer, A.; Richard, H.; Tröster, T.; Schwarze, D. Highly Anisotropic Steel Processed by Selective Laser Melting. *Metall. Mater. Trans. B* **2013**, *44*, 794–796. [CrossRef]

39. Garibaldi, M.; Ashcroft, I.; Simonelli, M.; Hague, R. Metallurgy of high-silicon steel parts produced using Selective Laser Melting. *Acta Mater.* **2016**, *110*, 207–216. [CrossRef]

40. Pham, T.Q.; Do, T.T.; Kwon, P.; Foster, S.N. Additive Manufacturing of High Performance Ferromagnetic Materials. In Proceedings of the 2018 IEEE Energy Conversion Congress and Exposition (ECCE), Portland, OR, USA, 23–27 September 2018; pp. 4303–4308. [CrossRef]

41. Cramer, C.L.; Nandwana, P.; Yan, J.; Evans, S.F.; Elliott, A.M.; Chinnasamy, C.; Paranthaman, M.P. Binder jet additive manufacturing method to fabricate near net shape crack-free highly dense Fe-6.5 wt.% Si soft magnets. *Heliyon* **2019**, *5*, e02804. [CrossRef] [PubMed]

42. Pham, T.Q.; Suen, H.; Kwon, P.; Foster, S.N. Characterization of Magnetic Anisotropy for Binder Jet Printed Fe93.25Si6.75. In Proceedings of the 2019 IEEE Energy Conversion Congress and Exposition (ECCE), Baltimore, MD, USA, 29 September–3 October 2019; pp. 745–752. [CrossRef]

43. Pham, T.Q.; Mellak, C.; Suen, H.; Boehlert, C.J.; Muetze, A.; Kwon, P.; Foster, S.N. Binder Jet Printed Iron Silicon with Low Hysteresis Loss. In Proceedings of the 2019 IEEE International Electric Machines Drives Conference (IEMDC), San Diego, CA, USA, 12–15 May 2019; pp. 1045–1052. [CrossRef]

44. Pham, T.Q.; Suen, H.; Kwon, P.; Foster, S.N. Reduction in Hysteresis Loss of Binder Jet Printed Iron Silicon. In Proceedings of the 2020 International Conference on Electrical Machines (ICEM), Gothenburg, Sweden, 23–26 August 2020; Volume 1, pp. 1669–1675. [CrossRef]

45. Goll, D.; Schuller, D.; Martinek, G.; Kunert, T.; Schurr, J.; Sinz, C.; Schubert, T.; Bernthaler, T.; Riegel, H.; Schneider, G. Additive manufacturing of soft magnetic materials and components. *Addit. Manuf.* **2019**, *27*, 428–439. [CrossRef]

46. Kang, N.; Mansori, M.E.; Guittonneau, F.; Liao, H.; Fu, Y.; Aubry, E. Controllable mesostructure, magnetic properties of soft magnetic Fe-Ni-Si by using selective laser melting from nickel coated high silicon steel powder. *Appl. Surf. Sci.* **2018**, *455*, 736–741. [CrossRef]

47. Gargalis, L.; Madonna, V.; Giangrande, P.; Hardy, M.; Ashcroft, I.; Galea, M.; Hague, R. 3D Printing as a Technology Enabler for Electrical Machines: Manufacturing and Testing of a Salient Pole Rotor for SRM. In Proceedings of the 2020 International Conference on Electrical Machines (ICEM), Gothenburg, Sweden, 23–26 August 2020; Volume 1, pp. 12–18. [CrossRef]

48. Plotkowski, A.; Pries, J.; List, F.; Nandwana, P.; Stump, B.; Carver, K.; Dehoff, R. Influence of scan pattern and geometry on the microstructure and soft-magnetic performance of additively manufactured Fe-Si. *Addit. Manuf.* **2019**, *29*, 100781. [CrossRef]

49. Electromagnetic Applications. Available online: https://www.gknpm.com/globalassets/downloads/powder-metallurgy/2018/electromagnetic-application_brochure.pdf (accessed on 20 December 2018).

50. Studnitzky, T.; Dressler, M.; Andersen, O.; Kieback, B. 3D Screen Printing-Mass Production for Metals, Ceramics and its Combinations. In Proceedings of the Fraunhofer Direct Digital Manufacturing Conference DDMC 2016, Berlin, Germany, 16–17 March 2016; pp. 9–13.

51. Kulkarni, D.P.; Rupertus, G.; Chen, E. Experimental Investigation of Contact Resistance for Water Cooled Jacket for Electric Motors and Generators. *IEEE Trans. Energy Convers.* **2012**, *27*, 204–210. [CrossRef]

52. Sagawa, M.; Fujimura, S.; Yamamoto, H.; Matsuura, Y.; Hiraga, K. Permanent magnet materials based on the rare earth-iron-boron tetragonal compounds. *IEEE Trans. Magn.* **1984**, *20*, 1584–1589. [CrossRef]

53. Tenaud, P.; Vial, F.; Sagawa, M. Improved corrosion and temperature behaviour of modified Nd-Fe-B magnets. *IEEE Trans. Magn.* **1990**, *26*, 1930–1932. [CrossRef]

54. Kramer, M.J.; McCallum, R.W.; Anderson, I.A.; Constantinides, S. Prospects for Non-Rare Earth Permanent Magnets for Traction Motors and Generators. *JOM J. Miner. Met. Mater. Soc.* **2012**, *64*, 752–763. [CrossRef]

55. Technologies, A.M. *Plastiform® Injection Molded Magnets*; Arnold Magnetic (Shenzhen) Ltd: Guangdong, China, 2014.

56. Li, L.; Tirado, A.; Nlebedim, I.C.; Rios, O.; Post, B.; Kunc, V.; Lowden, R.R.; Lara-Curzio, E.; Fredette, R.; Ormerod, J.; Lograsso, T.A.; Paranthaman, M.P. Big Area Additive Manufacturing of High Performance Bonded NdFeB Magnets. *Sci. Rep.* **2016**, *6*, 36212. [CrossRef] [PubMed]

57. Duty, C.E.; Kunc, V.; Compton, B.; Post, B.; Erdman, D.; Smith, R.; Lind, R.; lLoyd, P.; Love, L. Structure and mechanical behavior of Big Area Additive Manufacturing (BAAM) materials. *Rapid Prototyp. J.* **2017**, *23*, 181–189. [CrossRef]

58. Urban, N.; Huber, F.; Franke, J. Influences of process parameters on rare earth magnets produced by laser beam melting. In Proceedings of the 2017 7th International Electric Drives Production Conference (EDPC), Wuerzburg, Germany, 5–6 December 2017; pp. 1–5. [CrossRef]

59. Compton, B.G.; Kemp, J.W.; Novikov, T.V.; Pack, R.C.; Nlebedim, C.I.; Duty, C.E.; Rios, O.; Paranthaman, M.P. Direct-write 3D printing of NdFeB bonded magnets. *Mater. Manuf. Process.* **2018**, *33*, 109–113. [CrossRef]

60. Paranthaman, M.P.; Shafer, C.S.; Elliott, A.M.; Siddel, D.H.; McGuire, M.A.; Springfield, R.M.; Martin, J.; Fredette, R.; Ormerod, J. Binder Jetting: A Novel NdFeB Bonded Magnet Fabrication Process. *JOM* **2016**, *68*, 1978–1982. [CrossRef]

61. Li, L.; Tirado, A.; Conner, B.; Chi, M.; Elliott, A.M.; Rios, O.; Zhou, H.; Paranthaman, M.P. A novel method combining additive manufacturing and alloy infiltration for NdFeB bonded magnet fabrication. *J. Magn. Magn. Mater.* **2017**, *438*, 163–167. [CrossRef]

62. Li, L.; Post, B.; Kunc, V.; Elliott, A.M.; Paranthaman, M.P. Additive manufacturing of near-net-shape bonded magnets: Prospects and challenges. *Scr. Mater.* **2017**, *135*, 100–104. [CrossRef]

63. Li, L.; Jones, K.; Sales, B.; Pries, J.L.; Nlebedim, I.; Jin, K.; Bei, H.; Post, B.K.; Kesler, M.S.; Rios, O.; et al. Fabrication of highly dense isotropic Nd-Fe-B nylon bonded magnets via extrusion-based additive manufacturing. *Addit. Manuf.* **2018**, *21*, 495–500. [CrossRef]

64. Khazdozian, H.A.; Manzano, J.S.; Gandha, K.; Slowing, I.I.; Nlebedim, I.C. Recycled Sm-Co bonded magnet filaments for 3D printing of magnets. *AIP Adv.* **2018**, *8*, 056722. [CrossRef]

65. White, E.M.H.; Kassen, A.G.; Simsek, E.; Tang, W.; Ott, R.T.; Anderson, I.E. Net Shape Processing of Alnico Magnets by Additive Manufacturing. *IEEE Trans. Magn.* **2017**, *53*, 1–6. [CrossRef]

66. Kano, Y.; Terahai, T.; Kosaka, T.; Matsui, N.; Nakanishi, T. A new flux-barrier design of torque ripple reduction in saliency-based sensorless drive IPM motors for general industrial applications. In Proceedings of the 2009 IEEE Energy Conversion Congress and Exposition, San Jose, CA, USA, 20–24 September 2009; pp. 1939–1945.

67. Fei, W.; Luk, P.C.K.; Shen, J.X.; Xia, B.; Wang, Y. Permanent-Magnet Flux-Switching Integrated Starter Generator With Different Rotor Configurations for Cogging Torque and Torque Ripple Mitigations. *IEEE Trans. Ind. Appl.* **2011**, *47*, 1247–1256.

68. Valavi, M.; Nysveen, A.; Nilssen, R. Characterization of radial magnetic forces in low-speed permanent magnet wind generator with non-overlapping concentrated windings. In Proceedings of the 2012 XXth International Conference on Electrical Machines, Marseille, France, 2–5 September 2012; pp. 2943–2948.

69. Dyck, D.N.; Lowther, D.A. Automated design of magnetic devices by optimizing material distribution. *IEEE Trans. Magn.* **1996**, *32*, 1188–1193. [CrossRef]

70. Okamoto, Y.; Hoshino, R.; Wakao, S.; Tsuburaya, T. Improvement of Torque Characteristics For a Synchronous Reluctance Motor Using MMA-based Topology Optimization Method. *IEEE Trans. Magn.* **2018**, *54*, 1–4. [CrossRef]

71. Huber, C.; Abert, C.; Bruckner, F.; Pfaff, C.; Kriwet, J.; Groenefeld, M.; Teliban, I.; Vogler, C.; Suess, D. Topology optimized and 3D printed polymer-bonded permanent magnets for a predefined external field. *J. Appl. Phys.* **2017**, *122*, 053904. [CrossRef]

72. Wang, S.; Kang, J. Topology optimization of nonlinear magnetostatics. *IEEE Trans. Magn.* **2002**, *38*, 1029–1032. [CrossRef]

73. Bendsoe, M.; Sigmund, O. *Topology Optimization: Theory, Methods, and Applications*; Engineering Online Library; Springer: Berlin/Heidelberg, Germany, 2003.

74. Lee, J.; Dede, E.M.; Nomura, T. Simultaneous Design Optimization of Permanent Magnet, Coils, and Ferromagnetic Material in Actuators. *IEEE Trans. Magn.* **2011**, *47*, 4712–4716. [CrossRef]

75. Jung, J.; Lee, B.; Kim, D.; Hong, J.; Kim, J.; Jeon, S.; Song, D. Mechanical Stress Reduction of Rotor Core of Interior Permanent Magnet Synchronous Motor. *IEEE Trans. Magn.* **2012**, *48*, 911–914. [CrossRef]

76. Guo, F.; Salameh, M.; Krishnamurthy, M.; Brown, I.P. Multimaterial Magneto-Structural Topology Optimization of Wound Field Synchronous Machine Rotors. *IEEE Trans. Ind. Appl.* **2020**, *56*, 3656–3667. [CrossRef]

77. Jung, S.; Ro, J.; Jung, H. A Hybrid Algorithm Using Shape and Topology Optimization for the Design of Electric Machines. *IEEE Trans. Magn.* **2018**, *54*, 1–4. [CrossRef]

78. Lee, J. Structural Design Optimization of Electric Motors to Improve Torque Performance. Ph.D. Thesis, University of Michigan, Ann Arbor, MI, USA, 2010.

79. Yan, Y.; Liu, L.; Ding, C.; Nguyen, L.; Moss, J.; Mei, Y.; Lu, G. Additive Manufacturing of Magnetic Components for Heterogeneous Integration. In Proceedings of the 2017 IEEE 67th Electronic Components and Technology Conference (ECTC), Orlando, FL, USA, 30 March–2 June 2017; pp. 324–330. [CrossRef]

80. Pham, T.Q.; Foster, S.N. Additive Manufacturing of Non-homogeneous Magnetic Cores for Electrical Machines Opportunities and Challenges. In Proceedings of the 2020 International Conference on Electrical Machines (ICEM), Gothenburg, Sweden, 23–26 August 2020; Volume 1, pp. 1623–1629. [CrossRef]

81. Dyck, D.N.; Lowther, D.A. Composite microstructure of permeable material for the optimized material distribution method of automated design. *IEEE Trans. Magn.* **1997**, *33*, 1828–1831. [CrossRef]

82. Liu, L.; Ding, C.; Lu, S.; Ge, T.; Yan, Y.; Mei, Y.; Ngo, K.D.T.; Lu, G. Design and Additive Manufacturing of Multipermeability Magnetic Cores. *IEEE Trans. Ind. Appl.* **2018**, *54*, 3541–3547. [CrossRef]
83. Waterman, J.; Clucas, A.; Costa, T.B.; Zhang, Y.; Zhang, J. Numerical modeling of 3D printed electric machines. In Proceedings of the 2015 IEEE International Electric Machines Drives Conference (IEMDC), Coeur d'Alene, ID, USA, 10–13 May 2015; pp. 1286–1291. [CrossRef]
84. Lemke, J.; Simonelli, M.; Garibaldi, M.; Ashcroft, I.; Hague, R.; Vedani, M.; Wildman, R.; Tuck, C. Calorimetric study and microstructure analysis of the order-disorder phase transformation in silicon steel built by SLM. *J. Alloy. Compd.* **2017**, *722*, 293–301. [CrossRef]
85. Sixel, W.; Liu, M.; Nellis, G.; Sarlioglu, B. Ceramic 3D Printed Direct Winding Heat Exchangers for Improving Electric Machine Thermal Management. In Proceedings of the 2019 IEEE Energy Conversion Congress and Exposition (ECCE), Baltimore, Maryland, USA, 29 September–3 October 2019; pp. 769–776. [CrossRef]
86. Urbanek, S.; Ponick, B.; Taube, A.; Hoyer, K..; Schaper, M.; Lammers, S.; Lieneke, T.; Zimmer, D. Additive Manufacturing of a Soft Magnetic Rotor Active Part and Shaft for a Permanent Magnet Synchronous Machine. In Proceedings of the 2018 IEEE Transportation Electrification Conference and Expo (ITEC), Long Beach, CA, USA, 13–15 June 2018; pp. 668–674. [CrossRef]
87. TUChemnitz. 3D-Multimaterialdruck. Available online: https://www.tu-chemnitz.de/etit/ema/AMMM/index.php (accessed on 25 December 2020)
88. Xue, S.; Acharya, V. Topology Optimization Empowers the Design of Interior Permanent Magnet (IPM) Motors. In Proceedings of the 2020 IEEE Transportation Electrification Conference Expo (ITEC), Chicago, IL USA, 23–25 June 2020; pp. 1–5. [CrossRef]

MDPI

St. Alban-Anlage 66

4052 Basel

Switzerland

Tel. +41 61 683 77 34

Fax +41 61 302 89 18

www.mdpi.com

Energies Editorial Office

E-mail: energies@mdpi.com

www.mdpi.com/journal/energies

9 783036 542850